AF772359

Regional Innovation Systems and Sustainable Development:
Emerging Technologies

Patricia Ordóñez de Pablos
University of Oviedo, Spain

W. B. Lee
Hong Kong Polytechnic University, Hong Kong

Jingyuan Zhao
Harbin Institute of Technology, China

Information Science REFERENCE

INFORMATION SCIENCE REFERENCE
Hershey · New York

Director of Editorial Content: Kristin Klinger
Director of Book Publications: Julia Mosemann
Acquisitions Editor: Lindsay Johnston
Development Editor: David DeRicco
Typesetter: Michael Brehm
Production Editor: Jamie Snavely
Cover Design: Lisa Tosheff

Published in the United States of America by
Information Science Reference (an imprint of IGI Global)
701 E. Chocolate Avenue
Hershey PA 17033
Tel: 717-533-8845
Fax: 717-533-8661
E-mail: cust@igi-global.com
Web site: http://www.igi-global.com

Copyright © 2011 by IGI Global. All rights reserved. No part of this publication may be reproduced, stored or distributed in any form or by any means, electronic or mechanical, including photocopying, without written permission from the publisher. Product or company names used in this set are for identification purposes only. Inclusion of the names of the products or companies does not indicate a claim of ownership by IGI Global of the trademark or registered trademark.

Library of Congress Cataloging-in-Publication Data

Regional innovation systems and sustainable development : emerging technologies / Patricia Ordonez de Pablos, W.B. Lee and Jingyuan Zhao, editors.
p. cm.
Includes bibliographical references and index.
Summary: "This book promotes scientific discussion on standards and practices of regional development, while also covering emerging research topics in regional innovation systems and sustained development"--Provided by publisher.
ISBN 978-1-61692-846-9 (hbk.) -- ISBN 978-1-61692-848-3 (ebook) 1. Technological innovations--Economic aspects. 2. Regional planning. 3. Economic development. I. Pablos, Patricia Ordonez de. II. Lee, W. B. III. Zhao, Jingyuan, 1968-
HC79.T4R434 2011
338'.064--dc22

2010016307

British Cataloguing in Publication Data
A Cataloguing in Publication record for this book is available from the British Library.

All work contributed to this book is new, previously-unpublished material. The views expressed in this book are those of the authors, but not necessarily of the publisher.

Table of Contents

Detailed Table of Contents

In a global world, where a lot of international trade treaties have been signed, we must face new threats such as influenza A (H1N1) or bioterrorism. In this context, to control the origins of food can save lives and big amounts of money. Meanwhile, new regulations in the European Union have arisen. In accordance with the provisions of Article 18 of European Regulation 178/2002, from 1(st) January 2005, all European companies in the food industry must have implemented a traceability system. However, it may sound strange that after all these years with this regulation in force some of the workers in this area don´t even know what traceability is. This document pretends to give some details on food traceability, and it will expose a real case in which cutting-edge technologies have been applied to solve the traceability in rural areas of northern Spain, by improving productivity and allowing small businesses to offer more competitive and safer products, and of higher quality.

The paper focuses on comparing and analyzing the development models of typical regional innovation systems in the world, discussing the relationship of regional innovation system and R&D centre growth, and points out that the emergence of R&D centre and growth are tight relative with the development of regional innovation system. Through researching on typical cases of India's Banglore, Singapore and Taiwan's Xinzhu, the paper summarizes the experience that establishing and perfecting regional innovation system will improve R&D centre growth. Using the experience for reference, some strategies to promote R&D centre are put forward as conclusions.

This chapter aims to present a conceptual model aimed to understand the Intellectual Capital-based (IC) characteristics of the regional innovation capacity. The proposed Regional Innovation Capacity Model (RICM) can be used for interpretative and normative purposes to analyse the innovation dynamics taking place at regional and territorial level. From an interpretative point of view, the model identifies the pillars grounding the innovation capacity of a local system. While, from a normative perspective, the model can inspire the definition of guidelines driving the design and the implementation of actions, projects and programmes aimed to stimulate and sustain regional development dynamics. The RICM adopts a knowledge-based perspective assuming that IC, in the forms of regional knowledge assets, and knowledge dynamics, in the form of knowledge transfer and learning processes, are the drivers of innovative processes and outputs. The chapter concludes proposing a future research agenda.

The link between KM and network theory lies in the cultivating of knowledge networks by viewing knowledge as connections and understanding it as the aggregation of connections across different entities albeit it people or organizations (Kadushin, 2004). The study explores primarily on the capabilities and capacity building of government-linked corporations (GLC) in Malaysia in relation to its evolution as a knowledge network. In this paper, we explore the overlap that exists in KM particularly in the entities and sharing of knowledge assets that undergoes a multiplying effect in its value propositions with the forming of interoganizational relationships. The findings of the study indicates that in order to benefit from the knowledge multiplying affects, organizations need to create a "network identity" and develop capabilities as a learning entity that would add value to the network itself. From the context of KM, majority of the GLCs are in the transitory stages in which issues of trust, competencies and communications have been key indicators in forming network relations. As such this paper would be useful in highlighting the significance of network relations and provides further proliferation in the study of social network theory and significance to KM.

This chapter discusses strategies aimed at regional-economic structural change in the German Ruhrgebiet. The Ruhrgebiet used to be the largest industrial area in Western-Europe. After the crisis in the coal and steel industry the region pursued re-industrialisation policies in the 1960s and 1970s. These attempts were largely unsuccesful. Therefore, since the 1980s the involved actors gradually adopted regional innovation strategies. Thus, they were able to break out from the region's lock-in situation. The re-orientation of the Ruhrgebiet towards innovation based on the industries' expertise and past (e.g. environmental technology, energy and industrial tourism) is more successful than its earlier re-industrialisation attempts. Also for other old industrial areas in Europe this kind of place-based renewal might be the future.

The biotechnology is a very complex sector and its growth depends on many variables like intellectual properties, venture capital, clusters, public policies, alliances, amongst others. In biotechnology, the complexity surrounding alliances can be observed. The market globalization, the exorbitant costs of R&D and the rapid changes in technology, are arguably in the midst of the principal reasons that push companies to establish alliances. Biotechnology companies use this instrument to develop external features in the search for resources and missing expertise. This chapter sets out to identify whether such alliances in biotechnology companies are an advantage in themselves, sufficient for the acquisition of new capabilities and whether they help the growth of these companies. For this approach, a private database of companies (900 companies including years 1996-2001) in the two most advanced countries in this sector: United States and United Kingdom will be used.

In theory, patents work by providing the inventor an incentive to invent in the first place and then to disclose. Disclosure to the public is rewarded by giving the inventor a monopoly. As product patent and higher patent protection has been advocated by Art 27.1 of the TRIPs agreement on the basis that for greater innovation through transfer of technology is a necessity in developing countries like India as it provides capital to fund expensive innovations, who are otherwise not be able to fund expensive innovations on its own. On the other hand, at the same time drugs are also related with the health of the people and to take care of the health of the people is the utmost priority of any Government and there are issues like accessibility with regard to strong patent protection to biopharma products and data exclusivity. Also as per Art 7 of the TRIPs transfer of technology has to occur to the developing countries in order to promote technological innovations, which is conducive to social and economic welfare. Therefore, striking the right balance between incentive and public access creates a tension is essential. This study suggests optimal policy (Patent and other regulations) to have a balance between biopharma drugs innovation and their access in India while complying with the provisions of the TRIPs agreement by broadly categorising variables such as (1) patent policy such as the scope of biotech patents and the extent of the right in terms of breadth and length; and (2) regulatory environment such as the taxation incentive, Investment policy, Government initiative for the development of this sector etc.

Many countries are attempting to develop themselves as intelligent nations, equipped and empowered by advanced info-communications technology (ICT). This paper describes and analyses how Singapore attempts to develop an intelligent nation under its umbrella vision and strategy of Intelligent Nation 2015 (iN2015) vision. It examines the substance of the vision and highlights some of the initiatives implemented to realise the vision. It also discusses the different roles of the government in implementing the iN2015 vision and some of the challenges ahead.

By introducing nonlinear technology gap into Jones (1995b), this paper constructs an R&D non-scale growth model that includes endogenous human capital and technological progress. The goal is to take the model's implications to the data to explaining the Chinese economic development experiences at period 1979-2004. Our model suggests that the technology gap has the block neck effect on the economic development. The market competitive equilibrium solution shows that R&D and technology spillover can enhance the steady state growth rate. The mode's transitional dynamics is also analyzed on the effects of human capital, capital, technological progress and intersectional labor movements on economic growth, technological progress has the most effect on the economic development, and more human capital shift into R&D sector from final goods sector.

Internet Cafés are the most popular locations for internet users to enjoy online game playing in China. At the beginning of this paper, the authors analyzed reasons for Internet cafes' popularity as well as listing difficulties the Internet Cafés are confronting. Applying the distribution strategy put forward by Kasper, the authors analyzed how five of the top ten most popular games are distributed in the Internet cafés by their game operators. In the conclusion, the authors suggest game operators of different market dominance should utilize different distribution strategies (intensive or selective one) to match the game product with the players' demand. At the end of this paper, some recommendations are put forward for the Internet cafés to take into account.

For nations to compete successfully in the long run, they must innovate; and innovating entails building new competencies, new capabilities and new knowledge. Starting with a brief review of the concepts of Science, Technology and Innovation, this chapter discusses the factors militating against the translation of scientific R&D outputs into innovations in Nigeria. The situation in selected Asian countries are reviewed to highlight the significance of specific factors such as institutional framework, adequate funding and knowledge pool in the process of building indigenous innovation capability for national development. Policies, strategies and structures that could facilitate the effective management of R&D and innovation, and enhance university-industry connections are suggested.

The marketing of new products for new markets is uncertain and often leads to confusion. One of the many causes is the wrong use of marketing techniques for mature products whose life cycles are different, although they are very familiar. Biotechnology, with a strong component of technological knowledge and innovative cross-refers not to an industry, but rather a set of applications of intricate characteristics mainly due to genetic manipulation, which represents a particular challenge unlike other more mature or traditional product or technology. For this reason, through an innovation based marketing for new product, new customers and new markets are essential to maximize the success of this technology.

The purpose of this study is to present a framework, which can be used to analyze the regional dimension of the institutional performance in higher education. This paper combines strategic planning and corporate social responsibility reporting. The strategic planning of a higher education institution defines strategic objectives, which take into account educational policy, national and regional networked strategies, and the regional demand for labor. The evaluation of the external impact of an institution means that a general conceptual framework must describe the outreach and engagement of the institution with the region. The framework presented in this study includes the elements of the networked knowledge society, strategic management, global reporting initiative and quality assurance. The empirical part of this study is based on the network study of the Turku University of Applied Sciences. The framework can be used in decision making when an institution is invited to join local, national and international networks.

This chapter focuses on the actors of knowledge, innovation and entrepreneurship, as comparative advantages towards regional sustainable development. The chapter describes the analytical framework in which technical change, innovation and entrepreneurship activities are considered as among the most important factors influencing sustainable economic development, both in national and regional level. Within this analytical framework, the chapter aims to provide a clear understudying of the main related questions summarized in the following: To which extent economic development depends on the presence of a network favorable to knowledge, innovation and entrepreneurship, based on the endogenous development capabilities? How could the level of regional innovation activities be enhanced? How could the effect of innovation activities on regional development be estimated? What is the relative position of a region as far as development and innovation competitiveness are concerned?

This chapter analyses the innovation activities and the way they contribute essentially to the regional dimension and growth. The chapter aims to investigate how technological infrastructure and innovation capabilities affect not only regional growth, but also the whole periphery and economy, as well as how technological innovation is essential if companies - and countries - are to maintain their competitiveness in global markets. Furthermore, this chapter describes the relative measures and indexes introduced by E.U. and O.E.C.D. during the last years, concerning Research and Development Expenditures, patents e.t.c., namely measuring innovation activities. However, there are a lot of underlying problems and questions regarding the measurement of innovation activities at a regional level. This chapter attempts to analyze the whole framework of innovation statistics and in particular to examine the measurement and also the statistical estimation of innovation activities in E.U.. Within this context, it is also aiming to emphasize and review the appropriate techniques, the most common methods and the particular related problems.

The chapter has measured the relationship between the intellectual property rights and Foreign Direct Investment in the context of Indian biopharma Industry in order to know the impact of TRIPs agreement of WTO on the biopharmaceutical industry of developing countries. The central issue in this study is the extent to which patent reform (after the imposition of the TRIPs agreement in 1995) affects In-

Preface

The regional development of society and economy are closely related with innovative capacities. Choosing suitable development model can give full play to regional advantages, integrate scientific and technological resources, and continually improve the regional innovative capacities. It is an important issue that finds the rules and experiences of establishing and developing innovative city based on economic theory, innovative theory and intellectual capital management theory, puts forward the framework of establishing innovative city according to the unique characteristics of urban development, researches on structural and systemic characteristics of innovative city, including innovative investment system, innovative service system, essential security system, innovative management system, innovative policy system. More again, the design principle of innovative city's goal system and evaluation system of achievement should be studied.

Regional Innovation System (RIS) is an emerging concept, and is a new field of world economy and geography. Many scholars pay much attention to the theoretical study of RIS, and score certain achievements in terms of the basic concept, content, formation mechanism and function of RIS. RIS focuses on a specific region, and its cross-organizational knowledge innovation mainly gets through the share of common tacit knowledge and social capital link, the knowledge transfer factor of RIS is the geographical and cultural adjacency and knowledge spillover. The study on RIS can help with the establishment of innovative city.

Government governance plays a critical role in the establishment of RIS and innovative city. The emphasis and model of government governance still need further exploration. Meanwhile, the issues that need further study are the mechanisms of innovation, business and creativity, information technology and Internet platform, infrastructure environment construction and investment, improvement of the education system and personnel training policies for social needs of innovative society, development countermeasures and suggestions of knowledge-intensive service industries, improvement of living environment and establishment of livable city.

The subject area is a combination of Innovation, Regional Analysis, Technology, Strategic Management of IT and Sustainable Development, and aims to be the leading source of information for all those interested in regional innovation systems and emerging technologies since it promotes scientific discussion on standards, and practice of regional development, as well as to cover the emerging research topics which are going to define the future of regional innovation systems and sustained development.

Furthermore, this book will demonstrate the capacity of regional innovation systems, information technology, management and sustainable development for the mutual understanding, prosperity and well being of all the citizens in the world.

Patricia Ordóñez de Pablos
W. B. Lee
Jingyuan Zhao
Editors
January 2010

Chapter 1
Traceability Systems for Sustainable Development in Rural Areas

Vicente García-Díaz
University of Oviedo, Spain

B. Cristina Pelayo G. Bustelo
University of Oviedo, Spain

Oscar Sanjuán Martínez
University of Oviedo, Spain

Juan Manuel Cueva Lovelle
University of Oviedo, Spain

Patricia Ordóñez de Pablos
University of Oviedo, Spain

ABSTRACT

In a global world, where a lot of international trade treaties have been signed, we must face new threats such as influenza A (H1N1) or bioterrorism. In this context, to control the origins of food can save lives and big amounts of money. Meanwhile, new regulations in the European Union have arisen. In accordance with the provisions of Article 18 of European Regulation 178/2002, from 1(st) January 2005, all European companies in the food industry must have implemented a traceability system. However, it may sound strange that after all these years with this regulation in force some of the workers in this area don't even know what traceability is. This document pretends to give some details on food traceability, and it will expose a real case in which cutting-edge technologies have been applied to solve the traceability in rural areas of northern Spain, by improving productivity and allowing small businesses to offer more competitive and safer products, and of higher quality.

DOI: 10.4018/978-1-61692-846-9.ch001

Copyright © 2011, IGI Global. Copying or distributing in print or electronic forms without written permission of IGI Global is prohibited.

INTRODUCTION

As it has been used for a very long time, traceability is not really something new. For example, Greek artists began to sign their work regularly in the seventh century, thanks to which we are familiar with great philosophers like Heraclitus or poets like Theognis of Megara. However, the names of Egyptian artists, who tend not to sign their works because of the set of mind of another culture, are not so familiar to us. Without doubt, the key of traceability is the ability to obtain information about the origins of a product. In the text we will show how we can enhance sustainable development, regional competitiveness and regional innovation by using traceability systems.

This project below is about a regional innovation system capability to answer to the needs of traceability systems and sustainable development in rural areas.

Food Traceability

The concept of Food traceability is relatively new, and because of that many employers and many consumers do not know what it is. In 2003, the OCU (Users and Customers Organization, 2009) undertook a study to 310 partners in Spain, asking them if they knew where their work has been done, and 95% of them answered that when they buy a product they would like to have as much information as possible about where it came from and if possible to even find out the exact origin of the raw materials from which is obtained. However, paradoxically, only 8% knew that that description was precisely the definition of food traceability.

More specifically, under Article 3 of European Regulation 178/2002 (The General Food Law Regulation (EC) 178/2002, 2002), food traceability means *the ability to trace and follow a food, feed, food-producing animal or substance intended to be, or expected to be incorporated into a food or feed, through all stages of production, processing and distribution.*

Food traceability was born to prevent food crises such as the mad cow (Ratzan, 1998). On one hand consumers demand each time more details about what they are eating, about their origin and the route that the product has followed from extracting raw materials through the entire evolutionary process undertaken. On the other hand, producers want to make sure that what they sell will not cause any problem to their customers, ensuring that they will not to damage their brand image, or that a food problem caused by a competitor company can disrupt them in any way. The right to a healthy diet, promoted by global agencies such as Food and Agriculture Organization of the United Nations (FAO) (*Right to Adequate Food*, 2002), shows that citizens can demand a control on the food they eat. The problem is that the production process is not always controlled, making traceability a very useful tool to control the entire food chain. Traceability is synonymous of *reliable information.*

Contributions of Food Traceability

Traceability by itself does not serve to make food safer, but as a tool to manage risk, it can avoid as far as possible food disasters. Briefly, one can say that with traceability key information can be obtained from products and knowing the background of such products you can find out if they are dangerous for people.

What we achieved with traceability is just to identify a product, providing data on raw materials, origin, destination, the way it was manufactured, the checks on the product, the results obtained and the dates when the whole process took place.

Figure 1 shows that there are two types of traceability. Backward traceability, which basically responds to the following questions: What was done? Who did it? Where? Forward traceability, on the other way, responds to the following questions: What is performed? Who performs it? Where?

As can be seen, two of the most efficient methods used to identify a product unmistakably,

Figure 1. Two types of traceability

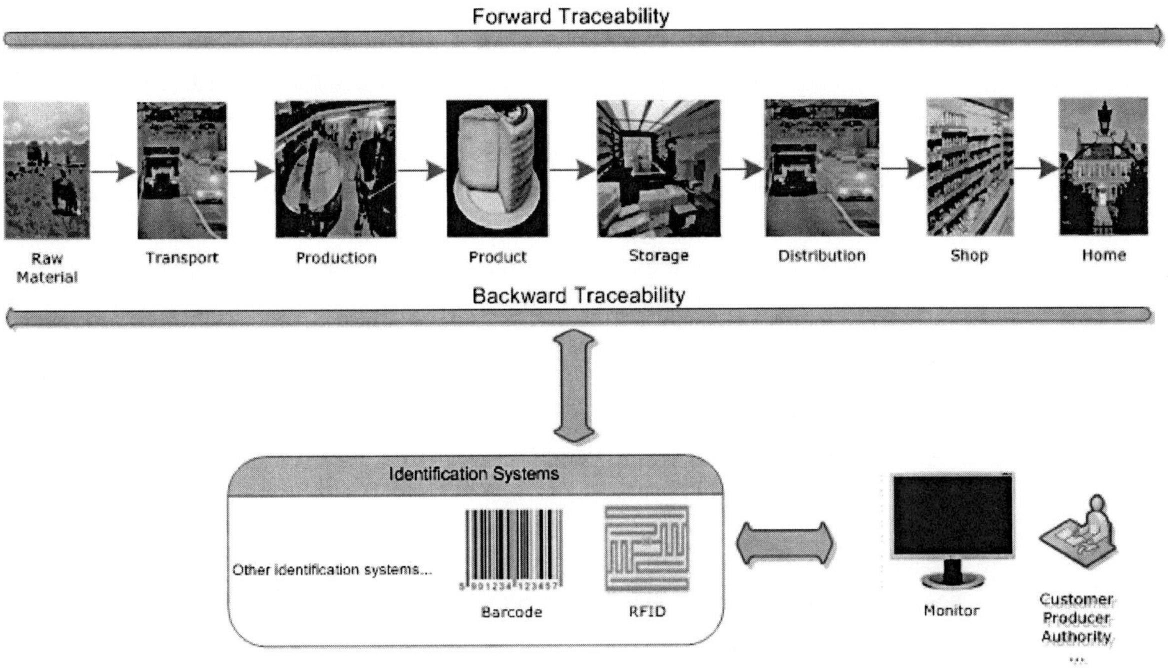

is through a barcode or an RFID tag. However, other methods can be used like manual labeling of dates and times.

REAL CASE

The origin of this project has two key elements. Firstly to comply with new legislation and secondly to ensure the quality and image of food in the region where the project was performed. This project was carried out in Asturias, a small region of northern Spain, where there are over 70 types of cheeses over all its villages.

When we began this work the first thing we did was an analysis of the different cheese factories and we realized that even small factories (with one or two employees) of the same type of cheese, do not have exactly the same manufacturing processes. We also appreciate that all the cheese factories have the same basic needs, i.e. all need to control the trace forward and backward, list products, make reports, etc.

We want to explain briefly some of the differences between cheeses in order to understand better the needs of each type of cheese. And more specifically, the differences between the five types of cheese we have focused on:

- **Cabrales.** It is the cheese from which we made the first prototype of all. It is a cheese made from milk of sheep, goat or cow (also can be any of the three combinations) that after a maturation period of approximately 60 days in the caves of the *Picos de Europa* is sold at a price that depends on its size. The treatment is done by units of cheese, so a cheese from the same batch may for instance be inserted on the site where they are going to dry before another. There are 21 different possible subtypes of cheese.
- **Casín.** It is a cheese produced quite faster than Cabrales cheese. For its production,

they used different raw material from those of Cabrales cheese such as calcium and ferments. It comprises three different types of kneading of products that are getting increasingly close to the cheese. The treatment is done by groups of cheese. All cheeses are sold at the same price and have the same size.

- **Afuega'l pitu.** It is similar to Casín cheese but it is different in some ways first in addition to cheese, also is prepared cottage cheese, secondly there are several different types of cheese depending on the type of paprika being used. Thirdly there are boxes that are marked in order to carry the cheese and not mixing batches.
- **Gamoneu.** The main difference between this cheese and Cabrales cheese is that the production of the dairies of Gamoneu is much greater in number than those of Cabrales, so it will be necessary to bring an organization using cheese batches instead of individual cheeses but at the time of packaging you need to weight each of the cheeses. There are 6 possible subtypes of different cheeses.
- **Beyos.** The Beyos cheese can be seen as a mixture of Afuega'l pitu cheese and Casín cheese with some changes in the various processes.

As it has been exposed, creating a traceability system that reflects the wide variety of cheese's assortments discussed briefly above is a complex process. The next Section will show the solution we have adopted to face this challenge.

ADOPTED SOLUTION

Works as (Sebestyen et al., 2008) shows that it is not a trivial process to develop a food traceability system. Although, nothing compels you to use the latest technology, undoubtedly the best alternative is to use a computerized system. Our solution is based on the use of the latest hardware and latest software technologies.

The *hardware* for each customer is made up of several advanced devices. A server with high capacity to host the software, one or more printers for printing reports, one or more labellers for product labels, one or more industrial terminals to make the data introduction easier even in unfriendly environments (high humidity, water, high temperatures, etc.), are just some of the hardware devices that could be used. In addition, producers are beginning to use non-optical codes to identify products. RFID (Radio Frequency Identification) (Glover & Bhatt, 2006), which is a technology that allows the use of electromagnetic or electrostatic coupling in the radio frequency of the electromagnetic spectrum to identify anything. Unlike barcodes, RFID does not require direct contact or lineal scanning.

Regarding the *software*, we applied MDE (Model-Driven Engineering) (Kent, 2002), the latest approach to software engineering has been proposed, being the fruit of countless scientific studies. Moreover, to focus on that technology is the goal of the most important software companies. MDE is based on raising the abstraction level of programming languages, enabling even non-specialized users to program when they know the concepts of the domain for which the language is designed.

The cheese traceability issue can be addressed as a product line to produce different systems tailored to individual needs of each client, but with much in common. In fact, the MDE initiatives tend to rely on the concept of SPL (Software Product Line) (van der Linden & Schmid, 2007). A SPL is a set of software products that are very similar but have slight differences between them. Thus, to develop the common points, an architecture that serves to avoid duplication and reusing code can be used. However, to develop the variable points among the different products a DSL is used, which serves to raise the level of abstraction of

Figure 2. Creating custom traceability systems

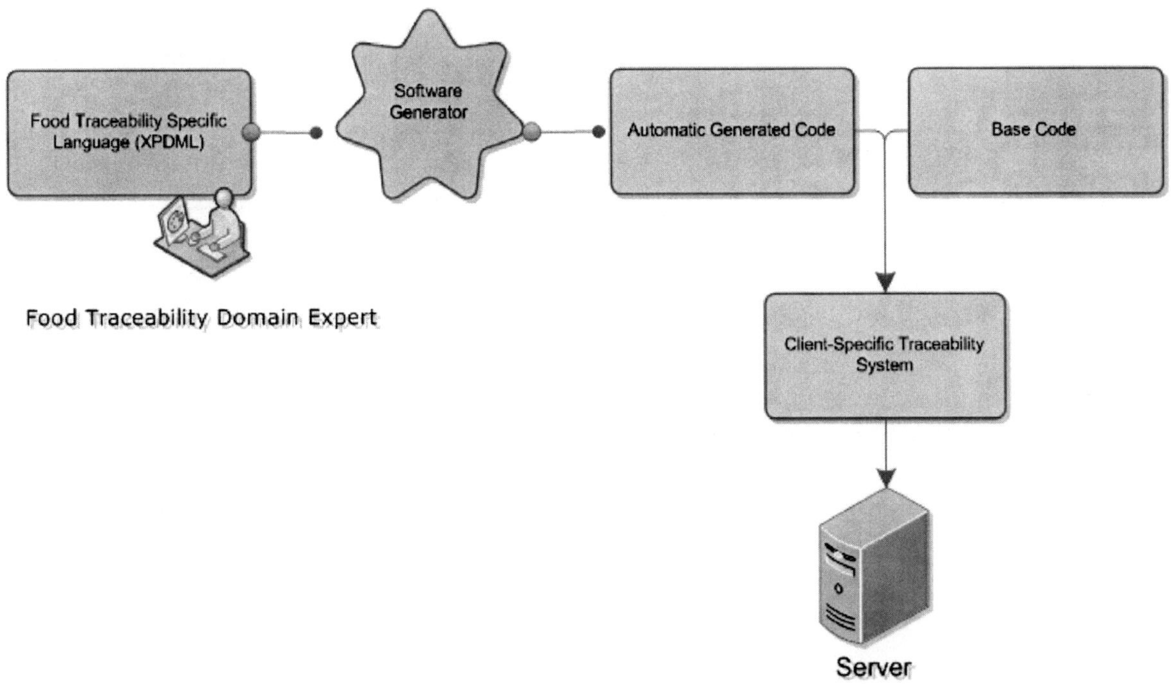

the language. We have created a language called XPDML (eXtensible Markup Process Definition Language), which allows customers to specify their own manufacturing process (see Figure 2). From that definition, using a code generator that we have also created, we will generate a specific code adapted to the needs of each customer without having to program based on the automatic processing of XPDML language.

Figure 3 shows how one customer could represent its manufacturing process to achieve creating the traceability system. It can be seen that no technical concepts are used, only concepts related to business. In addition, our system can generate easily traceability systems for other food industries different from cheese, since all follow the same basic principles.

Basically all the solutions need the same items, that is to say, the software generated, a SQL server data base, a computer, a scale, a printer, a label printer, RFID readers, industrial terminals, and lots of chips. But depending on the mode of manufacturing the number of items may change, or for example if the customer requires some other devices such as RFID printers or more scales. What never changes is that our software is the one which controls and manages all the above mentioned hardware. The software automatically generated from the specification of the characteristics of a customer, is responsible for managing all communication with hardware devices, to store information in a database and display the appropriate interface according to the user profile that is logged in. Figure 4 shows an idea of an orchestration of elements that interact in the food traceability system. More information about our system can be reached at (García-Díaz et al., 2008).

CONCLUSION AND FUTURE WORK

Traceability is a feature that has been imposed or required in a variety of products since time immemorial.

Figure 3. A sample of a manufacturing process

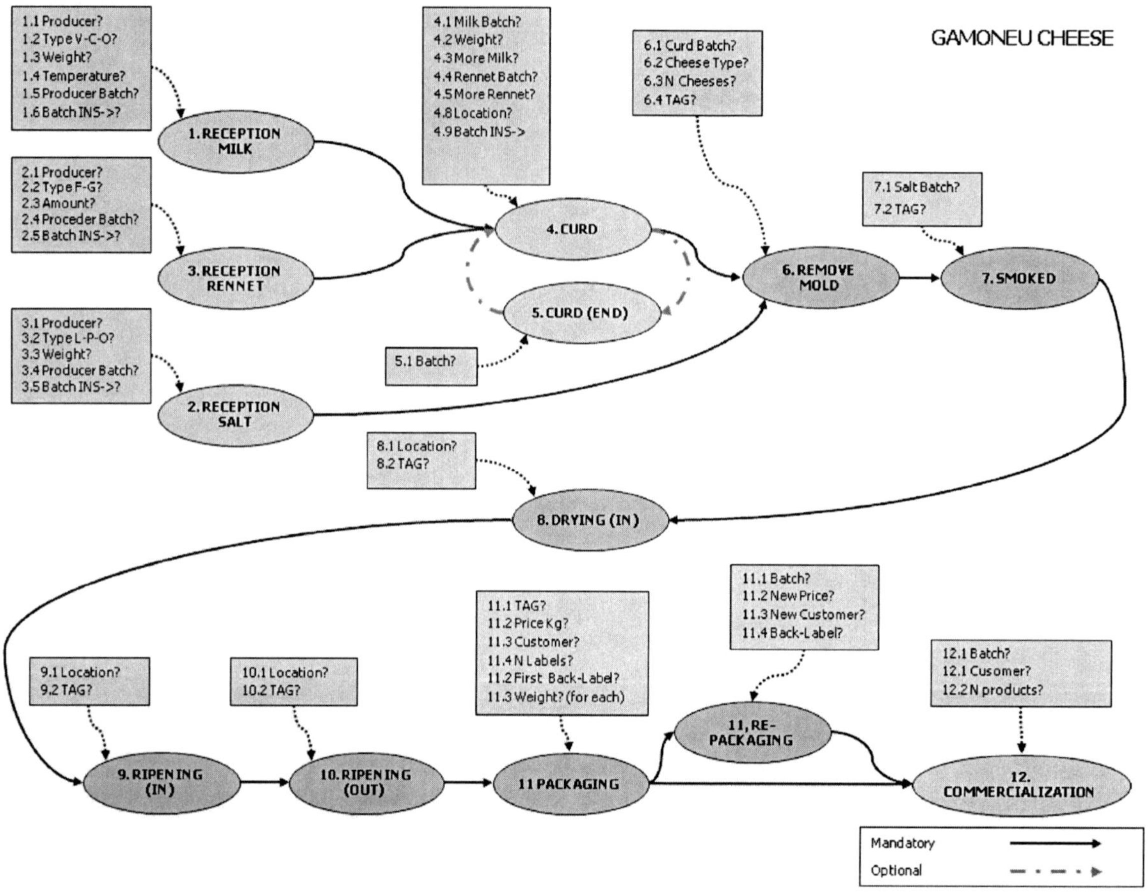

The traceability of food, understood as the ability to obtain a food history from its origins from any segment of the life cycle of a product (phases of manufacturing, marketing, distribution and consumption), is an important step forward to improve quality and increase the safety of food consumers. It could be defined as a process that ensures the producer, distributor and consumer, that the products have the guarantee of origin and have been processed using an auditable system. I.e., producers, manufacturers, traders, consumers and public authorities can track a particular object along all or part of its life.

It is also true that many companies use manual systems, causing a low efficiency and slowing the manufacturing processes by the additional tasks to control the traceability. When you use manual systems, even worse is the slow response from producers to potential food crises. In addition, the manual information is more difficult to store and to manage. Undoubtedly, it is best to use a computerized tracking system like the one we presented in this paper that is virtually transparent to its users.

We have developed a system that allows regional innovation in areas in which some types of cheese are produced, helping to create products with higher quality and promoting sustainable development.

Our future work focuses on further improving the current one. We will focus on: (1) add support for managing new hardware devices, (2) refine the abstract syntax of the domain-specific language, (3) do usability testing of the concrete syntax of

Figure 4. A simple of component organization

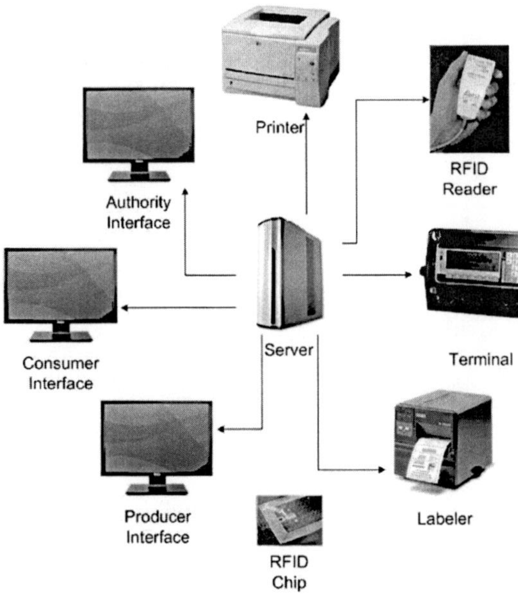

the language, (4) increase functionalities offered by the basic architecture developed, (5) perform automatic backups and send them to a central repository.

It is also worth noting that in our research group, Oviedo3-OOTLAB, we conduct several researches that may be related to the project outlined here. In (Fernández et al., 2009), we made a proposal to simplify the notation BPMN, allowing to simplify the way that could be defined manufacturing processes, In (Tolosa et al., 2009), we show the first steps towards meta-model interoperability of models through transformations, facilitating the generation of artifacts from models. Finally, other MDE tools can be seen in (Palacios-González et al., 2008).

REFERENCES

Fernández, H., Palacios-González, E., & García-Díaz, V. G-Bustelo, B.C.P., Sanjuan-Martínez, O., & Lovelle, J.M.C. (2009). SBPMN - An Easier Business Process Modeling Notation for Business Users. *Computer Standards & Interfaces.*

García-Díaz, V., Fernández-Fernández, H., & Palacios-González, E. G-Bustelo, B. C. P., & Lovelle, J. M. C. Intelligent traceability system of Cabrales cheese using MDA TALISMAN. In *Proceedings of the 2008 International Conference on Artificial Intelligence.* July 14-17, 2008, Las Vegas, Nevada, USA

Glover, B. & Bhatt, H (2006). RFID Essentials (Theory in Practice). *O'Reailly Media, Inc.*

Kent, S. (2002). *Model-Driven Engineering. Proceedings of the Third International Conference on Integrated Formal Method.* (pp. 286-298). London, UK: Springer.

Palacios-González, E., Fernández-Fernández, H., & García-Díaz, V., G-Bustelo, B.C.P., Lovelle, J.M.C., & Sanjuan-Martínez, O. (2008). General Purpose MDE Tools. *International Journal of Artificial Intelligence and Interactive Multimedia, 1*(1), 77–88.

Ratzan, S. (1998). *Mad Cow Crisis: Health and the Public Good.* New York: NYU Press.

Ratzan, S. (1998). *Mad Cow Crisis: Health and the Public Good.* New York: NYU Press.

Right to Adequate Food. (2002). Retrieved May 1, 2009, from Food and Agriculture Organization of the United Nations: http://www.fao.org/righttofood/principles_en.htm

Sebestyen, G., Nedevschi, S., Saplacan, G., Cerghizan, M., Todor, N., & Rusu, M. Towards a Traceability Solution on the Food Supply Chain. *The 8th International Conference on Technical Informatics.* June 5-6, 2008. Timisoara, Romania.

The General Food Law Regulation (EC) 178/2002 (2002). *The European Parliament.* Retrieved April 10, 2009, from http://europa.eu/eur-lex/pri/en/oj/dat/2002/l_031/l_03120020201en00010024.pdf

Tolosa, J. B., García-Díaz, V., Sanjuan-Martínez, O., Fernández-Fernández, H., & García-Fernández, G. (2009). Towards Meta-Model Interoperability of Models and Transformation Models. *International Symposium on Distributed Computing, Artificial Intelligence, Bioinformatics, Soft Computing and Ambient Assisted Living*(LNCS 5518, pp.315-322) June 10-12, 2009 Heidelberg, Germany: Springer.

Users and Customers Organization. *OCU*. Retrieved June 14, 2009, from: http://www.ocu.org van der Linden, F.J. & Schmid, K. (2007). Software Product Lines in Action. *The Best Industrial Practice in Product Line Engineering*. New York: Springer.

Chapter 2
R&D Centre Growth Based on Regional Innovation System

Jingyuan Zhao
Harbin Institute of Technology, China

ABSTRACT

The chapter focuses on comparing and analyzing the development models of typical regional innovation systems in the world, discussing the relationship of regional innovation system and R&D centre growth, and points out that the emergence of R&D centre and growth are tight relative with the development of regional innovation system. Through researching on typical cases of India's Banglore, Singapore and Taiwan's Xinzhu, the paper summarizes the experience that establishing and perfecting regional innovation system will improve R&D centre growth. Using the experience for reference, some strategies to promote R&D centre are put forward as conclusions.

1. INTRODUCTION

Since 1990s, overseas research and development (R&D) of multinational enterprises (MNEs) is speedily developing and growing. The objective that MNEs establish local R&D organizations or cooperate R&D with local enterprises is to effectively utilize local science and technology resource or to support the production and marketing of local subsidiary companies under the background of global competitive environment with expeditious science and technology development and change (Chiesa, 1996; Gerybadze & Reger,1999). The academy filed has the trend to research on the subject more and more, IEEE Transactions on Engineering Management and Research Policy discussed deeply on MNEs' overseas R&D in 1996 and 1999 as special issues, which means people have attached more importance to the subject.

MNEs usually choose international cities with active innovation, such as Silicon Valley, Boston, Austin, Seattle in the US, Tokyo, Singapore, India's Banglore, China's Beijing, and so on in Asia. Following the trend of global innovation, in another new round of competition, the stronger competitive power emerges in Asia-Pacific region,

DOI: 10.4018/978-1-61692-846-9.ch002

Copyright © 2011, IGI Global. Copying or distributing in print or electronic forms without written permission of IGI Global is prohibited.

Tokyo, Singapore, Korea, Banglore, as well as China's Hong Kong, Taiwan, Beijing, Shanghai, Shenzhen give prominence to innovational economy, and more and more R&D centres are emerging.

Cooke, Morgan (1994) researched on the theory and cases of Regional Innovation System (RIS) early and completely. Cook, Braczyk and Heidenreich (1996) particularly discussed regional innovation system. Some researchers discussed the conception of regional innovation system from different points (Asheim, 1997; Cassiolato, Lastres, 1999; Carlsson, et al., 1999). Liu, Xu (2002) summarized the international development of regional innovation system theory.

According to the theory and conception of regional innovation system, the regional innovation system is chiefly comprised with enterprises as innovation main body, science and research institutes and universities as technology innovation resource, agency and service organizations as communication means and important tache of knowledge flow, government as exterior environment builder for enterprise innovations. Enterprises are the main body of regional innovation, science and research institutes and universities are the resource of knowledge innovation, agency and service organizations are the tache of regional innovation, governments are the constructers and managers for regional innovation activities.

Then, what is the internal relationship between R&D centre growth and the development of regional innovation system? Which signification and effect does regional innovation system have on R&D centre growth? The study will analyze and compare different development models of important regional innovation systems in the world, research on the relationship between the development of regional innovation system and the emergence and growth of R&D centre. Through researching on typical cases of India's Banglore, Singapore and Taiwan's Xinzhu, the study discusses on how to push R&D centre growth by means of establishing and improving regional innovation system.

2. COMPARATIVE ANALYSIS ON REGIONAL INNOVATION SYSTEM

Zhongguancun Science Park Administration Commission and Great Wall Enterprise Stratagem Academy compare and analysis on Zhongguancun's benchmark and objective with world first level parks in 2006, the study finds that regional innovation system in the world is established through different development in terms of value chain tache and station in global industrial division, R&D model, cooperation model between academy and enterprises, shown as Table 1. Silicon Valley and Zhongguancun pay equal attention to basic research and application research; Cambridge focuses on basic research; Xinzhu, Banglore, Sophia Antipolis and Ireland mostly focus on application research.

In most of regional innovation systems of R&D centre, enterprises cooperate with universities and research institutes, for example, in Silicon Valley, enterprises' research organizations are united with universities and academies, in which Stanford is the delegate. Silicon Valley is the mirror of cooperation between enterprises and academy. Another example is Cambridge. Cambridge University and other universities play an extremely important role in technology innovation activities, and build a tight net of cooperation between academy and enterprises.

By comparing and researching on main regional innovation systems in the world, the study concludes that main value chain tache and industry division are different in different regional innovation systems, the trend of development, R&D model and cooperation model are different from each other, accordingly emerging R&D centre with different characteristic. Although there are different models of regional innovation systems,

Table 1. Comparing main regional innovation systems in the world

Name	Value chain tache and station in global industrial division	R&D model	Cooperation model between academy and enterprises
Silicon Valley	Priority on R&D, design and high-tech service; high-point station in global industrial chain, with advantage ofbrand and R&D	Pay equal attention to basic research, application research and experimentation; focus on original innovation.	Composed with university and research organizations in which Standford is the delegate, and enterprises' research organizations; the mirror of cooperation between enterprises and academy.
Xinzhu	IT R&D and manufacture; priority on Subcontracting	Focus on product exploitation of application technology, integration innovation and second innovation.	Scientific research institutes, particularly industry technology academe, pay a distinctly role in technology innovation
Banglore	R&D and coding of software; epiboly of software; software machining and expert	Priority on application software exploitation.	Tightly cooperation among India Institute technology and other university, scientific research institutes, enterprise and agency organization.
Cambridge	Priority on R&D	Priority on basic research and pay attention on transforming research conclusion.	Cambridge university and other universities play a key role in technology innovation, forming tight cooperative net of enterprises and academy.
Sophia Antipolis	Priority on innovation and R&D; superior station in value chain; comparatively lag on manufacture, Logistics, marketing	Facing application	Enterprises tightly cooperative with universities and scientific research organization, but lack tight relationship between enterprise.
Ireland	priority on software epiboly and Subcontracting	Focus on basic research.	Universities, scientific research institutes and enterprises are tightly cooperated.
Zhongguancun	Priority on R&D and service; advance level of industrial technology in China, but big gap from world advance level, focus on second exploitation of technology.	Focus on application software; pay equal attention to basic research and application research.	Universities, scientific research institutes and enterprises are tightly cooperated; academy organizations are the main force of technology; the station of enterprises as innovation main body is increased.

it can be seen that the emergence and growth of R&D centre are tightly relative with the development of regional innovation systems.

3. R&D CENTRE GROWTH CASES

MNEs that still give emphases of R&D internationalization on developed countries have leaned to developing countries and mid-developed countries, particularly Asian area. Some Asian countries are trying to attract MNEs to establish R&D organizations locally. China, India and Singapore give prominence to attract MNEs' R&D (Chen, 2005). According UN data, from 2005 to 2009, China is the most attractive base for MNEs' oversea R&D, which is on primacy, India is the third place, the next Asian countries/regions are Singapore, China's Taiwan, Malaysia, South Korea, Thailand, shown as Figure 1.

The study chooses India's Banglore, Singapore and China's Taiwan Xinzhu as typical cases to discuss the impact of regional innovation system development on R&D centre growth.

Figure 1. 2005-2009 Attractive R&D bases for MNEs (Resource: UN, 2005)

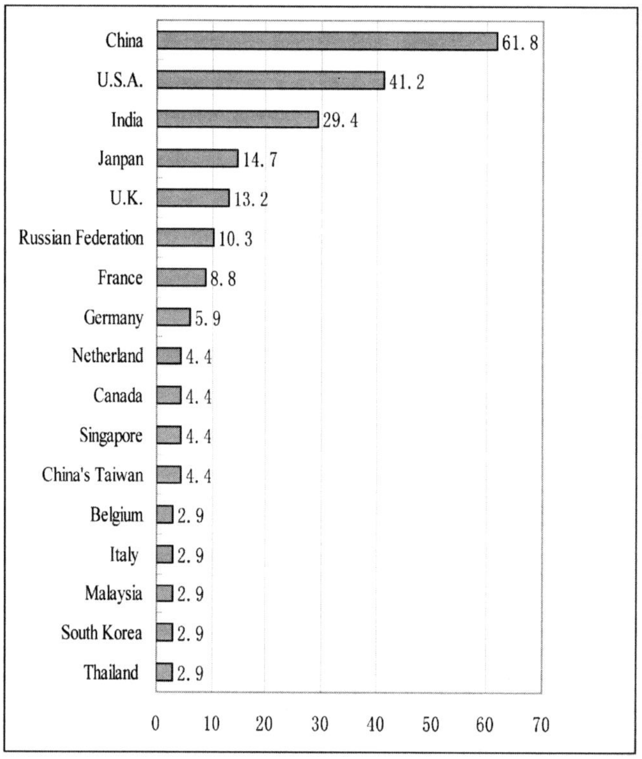

3.1 India's Banglore

In recent years India's Banglore attracts largely MNEs, particularly R&D investment of software industry, which is called Asian Silicon Valley. In 1991 India established the national level software technology park, which is Banglore. R&D organizations in Banglore have two characteristic: mainly focus on R&D of software industry; mother companies belong to America and Europe. The policy route of India is to enter the global innovation net by cooperation between local R&D organizations with MNEs R&D organizations, so as to improve internationalization level of local innovation, and make Banglore to be international R&D centre. We will discuss Banglore R&D centre from the point of view of technology innovation system, R&D fund, manpower resource and culture.

3.1.1 Technology Innovation System

By development for more than ten years, Banglore has built perfect science and technology R&D system, which centre is enterprises. There are many excellent software companies, there into, software companies with CMM5 (Capability Maturity Model 5) authentication are more than 30. Many software companies converge to bring cluster effect, which makes each individual in cluster renewedly optimize collocation for their Talents, science research fruits, funds and so on, so as to improve entire innovation ability. Banglore congregates many scientific research organizations, and has seven famous universities, such as India Technology Institute, several hundreds of high-level expert schools and vocational schools.

Banglore's IT industry cluster in software park forms a system that congregates fund embaying knowledge, organizes product element embay-

Figure 2. Bangalore technological innovation system

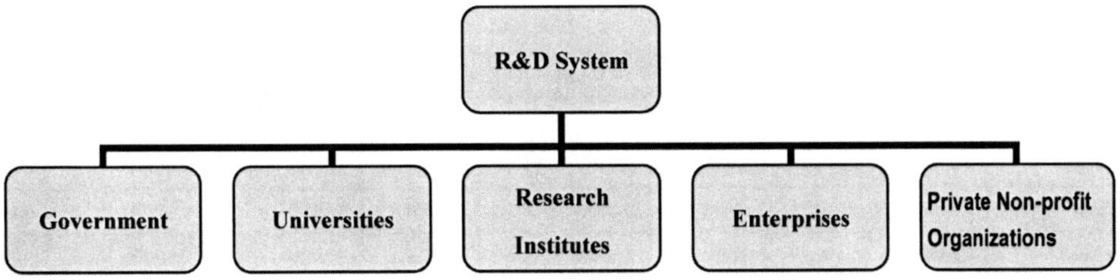

ing technology. In the system, central and local governments, venture capitalists, special technology personnel, as well as R&D organizations, periphery service organizations offer mechanism assurance and environment conditions. Enterprises that lie in core station continually create and invent to discover new technology, so as to impulse renovation and upgrade of entire cluster system. Personnel flow also results in technology spill, shown as Figure 2.

3.1.2 R&D Devotion and Policy Support

Indian government gradually increases R&D fund investment, and offers policy support. 85% R&D fund of science and technology are from central and local government. During the Eighth Five-Years Planning, science and technology R&D fund from all levels of government reaches 200,000 millions ruble, 1000 times than the first five-years planning.

For raising R&D fund, governments take policies as following: increase financial expenditure as scientific research funds, encourage cooperative innovation between scientific research organizations and enterprises, promote fruits of scientific research to realize industrialization and commercialization; take some special financial organizations as the centre, such as India Industry Exploiture Bank, India Technology Exploiture and Information Corporation, India Venture Capital and Technology Finance Corporation, to support financially high-new technology for commercialization and industrialization; constitute a serious of favorable policies; absorb energetically overseas Indian scientists (the main is American-Indian scientists), overseas Indian, as well as venture fund from multinational nation venture companies to IT, software technology and other high-new technology industry.

3.1.3 Manpower Resource

Banglore has perfect IT education and training system. Gai, Zhang and Lu(2004) discussed that Banglore's software talents be far more than any other city in Asia, there be some excellent technology and management organizations, 77 engineering institutes, giving 30 thousands of engineering technology talents, there into first third for IT persons. Furthermore, Banglore's evident advantage of software talents profits from a serious talents policies, which mainly consist that Science and Industry Committee and Industry Research Committee establish science and technology talents repertory; give various special treatments for homecoming scientists; implement policies of respecting knowledge and talents; create favorable working environment and support superior treatment; attract excellent science and technology talents back to homeland, in addition, make various rule to limit talents away overseas(Fu, 1993). Besides, talent cost in Banglore is very low, which attracts MNEs to establish R&D organizations here. Comparing with other countries with developed software industry in the world, such

Table 2. Wage total of software personnel in countries with developed software industry Unit: USD/ yearly salary

	Switzerland	USA	Canada	UK	Ireland	Greece	India
Project Manager	74000	54000	39000	39000	43000	24000	**23000**
Business Analyst	74000	38000	36000	37000	36000	28000	**21000**
System Analyst	74000	48000	32000	34000	36000	15000	**14000**
System Designer	67000	55000	36000	34000	31000	15000	**11000**
Project Developer	56000	41000	29000	29000	21000	13000	**8000**

Recourse: Beijing Committee on Science and Technology (2006)

as America, Switzerland, Canada, Ireland, and so on, India's wage of software industry personnel is the lowest, shown as table 2.

3.1.4 Culture

Culture of rule, cooperation and faith is Banglore's criterion for enterprises. Criterion is a kind of society capital, it can be explained to be derivable rule, formula and system in industry cluster. The enterprises in Banglore cluster are not dispersed, but a cooperative net under the background of faith. The faith is a bond that breeds relationship between interior organizations in cluster. The industry association in Banglore assures the establishment of credit mechanism, if some one enterprise destroy industry criterion, disobey promises, break faith easily, the enterprise will be strictly punished and collectively judged by industry association.

Put above together, the development of Banglore profits from following points: building technology innovation system; investing largely to R&D and support raising R&D fund; actualizing a serous of talents policies; establishing cultural environment of rule, cooperation and faith. Those experiences are worthy to use for reference.

3.2 Singapore

Among developing countries, Singapore is one of countries that attracts many R&D investment of MNEs (Amsden, Tschang, et. al, 2001).Singapore's advantage attracting R&D investment lies in internationalization level and integrated city management level. Besides, Singapore offers perfect foundation establishment and service system to create a attractive operating environment for investors. On the other hand, various encouraging policies make Singapore become MNEs' R&D centre for local area, even for the globe. In 2000, foreign companies' R&D expenditure accounted for 35.9% from Singapore R&D expenditure total, it was infrequent in the world (Du, 2003). From the point of view of development programming of science and technology, building R&D institutes, manpower resource and knowledge property right protection, we will analyze the experiences of Singapore R&D centre growth.

3.2.1 Programming of Science and Technology Development

According to national economical development level and MNEs' globalization status, Singapore continually adjusts science and technology programming and industry policy in order to adapt the need of science and technology advancement and industry upgrade. In the age of 1990, National Science and Technology Board constituted successively two national science and technology development programming: NTP (National Technology Plan: 1992-1995) and NSTP (National Science and Technology Plan: 1996-2000), which established the foundation for Singapore's science and technology development. In the programming, Singapore government promoted civilian organization's R&D; built efficient technology transfer and management system of knowledge property right; dig world extractive talents and trained local talents; exploited international cooperative net; 50% fund for government organizations' R&D, 30% fund for civilian organizations' R&D, 20% fund for talents exploiture. The two programming played an important role for supporting R&D in Singapore's regional innovation system.

3.2.2 Establishment of R&D Academy

Singapore attaches importance to build R&D academy, and makes great efforts to impel these R&D organizations to cooperate with MNEs, and offer R&D service to MNEs in order to attract MNEs' R&D and promote them upgrade. In Singapore's regional innovation system, high-level schools and enterprises pay attentions to basic research and experiment exploitation, science and technology academies mainly take hands to application research. For attracting MNEs' R&D investment, Singapore government energetically encourage and enhance R&D academies to cooperate with MNEs' R&D organizations, so as to increase science and technology academies' service for MNEs.

3.2.3 Manpower Resource

Enough talents with high education are necessary condition to attract MNEs' R&D. As a very small economy body, Singapore with only 3,200 thousands people always faces the puzzle of lacking high-level team. Therefore, Singapore government still put emphasis on exploiting manpower resource, and vigorously attracting international talents in past twenty years, Singapore has become one of minority countries that really implement entire opening manpower policy. For satisfying the need of R&D talents, Singapore adopts many policies, for example, enhancing training talents in existence so as to solve the problem of technology personnel lack; take loose immigrant policy to attract overseas talents.

3.2.4 Protecting Knowledge Property Right

Protecting effectively knowledge property is one of reasons why MNEs choose Singapore to do R&D for their products or to establish local science centre. Singapore pays attention to evaluating and protecting science fruits, and help exploiters or organizations to get patents. For example, Philip denoted that the reason why Philip cooperated with government lab be government wouldn't come down one knowledge property right (Amsden, Tschang, et al., 2001).

3.3 Taiwan's Xinzhu

Among many developing countries and area, China's Taiwan is such a location, which makes MNEs' R&D organizations deeply think a lot. Zhong (2001) discussed that the establishment of Xinzhu Science Park offered an industry congregation space for Taiwan to attract outside high-tech industry investment and transfer technology. By the development in past twenty years, Xinzhu gradually emerged six support industries, integrate circuit, computer and assistant equipment, communication, photoelectricity, exactitude engine, biology technology. We will analyze Taiwan's Xinzhu R&D centre growth from the point of view of technology innovation, policy support and manpower resource.

3.3.1 Technology Innovation

Xinzhu's technology innovation is the important element for R&D centre growth. Xinzhu's R&D investment proportion is high, so as to promote industry's technology content and product's international completive power. There are many technology patents in Xinzhu, which is one of

Science Parks of owning patents densely and most for per person. In 2003, Xinzhu had 3026 patent authorizations, increased 12.6% than last year. The average of per enterprise owning patents is 8, per thousand persons owning patents is 29.7 per 100 million R&D investment will come out 173 patents, which is higher than Silicon Valley. Science academies, particularly Industry Technology Institute plays a key role in Xinzhu's technology innovation. By science academies and institutes' cooperation and participation, Xinzhu's innovational power is rapidly increased, so as to drive entire level of Taiwan up.

3.3.2 Science and Technology Policy Support

Science and technology policy support plays an important role that should not be ignored. By inspiring policies, such as science and technology awards and funds support, Xinzhu encourage enterprises to invest to R&D, so as to impel Xinzhu to become one of R&D centres in the world, which innovations most congregate.

Taiwan Government setup various science and technology awards and funds to inspire enterprises in park to increase R&D investment and product innovation. In 2004, administration office of park imburse 36 enterprises in all, 39 projects, fund amount is 106 million new TAIB, account for 22.2% of project general cost. The innovation technology R&D planning award has set up for 10 years, and has become important resource for science and technology product to advance their completive power. For pushing high-tech industry to develop, government combines banks to offer lover interest loan for enterprises and establish R&D awards.

3.3.3 Manpower Resource

The characteristic of Xinzhu's manpower resource is that the personnel number is steadily creasing and the proportion of high educational level is high. In 2004, the personnel with Ph.d is 1.7%, with master degree is 28.4%, with bachelor degree is 36.1%, total account for 66.1% of entire personnel.

The advantage of Xinzhu's manpower resource profits from Taiwan's emphases on using and training local talents, and cultivate a great deal science and technology personnel with better R&D ability and skilled worker with high quality. In the mean time, Xinzhu thinks much of attracting international R&D talents to advance Taiwan's level of R&D talents and degree of internationalization. Taiwan's economy growth, talent policy finally stimulate Brain Channel to flow back. Taiwan's case shows that some one area wants to be one part of global technology net, at least should improve technology education and train, so as to create support foundation for enterprises, and build multinational nation relationship of society and technology, which will attract overseas investment. Taiwan's talents policy makes Xinzhu establish a relationship system with Silicon Valley, which is an important element that Xinzhu is speedily developed to be R&D centre, and worthy to use for reference.

4. CONCLUSION

By comparing cases of regional innovation system, the paper discusses the characteristic of regional innovation systems, and draws the conclusion that different models of regional innovation system emerge different models of R&D centre, the growth of R&D centre is tightly relative with regional innovation system, the development of regional innovation system promotes R&D centre growth. Therefore, we can develop regional innovation system to promote R&D centre growth.

Choosing India's Banglore, Singapore and Taiwan's Xinzhu as typical cases with much attraction for MNEs' overseas R&D, we more research on their regional innovation systems, consequently get the conclusion to develop regional innovation system to promote R&D centre growth. Some experiences are worthy to use for reference as following:

- Government's science and technology planning, policy and mechanism are the guidance and guarantee for R&D centre growth.
- Improving technology innovation and emerging perfect technology innovation system will push R&D centre growth.
- Talents are the first resource of R&D centre development, talents policy is the key of R&D centre growth.
- Culture of standard, cooperation and faith offers favorable environment for R&D centre growth.

REFERENCES

Asheim, B. T., & Isaksen, A. (1997). Location, Agglomeration and Innovation: Towards Regional Innovation Systems in Norway? *European Planning Studies*, *5*(3), 299–330.

Carlsson, B., Jacobsson, S., Holmen, M., & Rickne, A. (2002). Innovation Systems: Analytical and Methodological Issues. *Research Policy*, *2*, 233–245. doi:10.1016/S0048-7333(01)00138-X

Cassiolato, J. E., & Lastres, H. M. (1999). *National and Regional Systems of Innovation in the Mercosur*. Prepared for the DRUID Summer Conferences on National Innovation.

Chen, X. (2005) R&D Internationalization and Two-Shore's High-tech Industry Competition and Cooperation. *Report of China Economy Institute*. Retrieved from http://www.docin.com/p-13289093.html

Chiesa, V. (1996). Managing the Internationalization of R&D Activities. *IEEE Transactions on Engineering Management*, (5): 7–23. doi:10.1109/17.491264

Cooke, P., & Braczyk, H. J. (1996). *Regional Innovation Systems: the Role of Government in the Globalized World* (Heidenreich, M., Ed.). London: UCL Press.

Cooke, P., & Morgan, K. (1994) The creative Milieu: A regional Perspective on Innovation. In M. Dodgson & R. Rothwell(Eds.). *The Handbook of Innovation*, 25-32.

Fu, D. (1993). *Analysis on Foreign High Technology and Its Industry Policy*. Bejing: Earthquake Press. Amsden, A., Tschang, T., et al. (2001). Do *Foreign Companies Conduct R&D in Developing Countries?* Asian Development Bank Institute Working Paper. Du, D., et al. (2003). *Internationalization of MNEs' R&D and reply strategy research*. The Subject Report of Shanghai Municipal Science & Technology Commission.

Gai, W., Zhang, H., & Lu, W. (2004). Compare and Analysis on International Typical High-Tech Industry Cluster and Experience Inspire. *China Soft Science*, *2*, 102–108.

Gerybadze, A., & Reger, G. (1999). Globalization of R&D: Recent Changes in Management of Innovation in Transnational Corporations. *Research Policy*, *28*(2/3), 251–274. doi:10.1016/S0048-7333(98)00111-5

Liu, S., & Xu, S. (2002). Research Summarization on International Development of Regional Innovation System. *China Science and Technology Forum*, *5*, 33–37.

Zhong, J. (2001). *System Analysis on World Silicon Models*. Bejing: China Society Science Press.

Chapter 3
The Regional Innovation Capacity Model

Antonio Lerro
University of Basilicata, Italy

Giovanni Schiuma
University of Basilicata, Italy

ABSTRACT

This chapter aims to present a conceptual model aimed to understand the Intellectual Capital-based (IC) characteristics of the regional innovation capacity. The proposed Regional Innovation Capacity Model (RICM) can be used for interpretative and normative purposes to analyse the innovation dynamics taking place at regional and territorial level. From an interpretative point of view, the model identifies the pillars grounding the innovation capacity of a local system. While, from a normative perspective, the model can inspire the definition of guidelines driving the design and the implementation of actions, projects and programmes aimed to stimulate and sustain regional development dynamics. The RICM adopts a knowledge-based perspective assuming that IC, in the forms of regional knowledge assets, and knowledge dynamics, in the form of knowledge transfer and learning processes, are the drivers of innovative processes and outputs. The chapter concludes proposing a future research agenda.

INTRODUCTION

The emergence of the knowledge economy, intense global competition and considerable technological advance has seen innovation becoming increasingly central to regional development (Cooke, 2008). As regional systems become increasingly focused on innovation, they must systematically invest and nurture innovation capabilities, from which they execute effective innovation processes, leading to innovations in new regional processes and sustainable development results (Chaminade & Vang, 2008). In this context, knowledge resources represent fundamental drivers to best enhance and support innovation dynamics, the maintenance of competitive strength and ultimately regional value creation (Asheim & Coenen, 2005).

Therefore, managers and policy-makers need to understand how knowledge resources are linked and affect innovation and regional development.

DOI: 10.4018/978-1-61692-846-9.ch003

Copyright © 2011, IGI Global. Copying or distributing in print or electronic forms without written permission of IGI Global is prohibited.

This involves the need to define and use models, approaches and tools able both to improve the understanding about the region's innovation capacity and to provide useful insights for maintaining and developing this crucial ability. From a research perspective, the challenge for managerial and regional science research lies in formulating and extending theoretically grounded models and empirical applications in order to develop more rigorous conceptualisations, useful for regional innovation management.

Although some theoretical contributions have stressed the strategic importance as well as the role of the knowledge resources as key value-drivers for regional systems' innovation dynamics and territorial excellence, there is still a need for a better understanding of the approaches for the identification, development and deployment of the knowledge resources for the improvement of innovation performance.

The main objective of this chapter is to present a conceptual model aimed to understand the Intellectual Capital-based (IC) characteristics of the regional innovation capacity. The proposed Regional Innovation Capacity Model (RICM) can be used for interpretative and normative purposes to analyse the innovation dynamics taking place at regional and territorial level. From an interpretative point of view, the model identifies the pillars grounding the innovation capacity of a local system, such as the innovation sources, innovation capacity, innovation processes and innovation results. While, from a normative perspective, the model can inspire the definition of guidelines driving the design and implementation of actions, projects and programmes aimed to stimulate and sustain regional development dynamics. The RICM adopts a knowledge-based perspective assuming that IC, in the forms of regional knowledge assets, and knowledge dynamics, in the form of knowledge transfer and learning processes, are the drivers of innovative processes and outputs. Specifically, we adopt the notion of Intellectual Capital (IC) as umbrella concept to identify the main knowledge assets affecting innovation and regional development dynamics (Carlucci & Schiuma, 2007; Bounfour & Edvinsson, eds., 2005; Lerro & Carlucci, 2007; Bradley, 1997). Knowledge assets are those critical and strategic knowledge resources affecting the performance and the value creation dynamics of a specific territorial system (Teece, 1998; Marr and Schiuma, 2001).

Following, the Regional Innovation Capacity Model (RICM) is presented. Then, key results and research and policy implications are discussed. Finally, limitations and suggestions for future research are provided.

THE REGIONAL INNOVATION CAPACITY MODEL

Innovation and Regions: Conceptual Foundations

Over the past two decades, social scientists and policy-makers have been paying more and more attention to regions as designated sites of innovation and competitiveness in the globalising economy (Chaminade & Vang, 2008; Gu & Lundvall, 2006; Vang & Asheim, 2006). The popularity of this issue can be traced back to various empirical studies of regional success stories, such as the fast economic growth of networked SMEs in the Italian industrial districts (Albino & Schiuma, 2003; Piore & Sabel, 1984), the industrial system of Silicon Valley (Saxenian, 1994) as well as other examples of successful regional clustering in most developed and developing economies (Porter, 1990). These studies all drawn on the common rationale that territorial agglomeration provides the best context for an innovation-based globalising economy because of localized learning processes and "sticky" knowledge grounded in social interaction (Asheim & Coenen, 2005; Asheim, 2002; Asheim and Isaksen, 2002; Gertler, 2004). They have emphasized the significance of the regional

Figure 1.

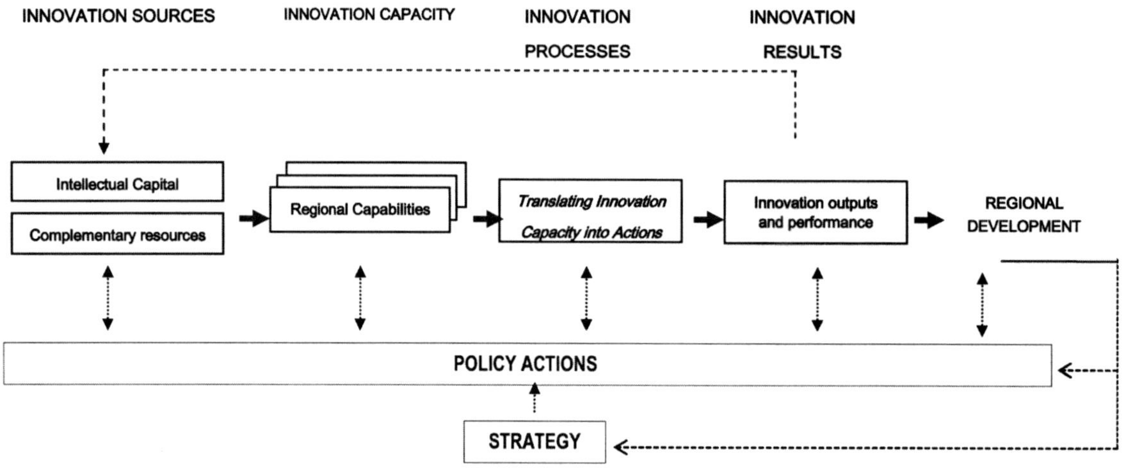

level in economic development in addition and sometimes over the national level (Dolfsma, 2008; Carrillo, 2006; Bounfour and Edvinsson, eds., 2005; Budd and Hirmis, 2004; Dakhli and De Clercq, 2004; European Commission, 2005; 1999; Gardiner et al., 2004; Huggins and Izushi, 2007; Kitson et al., 2004; Pike et al., 2006).

However, although in the last years many studies have investigated the knowledge factors grounding innovation at local level (Sanchez Medina et al., 2007; Bontis, 2004; Pasher & Shachar, 2005; Pulic, 2005; Smedlund & Poyhonen, 2005), there is still a need for the definition of the theory pillars of a knowledge-based perspective of innovation dynamics at regional level. In particular, there have been few attempts using a knowledge-based approach to generate an effective conceptual model of regional innovation capacity. Moreover, despite this 'innovation fad' at regional level, many region's decision-makers frequently embark on policies, plans, programmes, actions to sustain their own region's innovation dynamics without a clear framework explaining the dynamics grounding the innovation capacity of a region.

In order to answer to these needs, the Regional Innovation Capacity Model (RICM) is proposed (Figure 1). Taking and expanding upon previous categorizations and drawing upon the various approaches in innovation research, we provide the RICM as a conceptual model adopting a knowledge-based view of the regional innovation capacity. Our view of the RICM comprises four main pillars – innovation sources, innovation capacity, innovation processes and innovation results. Specifically, RICM models the recursive process through which regional systems source the knowledge they need to undertake innovation, transform this knowledge into an innovation capacity driving innovation processes and then exploit innovation results to generate growth and development. The effectiveness of the regional systems in translating innovation sources into innovation results driving regional value creation dynamics is also linked to the strategy of development as well as to the policy actions defined and implemented by the regional governments. The RICM pillars are deeply analysed in the following sub-sections.

Innovation Sources

Political and institutional changes and economic restructuring at global level have forced the regional systems to review the sources of their innovation dynamics, competitiveness and development patterns. Both traditional assets (e.g., infrastructures, financial capital, raw materials) and knowledge resources (people, relationships, practices) play an important role in creating value for the regional systems. However, in the last decades, strategic approaches such as the Resource-based View, the Competence-based View and the Knowledge-based View have largely stressed that competitiveness is increasingly based on the possession and development of intangible and knowledge resources. This involves that, although traditional assets still have a critical role, the regional policies have increasingly to take into account the knowledge-base dimensions of their development if they want to prosper in today's knowledge economy (Lerro & Schiuma, 2008).

Huggins and Izushi (2007) suggest that a region's superior development depends on its ability to defend and use the knowledge resources and assets it creates. According to Sanchez Medina et al. (2007, p.474), "this type of resources will be one of the most important factors for social and economic development. It will be the countries that are better equipped with (intellectual) capital that will make most progress". Drawing on these insights, according to which the knowledge assets have strategic importance for the innovation dynamics and development of the regional systems, in the following the knowledge assets categories characterizing Intellectual Capital of a region are discussed: human capital, relational capital, structural capital and social capital (Schiuma et al., 2007; Schiuma & Lerro, 2008). The human capital comprises the know-how characterizing the different actors operating within the region. Know-how involves knowledge that is tacit, complex, causally ambiguous and difficult to codify as well as the knowledge mainly held and exercised by individuals and 'collectively' by the different regional stakeholders. Thus, it can be argued that know-how is the basis of capabilities, whether conceptualized as routines or specific regional-level activities. Given the above discussion, human capital may be best understood as those factors that are built upon or are reflective of know-how, both tacit and explicit, which individuals and stakeholders posses and exercise. In some cases, the know-how may reside in the individuals; in other cases, the know-how may be reflective of region's stakeholders. The relational capital indicates the group of the knowledge resources linked to the relationships characterizing the regional systems. This includes fundamentally relationships established and maintained among regional stakeholders driving innovation dynamics at regional level. It represents a collective effort of the know-how of many stakeholders in a variety of contexts and situations. The structural capital includes all those assets tangible in nature relevant for the development, acquisition, management and diffusion of knowledge as well as all the components linked to structural features of a region. Within this category it is possible to consider two sub-categories: the structural tangible assets and the structural intangible assets. The first include all region's infrastructure which can be tangible which support knowledge development and management. The second comprise basically intellectual property assets largely derived from the intellectual capacity of human know-how. The social capital comprises the assets having soft nature and contributing order, stability and quality to a region, such as regional culture, history, attitudes, norms, values, behaviours, image, and other cultural dimensions characterizing the regional systems. These particular regional assets provide the strength and cohesion between 'high-order' resources, such as know-how, and other tangible and intangible resources. Indeed, such assets are an important link between what the region does and how it does it. Besides knowledge resources, regions have to exploit also complementary resources which can be defined 'in negative' as the resources which are not made or incorporating knowledge (Teece et al., 1997) such as financial assets, cash, raised financial capital, financial investments, buildings, land and other regions' properties unable to incorporate or provide knowledge. (Dougherty & Hardy, 1996).

Innovation Capacity

The concept of innovation capacity is recently emerged in the academic and policy debate as a meta-concept to denote the real and potential capabilities of a system to convert knowledge into innovation able to drive long term economic growth and wealth creation (Freeman, 1995; Furman et al., 2002; Lundvall & Johnson, 1994; Nelson, 1993). It has been introduced and adopted by different scholars interested to investigate and understand the factors and the determinants at the basis of the innovation dynamics and capabilities grounding regional and local development (Furman et al., 2002; Howells, 2005; Tura & Harmaakorpi, 2005).

Cooke and Memedovic (2003, p.8) highlight that "there is a growing awareness among regional authorities that the economic growth and competitiveness of their regions depend largely on the capacity of indigenous firms to innovate. Offering the appropriate support to indigenous firms to become more competitive through innovation is a rising star on the regional policy agenda". According to this issues, Archibugi and Michie (1995, p. 1) state that "the production and use of knowledge is at the core of value-added activities, and innovation is at the core of firms and nations' strategies for growth". Tura and Harmaakorpi (2005), reviewing a number of theoretical frameworks and notions about regional innovation, have highlighted the importance of building and enhancing regional innovativeness by developing an environment characterized by networked systems, such as regional innovation systems (Cooke, 2008; Cooke et al., 1997; Doloreux, 2002), innovative milieu (Aydalot & Keeble, 1988; Camagni, 1991; Crevoisier & Maillat, 1991), industrial districts (Marshall, 1916; Piore & Sabel, 1984; Becattini, 1990), new industrial spaces (Storper and Scott, 1992; Storper, 1995), and learning regions (Florida, 1995; Asheim, 1996). While, from a policy perspective, the European Commission (2005) has underlined the relevance of the regional context to develop innovation capabilities. In particular, it has been argued that innovation requires long-term cooperation between investors, entrepreneurs, researchers, firms, public authorities and consumers. Such relations flourish more easily at regional level, enjoying benefits from short distances that facilitate both formal and informal contacts. Regional-based cooperation networks are ideal knowledge sources and entry points to exchange information and to set up new ideas (DTI, 1998).

Furman et al. (2002, p. 899) specifically refer to the innovation capacity as "the ability of a geographical area to produce and commercialize a flow of innovative technology over the long term" and underline that it depends on the strength of the common innovation infrastructure, on the environment for innovation and on the strength of linkages between these two. Rondè and Hussler (2005) have highlighted that a regional innovation capacity requires also to deeply understand the nature of the competences and capabilities that have to be mastered in order to be innovative. In accordance with Teece and Pisano (1998) an "innovative capability" is defined as an actor's ability to sense the changes in the environment and to be able to exploit existing resources and competencies in order to create competitive advantage by innovation activities.

We adopt the concept of "regional innovation capacity" to refer to the ability of any regional stakeholders to deal with changes both in accordance to an adaptive and pro-active behaviour, in order to enhance innovation dynamics taking place at regional level driving regional development.

On the basis of the literature about regional innovation systems (Cooke, 2004; Cooke & Schienstock, 2000; Lundvall & Johnson, 1994; Moulaert & Sekia, 2003), it seems possible to identify three main factors affecting the regional innovation capacity, these are: (1) the regional stakeholders, i.e. the actors operating at local level, (2) the networking, i.e. the synergistic relationships linking the stakeholders both among them, within a region, and with external innovation players, and (3) the local context, i.e. the regional space and the related resources in which the stakeholders and their relationships take place. They are interrelated and integrated dimensions which define the innovation capacity of a region.

Regional stakeholders play a fundamental role as "bridges" that allow to reach external and complementary knowledge sources in order to cover 'structural holes' (Burt, 1992).

Chaminade and Vang (2008) underline the importance of two main kinds of actors. The first type of actors are the companies, including their customers and suppliers. The second type of actors include research and higher education institutes – universities, R&D institute – technology transfer agencies, vocational training organizations, business associations, finance institutions (Asheim & Coenen, 2005). Regional stakeholders since are generally specialized on one knowledge area, rarely have all the required heterogeneous resources to successful innovate. Therefore, they need to create networks in order to acquire knowledge sources from the local and the external environment, activating knowledge spillovers (Audretsch & Feldman, 1996; Camagni, 1991; Cooke, 2001; Katz & Kahn, 1996; Maillat, 1995; Owen-Smith & Powell, 2004). In order to have access to complementary knowledge resources and capabilities for improving skills, competencies and routines McEvily and Zaher (1999) highlight the importance of establishing non redundant relationships as well as of involving the regional institutions in the networks. The relationships taking place at regional level spur and sustain localized regional learning dynamics which tend to idiosyncratically characterize the local context. Consequently, innovation assumes the traits of a locally embedded process taking place within a regional innovation environment. This environment mainly consists of innovation networks (Cooke & Wills, 1999) aiming to increase the innovation capability of a regional system. These networks may take different forms defined by, for example, the origin, size, structure and objective of the networks (Morgan & Neuwelaer, 1998).

Innovation Processes

Traditionally, innovation process research addresses how and why innovations emerge and grow to drive regions towards sustainable development. In our model, process is interpreted as a path-dependent phenomenon (King, 1992; Koput, 1997; Schroeder et al., 1989), i.e. a collection of tasks and/or activities and an integration and exploitation of capabilities which together transform inputs into outputs. Innovation is clearly not just about technical research and development, nor it is something that can be successfully performed in innovation sectors. Rather, it pervades all aspects of a region's existence, from the core value system to the measures and behaviors that are manifested on a daily basis. Specifically, our interest is the processes through which regions source, transform and exploit new and potentially preexisting knowledge through innovation. These region-specific processes can be seen as part of a broader evolutionary dynamics in which tangible and intangible resources are steadily refined and occasionally transformed.

The RICM provides great importance to two main innovation processes at regional level: the *regional learning dynamics* and the *knowledge transformation*.

The knowledge transformation process involves the dynamics of the transformation of the stakeholders' knowledge sources into innovation outputs. The effectiveness of a region's knowledge transformation is affected by various factors, such as enterprises characteristics, the strength of the firm's resources-base, financial, managerial and organizational capabilities of the firms as well as of the public administrations (Griliches, 1992; Love & Roper, 1999).

The regional learning dynamics emphasise the importance of the wider knowledge eco-system as well as the potential benefits of operating in an environment characterised by rich knowledge sources and extensive networking opportunities (Albino & Schiuma, 2003; Albino et al., 1998; Iansiti & Levien, 2004). Regional learning dynamics can be defined as the acquisition of knowledge and competences through interactive collaboration with firms and knowledge providers. It is considered a function of the soft infrastructure – increased qualifications of the human resources, organizational and social capital and so on (Chaminade & Vang, 2006; 2008). In particular, regional learning benefits from physical proximity (Albino and Schiuma, 2003; Morgan, 2004). Physical proximity implies that the firms are embedded in the same institutional setting and thus share certain norms, conventions and mindsets. Hence, a shared institutional setting facilitates the transmission of complex tacit knowledge needed for innovation. Regional learning dynamics and thus innovation will most probably take place when both IC is in place (Chaminade & Vang, 2008). Furthermore, physical proximity might in some cases support the emergence of social capital, which sets the ground for regional learning. Regional learning dynamics also take place when firms collaborate with universities and research centres and other institutions like business services, entrepreneurial associations and so on.

Universities represent the engines of growth, particularly at regional level. Yet, despite the large amount of research on university-industry interactions (Mansfield, 1991; 1998; Salter & Martin, 2001), the direct effects of the universities and research centres on the innovative performance of firms has proved difficult to trace (Larsen and Salter, 2004). Universities play a double role as the providers of qualified human resources and as the providers of R&D. As R&D providers, universities and research centres are believed to be especially relevant for certain sectors (Pavitt, 1984), such as the pharmaceutical industry (Mansfield, 1995) or biotechnology (Asheim & Coenen, 2005) while appearing to be almost irrelevant for some others such as furniture (Klevorick et al., 1995; Larsen & Salter, 2004).

Innovation Results and Regional Development

In terms of innovation results, the RICM emphasises the importance of examining both regional innovation outputs and outcomes. Knowledge from different sources and different processes may have substantial differential effects. This suggests the possibility of different routes through which knowledge of different types might influence various aspects of regions' innovation activity and hence sustainable regional development. Specifically, regional innovation outputs is about knowledge exploitation, i.e. the process by which regional performance are influenced by innovation (Gerosky et al., 1993). At this point, we envisage that regions' acquired knowledge has been codified into specific output variables defined by policy-makers together with key-stakeholders. These defined variables might drive an advancement of the sustainable regional development patterns. The strength of this linkage, however, will depend on regions' ability to appropriate the full rent from their innovations.

Finally, the ability to appropriate innovation returns has to drive sustainable regional development patterns. The issue of the regional development has been usually analysed in terms of regional competitiveness, and as the result of economic value creation (Behrens and Thisse, 2007; Brooksbank and Pickernell, 1999; European Commission, 1999; Porter, 1990). Nowadays, this interpretation is enriched and integrated according to a wider view of the value dimensions of regional development (Pike et al., 2006; Sanchez Medina et al., 2007). It implies the integration of the economic, social, and environmental dimensions, and requires that economic growth supports cognitive and social progress and respects the environment, that social policy underpins economic performance and that environmental policy is cost-effective. In turn, value creation capacity is strongly based on the ways as well as the effectiveness according to which the region's resources and assets nurture regional innovation capacity. It determines that decision-makers should systematically identify, manage and exploit the spectrum of resources and assets really or potentially grounding innovation and regional development. In particular, they have to identify what kind of resources does the region needs to put in place to allow it to operate its processes now and in the future, and how these resources, separately and interactively, enhance region value delivering.

Strategy

In a globalising economy, regional development and competitive advantage are based on the exploitation of unique competencies and resources driving innovation dynamics. A strategic perspective should focus on how to develop such unique competencies and resources in order to foster innovation and competitiveness (Porter, 1990). Strategy determines the configuration of resources, products, processes and systems that regions adopt to deal with the uncertainty existing in the today's economic scenario. It requires that regions make decisions about what businesses and functions they should be performing and in what sectors. Successful innovation requires a clear articulation of a common vision and the regional expression of the strategic direction. This is a critical step in institutionalizing innovation at regional level. In particular, the governments' policy should be driven by a vision and in order to activate and support the regional development dynamics based on the exploitation of the innovation capacity grounding on regional knowledge assets.

Regional Innovation Policies

Innovation dynamics and regional development are increasingly based on the way and the effectiveness with which a region's resources are employed. It depends on one side on the aims of the different stakeholders as well as on the decisions and the processes they deliver to reach their aims. Of course, the way to intend innovation and development and the level of priority to which any single stakeholders is interested to develop and gain benefits can be very different. For this reason, it is necessary to integrate and coordinate the single actions and decisions of the stakeholders through different kinds of policy actions aimed to balance potential disequilibria created. Policy-making decisions and actions play a fundamental role. Historically, the policy-making decisions and actions have been grounded on the traditional model of the "government". It is based on the traditional logic to define and adopt rules hampering or constraining a determined set of actions and often implies a negative role of the public authorities in managing innovation within regions. The RICM, instead, calls to a pro-active role of the regional decision-makers who contribute or try to contribute to achieve targets and performance improvement through the definition, the support, the implementation and the control of a panel of actions in function of the stakeholders aims as well as of the whole region's aims. The regional governments could pursue different initiatives to stimulate the development of systematic propensities within regional systems, focusing specifically on the weak pillars and related links in the system. For example, they could define policies aiming at stimulating vertical and horizontal collaboration among indigenous firms, or might play a role in stimulating the collaboration between knowledge providers and firms (Chaminade & Vang, 2008), or by using public procurement as a tool to enhance experimentation and innovation in the local firms (Arora & Gambardella, 2004; 2005). Public procurement, in fact, might be very important to create local markets and give the right incentives to the indigenous firms to use their capabilities for innovation. Additionally, there could be the need for developing policy measures and evaluating tools taking into account the position of the regional system in global competition. Policies could focus, for example, on competency specialization, on fostering interactive learning, on the structure of the networks, on the quality of human capital and so on.

DISCUSSION AND CONCLUSION

The principal objective of this research has been to present a conceptual model aimed to understand the Intellectual Capital-based (IC) characteristics of the regional innovation capacity to be used for interpretative and normative purposes to analyse the innovation dynamics taking place at regional and territorial level. The four interrelated pillars of the RICM - innovation sources, innovation capacity, innovation processes and innovation results – have been deeply analysed and discussed. In research terms, our RICM analysis has two main implications. First, we are able to clearly identify the drivers of region-level innovation and development, and in particular to highlight the complementary role of resources, capabilities, processes and output. This provides a clear evidence that each of these main elements is important in influencing innovation and regional development both through their direct effect but also potentially through complementary effects with other innovation drivers. The RICM also exposes the mechanisms through which these elements influence innovation performance, providing a potential structure for the evaluation of future regional policy initiatives. Secondly, through the RICM we are able to identify the drivers of innovation behaviour itself, emphasising the role of the knowledge assets as key-sources for innovation. The implication is that policy intervention to strenghten knowledge exploitation may have direct benefits for innovation, but may also have indirect benefits through complementary relationships with other innovation drivers.

Our current approach has some limitations, however, which could usefully be addressed in future work. This research has highlighted the need for further rigorous investigation of innovation and its antecedent variables and illustrated the importance of adopting a region-wide approach to the management of innovation through a better identification and exploitation of the intellectual capital of the regional system. The concept of innovation capacity identified in this paper can be refined, validated and testing using other research methods, including case studies and surveys. Through the application of scientific methods of investigation to this concept, further progress will be made in unlocking and analysing the complexities of regional innovation processes and the regional development patterns that results. Detailed exploratory case studies can provide richer, more textual background into innovation variables. Given that practitioners are often ahead of academic practice, this provides opportunity for new data and raises the basis of new research questions. Further research should be directed at identifying and refing variables impacting and influencing innovation management at regional level.

REFERENCES

Albino, V., Carbonara, N., & Schiuma, G. (1998). Le strategie di sviluppo dei network locali. *Sviluppo & Organizzazione*, *170*, 33–46.

Albino, V., & Schiuma, G. (2003). New Forms of Knowledge Creation and Diffusion in the Industrial District of Matera-Bari. In Belussi, F., Gottardi, G., & Rullani, E. (Eds.), *The Net Evolution of Local Systems – Knowledge Creation, Collective Learning and Variety of Institutional Arrangements*. Dordrecht, the Netherlands: Kluwer.

Archibugi, D., & Michie, J. (1995). Technology and innovation: an introduction. *Cambridge Journal of Economics*, *19*, 1–4.

Arora, A., & Gambardella, A. (2004). The globalization of the software industry: perspectives and opportunities for developed and developing countries. *NBER Working paper, series 10538*.

Arora, A., & Gambardella, A. (2005, eds.). *From Underdogs to Tigers: The Rise and Growth of the Software Industry in Brazil, China, India, Ireland and Israel*. New York: Oxford University Press

Asheim, B. T. (1996). Industrial Districts as 'learning regions': a condition for prosperity? *European Planning Studies*, *4*, 379–400.

Asheim, B. T. (2002). Temporary organizations and spatial embeddedness of learning and knowledge creation. *Geografiska Annaler. Series B, Human Geography*, *84*, 111–124. doi:10.1111/j.0435-3684.2002.00117.x

Asheim, B. T., & Coenen, L. (2005). Knowledge bases and regional innovation systems: Comparing Nordic clusters. *Research Policy*, *34*, 1173–1190. doi:10.1016/j.respol.2005.03.013

Asheim, B. T., & Isaksen, A. (2002). Regional innovation systems: the integration of local sticky and global ubiquitous knowledge. *The Journal of Technology Transfer*, *27*, 77–86. doi:10.1023/A:1013100704794

Audretsch, D. B., & Feldman, M. (1996). R&D spillovers and the geography of innovation and production. *The American Economic Review*, *86*, 630–640.

Aydalot, P., & Keeble, D. (1998). *High Technology, Industry and Innovation Environments: The European Experience*. London: Routledge.

Becattini, G. (1990). The Marshallian industrial district as a socio-economic notion. In Pyke, F., Becattini, G., & Sengenberger, E. (Eds.), *Industrial* (pp. 37–51). Geneva: Districts and Interfirm Cooperation in Italy.

Behrens, K., & Thisse, J. F. (2007). Regional economics. A new economic geography perspective. *Regional Science and Urban Economics*, *37*, 457–465. doi:10.1016/j.regsciurbeco.2006.10.001

Bontis, N. (2004). National Intellectual Capital Index – A United Nations initiative for the Arab region. *Journal of Intellectual Capital*, *5*(1), 13–39. doi:10.1108/14691930410512905

Bounfour, A., & Edvinsson, L. (2005, Eds). *Intellectual Capital for Communities*. Boston, MA: Elsevier Butterworth-Heinemann.

Bradley, J. (1997). Intellectual capital and the new wealth of nations. *Business Strategy Review*, *8*, 53–62. doi:10.1111/1467-8616.00007

Brooksbank, D. J., & Pickernell, D. G. (1999). Regional competitiveness indicators – a reassessment of method. *Local Economy*, *13*, 310–326. doi:10.1080/02690949908726459

Budd, L., & Hirmis, S. A. (2004). Conceptual Framework for regional competitiveness. *Regional Studies*, *38*(9). doi:10.1080/0034340042000292610

Burt, R. S. (1992). *Structural Holes: The Social Structure of Competition*. Cambridge, MA: Harvard University Press.

Camagni, R. (1991). Local 'milieu', uncertainty and innovation networks: towards a new dynamic theory of economic space. In Camagni, R. (Ed.), *Innovation Networks: Spatial Perspectives*. London: Belhaven.

Carlucci, D., & Schiuma, G. (2007). Exploring intellectual capital concept in strategic management research. In Joia, L. (Ed.), *Strategies for Information Technology and Intellectual Capital: Challenges and Opportunities*. Hershey, PA: Idea Group Inc.

Carrillo, F. J. (2006). *Knowledge Cities* (1st ed.). Boston, MA: Elsevier Butterworth-Heinemann.

Chaminade, C., & Vang, J. (2008). Globalisation of knowledge production and regional innovation policy: Supporting specialized hubs in the Bangalore software industry. *Research Policy*, *37*, 1684–1696. doi:10.1016/j.respol.2008.08.014

Cooke, P. (2001). Regional innovation systems, clusters, and the knowledge economy. *Industrial and Corporate Change*, *10*, 945–974. doi:10.1093/icc/10.4.945

Cooke, P. (2004). Evolution of regional innovation systems – emergence, theory, challenge for action. In Cooke, P. (Ed.), *Regional Innovation Systems*. London: Routledge.

Cooke, P. (2008). Regional Innovation Systems, Clean Technology & Jacobian Cluster-Platform Policies. *Regional Science Policy and Practice*, *1*(1), 23–45. doi:10.1111/j.1757-7802.2008.00002.x

Cooke, P., Gomez Uranga, M., & Extebarria, G. (1997). Regional innovation systems: Institutional and organizational dimensions. *Research Policy*, *26*, 475–491. doi:10.1016/S0048-7333(97)00025-5

Cooke, P., & Memedovic, O. (2003). *Strategies for Regional Innovation Systems: Learning Transfer and Application Policy Papers*. Wien: UNIDO.

Cooke, P., & Schienstock, G. (2000). Structural competitiveness and learning regions. *Enterprise & Innovation Management Studies*, *1*, 265–280. doi:10.1080/14632440010023217

Cooke, P., & Wills, D. (1999). Small firms, social capital and enhancement of business performance through innovation programmes. *Small Business Economics*, *13*, 219–234. doi:10.1023/A:1008178808631

Crevoisier, O., & Maillat, D. (1991). Milieu, industrial organizations and territorial production system: towards a new theory of spatial development. In Camagni, R. (Ed.), *Innovation Networks: Spatial Perspectives*. London: Belhaven.

Dakhli, M., & De Clerq, Q. D. (2004). Human capital, social capital, and innovation: a multicountry study. *Entrepreneurship & Regional Development*, *16*, 107–128. doi:10.1080/08985620410001677835

Dolfsma, W., Chaminade, C., & Vang, J. (2008). Knowledge dynamics out of balance: Knowledge biased, skewed and unmatched. *Research Policy*, *37*, 1657–1658. doi:10.1016/j.respol.2008.08.007

Doloreux, D. (2002). What should we know about regional systems of innovation. *Technology in Society*, *24*, 243–263. doi:10.1016/S0160-791X(02)00007-6

Dougherty, D., & Hardy, C. (1996). Sustained product innovation in large, mature organizations: overcoming innovation-to-organization problems. *Academy of Management Journal*, *39*(5), 1120–1153. doi:10.2307/256994

DTI - Department of Trade and Industry. (1998). Our Competitive Future: Building the Knowledge Driven Economy. London: Competitiveness White Paper DTI.

(1999). *European Commission*. Brussels: Sixth Periodic Report on the Social and Economic Situation of Regions in the EU.

European Commission. (2005). Measuring progress towards a more sustainable Europe – Sustainable development indicators for the European Union. Luxembourg.

Florida, R. (1995). Towards the learning region. *Futures*, *27*, 527–536. doi:10.1016/0016-3287(95)00021-N

Freeman, C. (1995). The national systems of innovation in a historic perspective. *Cambridge Journal of Economics*, *18*, 5–24.

Furman, J. L., Porter, M. E., & Stern, S. (2002). The determinants of national innovative capacity. *Research Policy*, *31*, 899–933. doi:10.1016/S0048-7333(01)00152-4

Gardiner, B., Martin, R., & Tyler, P. (2004). Competitiveness, productivity and economic growth across the European regions. *Regional Studies, 38*(9).

Geroski, P., Machin, S., & Van Reenen, J. (1993). The profitability ofinnovating firms. *The Rand Journal of Economics, 24*, 198–211. doi:10.2307/2555757

Gertler, M. (2004). *Manufacturing Culture: The Institutional Geography of Industrial Practice*. Oxford: Oxford University Press.

Griliches, Z. (1992). The search for research-and-development spillovers. *The Scandinavian Journal of Economics, 94*, S29–S47. doi:10.2307/3440244

Gu, S., & Lundvall, L. (2006). Policy learning as a key process in the transformation of China's innovation systems. In Lundvall, B.-A., Intakumnerd, P., & Vang, J. (Eds.), *Asian Innovation Systems in Transitions*. Cheltenham: Edward Elgar.

Howells, J. (2005). Innovation and regional economic development: a matter of perspective? *Research Policy, 34*, 1220–1234. doi:10.1016/j.respol.2005.03.014

Huggins, R., & Izushi, H. (2007). *Competing for Knowledge. Creating, connecting, and growing*. London, New York: Routledge.

Iansiti, M., & Levien, R. (2004). *The Keystone Advantage*. Cambridge, MA: Harvard Business School Press.

Katz, D., & Kahn, R. (1996). *The Social Psychology of Organizations*. New York: Wiley.

King, N. (1992). Modelling the innovation process: an empirical comparison of approaches. *Journal of Occupational and Organizational Psychology, 65*(2), 89–100.

Kitson, M., Martin, R., & Tyler, P. (2004). Regional Competitiveness: An Elusive yet Key Concept? *Regional Studies, 38*(9), 991–999. doi:10.1080/0034340042000320816

Klevorick, A. K., Levin, R. C., Nelson, R. R., & Winter, S. G. (1995). On the sources and significance of interindustry differences in technological opportunities. *Research Policy, 24*, 185–205. doi:10.1016/0048-7333(93)00762-I

Koput, K. (1997). A chaotic model of innovative search: some answers, many questions. *Organization Science, 8*(5), 528–542. doi:10.1287/orsc.8.5.528

Larsen, K., & Salter, A. (2004). Searching high and low: waht types of firms use universities as a source of innovation. *Research Policy, 33*, 1201–1215. doi:10.1016/j.respol.2004.07.004

Lerro, A., & Carlucci, D. (2007). Intellectual Capital and Regions: Origins, Theoretical Foundations and Implications for Decision-makers. *International Journal of Learning and Intellectual Capital, 4*(4), 357–376. doi:10.1504/IJLIC.2007.016333

Love, J. H., & Roper, S. (1999). The determinants of innovation: R&D, technology transfer and networking effects. *Review of Industrial Organization, 15*, 43–64. doi:10.1023/A:1007757110963

Lundvall, B. A., & Johnson, B. (1994). The learning economy. *Journal of Industry Studies, 1*, 23–42.

Maillat, D. (1995). Territorial dynamic, innovative milieu and regional policy. *Entrepreneurship & Regional Development, 7*, 157–165. doi:10.1080/08985629500000010

Mansfield, E. (1991). Academic research and industrial innovation. *Research Policy, 20*, 1–12. doi:10.1016/0048-7333(91)90080-A

Mansfield, E. (1995). Academic research underlying industrial innovation: sources, characteristics, and financing. *The Review of Economics and Statistics, 77*(11), 55–65. doi:10.2307/2109992

Mansfield, E. (1998). Academic research and industrial innovation: an update of results. *Research Policy*, *26*, 773–776. doi:10.1016/S0048-7333(97)00043-7

Marr, B., & Schiuma, G. (2001). Measuring and Managing Intellectual Capital and Knowledge Assets in New Economy Organisations. In Bourne, M. (Ed.), *Performance Measurement Handbook*. London: GEE Publishing Ltd.

Marshall, A. (1916). *The Principles of Economics*. London: Macmillan.

McEvily, B., & Zaheer, A. (1999). Bridging ties: a source of firm heterogeneity in competitive capabilities. *Strategic Management Journal*, *20*, 1133–1156. doi:10.1002/(SICI)1097-0266(199912)20:12<1133::AID-SMJ74>3.0.CO;2-7

Morgan, K. (2004). The exaggerated death of geography: learning, proximity and territorial innovation systems. *Journal of Economic Geography*, *4*, 3–22. doi:10.1093/jeg/4.1.3

Morgan, K., & Neuwelaer, C. (1998). *Regional Innovation Strategies: The Challenge for Less Favoured Regions*. London: Jessica Kingsley.

Moulaert, F., & Sekia, F. (2003). Territorial Innovation Models: A Critical Survey. *Regional Studies*, *37*(3), 289–302. doi:10.1080/0034340032000065442

Nelson, R. R. (1993). *National Innovation Systems: A Comparative Analysis*. New York.

Owen-Smith, J., & Powell, W. W. (2004). Knowledge networks as channels and conduits: the effects of spillovers in Boston biotechnology community. *Organization Science*, *15*, 2–21. doi:10.1287/orsc.1030.0054

Pasher, E., & Shachar, S. (2005). The Intellectual Capital of the State of Israel. In Bounfour, A., Edvinsson, L. (2005, Eds.), *Intellectual Capital for Communities*, Boston: Elsevier Butterworth-Heinemann.

Pavitt, K. (1984). Sectoral patterns of technological changes: towards a taxonomy and a theory. *Research Policy*, *13*, 343–347. doi:10.1016/0048-7333(84)90018-0

Pike, A., Rodriguez-Pose, A., & Tomaney, J. (2006). *Local and Regional Development* (1st ed.). London: Regional Studies Association-Routledge.

Piore, M., & Sabel, C. (1984). *The Second Industrial Divide*. New York: Basic Books.

Porter, M. E. (1990). *The Competitive Advantage of Nations*. Basingstoke: Macmillan.

Pulic, A. (2005). Value Creation Efficiency at National and Regional Levels: Case Study-Croatia and the European Union. In Bounfour, A., Edvinsson, L. (Eds.) *Intellectual Capital for Communities*. Boston, MA: Elsevier Butterworth-Heinemann, Rondè, P. & Hussler, C. (2005). Innovation in regions: What does really matter? *Research Policy*, 34, 1150-1172.

Salter, A., & Martin, B. (2001). The economic benefits of publicly funded basic research: a critical review. *Research Policy*, *30*, 509–532. doi:10.1016/S0048-7333(00)00091-3

Sanchez Medina, A. J., Melian Gonzales, A., & Garcia Falcon, J. M. (2007). Intellectual Capital and Sustainable Development on Islands: An Application to the Case of Gran Canaria. *Regional Studies*, *41*(4), 473–487. doi:10.1080/00343400600928327

Saxenian, A. (1994). *Regional Advantage: Culture and Competition in Silicon Valley and Route 128*. Cambridge, MA: Harvard University Press.

Schiuma, G., & Lerro, A. (2008). Knowledge-based Capital in Building Regional Innovation Capacity. *Journal of Knowledge Management*, *12*(5), 121–136. doi:10.1108/13673270810902984

Schiuma, G., Lerro, A., & Carlucci, D. (2007). The Knoware Tree and the Regional Intellectual Capital Index: An Assessment within Italy. *Journal of Intellectual Capital*, *9*(2), 283–300. doi:10.1108/14691930810870346

Schroeder, R. G., Van de Ven, A. H., Scudder, G. D., & Polley, D. (1989). The development of innovation ideas. In Van de Ven, A. H., Angle, H. L., & Poole, M. (Eds.), *Research on the management of innovation: the Minnesota studies*. New York: Harper & Row.

Smedlund, A., & Poyhonen, A. (2005). Intellectual Capital Creation in Regions: A Knowledge System Approach. In Bounfour, A., Edvinsson, L. (Eds.), *Intellectual Capital for Communities*. Boston, MA: Elsevier Butterworth-Heinemann., Storper, M. & Scott, A.J. (1992, Eds). *Pathways to Industrialization and Regional Development.* London: Routledge.

Storper, M. (1995). Competitiveness policy options: the technology-regions connection. *Growth and Change*, (Spring): 285–308. doi:10.1111/j.1468-2257.1995.tb00172.x

Teece, D., & Pisano, G. (1998). The dynamic capabilities of firms: an introduction. In G. Dosi, D. Teece, G. & Silverberg (Eds), *Technology, Organization, and Competitiveness: Perspectives on Industrial and Corporate Change*. Oxford, UK: Oxford University Press.

Teece, D. J. (1998). Capturing Value From Knowledge Assets: The New Economy, Markets for Know-How, and Intangible Assets. *California Management Review*, *40*(3).

Teece, D. J., Pisano, G., & Shuen, A. (1997). Dynamic capabilities and strategic management. *Strategic Management Journal*, *18*(7), 509–533. doi:10.1002/(SICI)1097-0266(199708)18:7<509::AID-SMJ882>3.0.CO;2-Z

Tura, T., & Harmaakorpi, V. (2005). Social Capital in Building Regional Innovative Capability. *Regional Studies*, *39*(8), 1111–1125. doi:10.1080/00343400500328255

Vang, J., & Asheim, B. T. (2006). Regions, absorptive capacity and strategic coupling with high-tech TNCs: lessons from India and China. *Science, Technology & Society*, *11*(1), 39–66. doi:10.1177/097172180501100103

Chapter 4
Understanding Knowledge Networks via Social Network Analysis (SNA)

Asleena Helmi
Swinburne University of Technology, Malaysia

Zainal Ariffin Ahmad
University Sains Malaysia, Malaysia

Daisy Kee Mui Hung
University Sains Malaysia, Malaysia

ABSTRACT

The link between KM and network theory lies in the cultivating of knowledge networks by viewing knowledge as connections and understanding it as the aggregation of connections across different entities albeit it people or organizations (Kadushin, 2004). The study explores primarily on the capabilities and capacity building of government-linked corporations (GLC) in Malaysia in relation to its evolution as a knowledge network. In this chapter, the authors explore the overlap that exists in KM particularly in the entities and sharing of knowledge assets that undergoes a multiplying effect in its value propositions with the forming of interoganizational relationships. The findings of the study indicates that in order to benefit from the knowledge multiplying affects, organizations need to create a "network identity" and develop capabilities as a learning entity that would add value to the network itself. From the context of KM, majority of the GLCs are in the transitory stages in which issues of trust, competencies and communications have been key indicators in forming network relations. As such this chapter would be useful in highlighting the significance of network relations and provides further proliferation in the study of social network theory and significance to KM.

DOI: 10.4018/978-1-61692-846-9.ch004

Copyright © 2011, IGI Global. Copying or distributing in print or electronic forms without written permission of IGI Global is prohibited.

1 INTRODUCTION

The KM phenomenon as a development of business studies has proven to be more than a management fad and intrigued both theorists and practioners. This ambiguity has an almost intuitive appeal due to its cross-industrial and cultural nature. Drucker (1998) predicted that the organization of the future would be knowledge-based and would comprise of specialists that form self-managed teams and performs through organized feedback from colleagues, customers and headquarters. There are evidences of viewing organizations and companies alike as a "unit of value creation" (Muller & Valikangas, 2002). A case in point is when BP developed its "epiphanies of scale" in knowledge by encouraging the formation of peer groups across different business units focused on areas of shared interest. The effect was the growth of hundreds of networks and subnetworks (Hansen & Oetinger, 2001).

Hence today a company is valued based on collective knowledge and technical capabilities of the organization including its people, equipment and systems. In other words, it is imperative for companies to be able to be innovative and leverage on intangible knowledge (example ideas, experience) to sustain performance within a business-web (b-webs) environment. Tapscott (2005) defined b-webs as an internetworked set of contributors which are either highly structured or amorphous that collaborate together to create value for customers and wealth for their shareholders. Consequently the notion of b-webs have invented new value propositions, transformed the rules of competition, and mobilizing people and resources to unprecedented levels of performance (Tapscott, Ticoll & Lowy, 2000).

When firms can manage the learning process better which is the more effective way to pass on tacit understandings of how things operate then clearly firms can become more efficient. Muller and Valikangas (2004) purported that a firms role as a unit of value creation can only be heightened and achieved which required firstly organizations form alliances by sharing resources and competencies in mature industries (example: joint ventures) and secondly engage in cross-industrial collaborations in new industries (example: research and development agreements). The challenge therein lies in understanding the impetus in developing interorganizational relations. Therefore this paper aims to offer a discussion on using a 'knowledge network' approach in the formation of relationships between firms and organizations alike. For the purpose of this paper the notion of a knowledge network denotes the firm's value as a network partner to gain further network-specific knowledge that would enhance the efficiency of the network itself (Olkkonen, 2001). This approach would also relate to network positioning (in relation to the level of interdependence) and learning capability (from a KM context). Thus the network theory will be discussed in the paper to describe the nature and development of KM and its significance in interorganizational relationships. The three main propositions of the research are (i) to explore the knowledge network among organizations and understanding the (ii) interorganizational relations (IOR) creates new value networks.

2 INNOVATION AND KM INITIATIVES IN MALAYSIA

According to Muller and Valikangas (2002), the following are factors of innovation in KM initiatives:

- reduced transaction costs with focus on innovation capabilities
- recognition of the value of intangible assets
- software tools for KM and collaborative working
- when firms form organization ties, partnership and alliances are more likely to innovate

- the growth of cross-industry knowledge networks and
- the speed with which value chains migrate, which sometimes necessitate a rapid recombination of assets by established companies

In this paper, the IOR or knowledge network was analyzed in the recognition of tacit knowledge rather than explicit values such as investments and costs. As such the study would be focused on company's capability and willingness in sharing intangible assets, collaborative working and recombination of assets. The inception of KM developments and practices among the GLCs and within Malaysia can be linked with the implementation of the Government Link Corporation Transformation Programme (GLCT) in 2005. Under the GLC Transformation programme (2005), among the more important of the ten initiatives is to intensify performance management in all GLCs to ensure focus on the key business priorities-particularly value creation-and to retain knowledge of key experts (PCG, 2005). The core areas of the programme were particularly on KM framework for the public administrative sectors, evaluation on system capabilities to support knowledge diffusion and KM strategy. Hence in the study, a social network analysis (SNA) was incorporated to analyze the extent of collaborative knowledge exchange within and across various stakeholders and the GLC.

The champions of KM in Malaysia have been local institutions and government-linked companies are known to have set-up KM initiatives (Chowdhury, 2006).

Nonetheless according to Woods (2005), the most common KM approaches in Malaysia are:

- 'imported' KM applications through the multinational companies (such as HP, Microsoft, etc.) where their existing global KM approaches are adopted by the Malaysian division;
- "adapted" KM applications for local use through multinational corporation –example: Motorola Corporate University as a model being considered by Malaysian institutions, Siemens "Sharenet" modified for use in an academic environment at Multimedia University and
- "homegrown" applications such as the KM blueprints of various government ministries.

These KM awareness and initiatives have indicated that both the private and public are actively contributing in developing their KM framework and aware of KM benefits to increase organizational performance and effectiveness. As such this study can contribute to the development of the Asian "network identity" which explores the opportunities of knowledge as a strategic resource and basis in developing successful and sustainable interorganizational ties.

2.1 Linking the Resource-Based View (RBV) and Network Theory

According to Zack (1999), the RBV allows enterprise a long-term view as it focuses on competitive advantage based on resources and capabilities which is often more sustainable compared to the traditional approach based on product and market positioning. The interfirm (re)formation is initiated by exogenous factors, such as the distribution of capital and technological resources and the pattern of resource dependence. From a resource dependence perspective, an organization will engage in a relationship with other business counterparts only if it considers itself dependent on them (Raak & Paulus, 2001). Organizations create ties with organizations that have resources and capabilities that can help them cope with uncertain environments and satisfy their resource needs (Baum & Ingram, 2000).

Nonetheless Godfroij (1989) argued that dependency between organizations may be competitive in relation to one goal and symbiotic with another and so may their relations (Raak & Paulus, 2001). In other words, the way in which resources are shared determines the level of interdependence in relationships among firms. The antecedent to the RBV was provided by Priem and Butler (2001) in which they argued that "how can knowledge as a strategic resource be valued and obtained; how does it contribute to competitive advantage and how does it interact with other resources?" (Choo & Bontis, 2002). According to Kang (2007), one of the key steps in competitive positioning is differentiation by identifying new value proposition to specific stakeholders (such as customers or employees). *Value proposition* is defined as measures in which KM can increase organizational performance (McManus, Fredericksen, Wilson & Snyder, 2004). As such the second proposition of the paper lies on the basis of knowledge as a strategic resource and network formation.

2.2 Boundaries in Interorganizational Relations (IOR)

The simplest definition to organizational boundaries is explained as "entities that separate different activities, people and process" (Ashford et. al, 2000). Yan and Louis (1999) defined boundary as "possible patterns that define how organizations structure and relate to one another" whereas Morash et. al (1997) purported the concept as the " context, emerging and shifting with the pace of organizational change" (McCarthy, Garavan & O' Toole (2003). Despite the different definitions, the common theme can be argued to be extending towards *relationship mapping on the dynamics or flow of organizational entities* (albeit implicit or explicit resources) based on either forms of intra or interorganizational relations. The boundary and dependence parities brings about the significance of an organization to associate itself to a network within its industrial constituents because the notion of a "network identity" is meant to capture the perceived attractiveness of an actor as an exchange partner due to its own resources and its unique set of connected relations with other actors, links to their activities and ties with their resources (Olkkonen, 2001). Hence today, an organization's role as a unit of value creation plays a significant factor in the setting of boundaries in a knowledge network as it enables the proliferation of new and unique knowledge. An example that illustrates the value of network is the Toyota Production System which inculcates collaboration across a complex network of value partners in its supply-chain that consequently elevated Toyota's competitive positioning over US automakers (Muller et. al, 2001). Studies conducted by Buttery, Fulop and Kelly (1997) in McCarthy et. al (2003) maintained that effective management of boundaries and networks consists of three elements mainly: trust- especially in new relationships and emerging industries; secondly, competencies- the failure to communicate in speedy manner or withholding information can adversely affect the effectiveness of interface and thirdly-the flow and frequency in communication to strengthen organizational ties. Thus organizations need to incorporate an embedded perspective in order to benefit various value propositions in a knowledge network environment.

3 RESEARCH METHODOLOGIES

The sample population of the study will comprised of approximately 58 government-linked corporations (GLCs) in Malaysia. *GLCs are defined as companies that have a primary commercial objective and in which the Malaysian Government has a direct controlling stake* (Putrajaya Committee, 2005). The GLCs were chosen on the basis that they serve as custodians of stakeholders investments in terms of high market value worth in the region of billions of ringgit (RM). The questionnaire was distributed across different segments of the organization which comprised of both middle management and the subject matter experts involved in the work processes. It is also important to note that the GLCs are operating in mature industry which each of the firm having long-standing relations with the government and been in operation for more than 5 till 7 years in their respective industries such as plantations, property, construction and investment. In today's challenging global business environment, government-linked corporations (GLCs) in Malaysia are compelled to transform into high-performing corporations in order to compete and achieve long-term sustenance. Size factor is a particularly salient factor for the GLCs because their performance would be obvious. More importantly, the larger the firm, the more organizational capital it possesses (Levinthal, 1991) in terms of financial resources legitimacy and goodwill (Baum, 1996). Consequently due to high accountability, large firms such as the GLCs face strong pressures to perform. The ongoing restructuring process of GLCs to achieve these expectations is clearly a challenging task. As such the study enforces the significance in understanding the impetus in developing ties among the GLCs. The Malaysian Deputy Prime Minister, Datuk Seri Najib Tun Razak, said "knowledge management models developed by the private sectors should be formulated and implemented in accordance with the country's culture and social norms" (News Straits Times, November, 2007).

As such the transition of the GLCs into world-class corporations represents a *knowledge strategy gap* that needs to be addressed and the study is important to understand and further develop effective strategies in strengthening organizational ties to encourage knowledge diffusion and sustain competitive advantage strategies pertaining to Asian business scenarios and cultures. According to Shipilov, Rowley and Aharonson (2006), because opportunities and constraints are unevenly distributed across networks, social structures are sources of competitive (dis)advantages depending on organizations' positions in their networks. The findings of the study would be also be incorporating a social network analysis (SNA) via mapping or "sociogram" (Sharma, 2005) in order to analyze the relationships and flows between the people, groups, and other knowledge processing entities of the GLCs within the context of inter-organizational learning and relationships. SNA is not a replacement for traditional KM tools such as knowledge databases or portals, but it can provide companies with a starting point for how best to proceed with KM initiatives (Levinson, 2008). However for the purposes of the study the cluster concept will be adapted onto a strategic knowledge map developed by Zack (1999). According to Zack (1999), this strategic knowledge framework offers the ability to take a snapshot of where the firm is today vis-à-vis its desired strategic knowledge profile (to assess its internal gaps) and vis-à-vis its competitors (to assess its external knowledge gaps). The primary data collection is based upon a survey based upon the Likert's summative scale approach in which the final score for the respondent on the scale is the sum of their ratings for all of the items (Trochim, 2006). The summative technique in rating the items was chosen in order to illustrate the multidimensional categories of knowledge strategy or positioning labeled by Zack (1999) as core, advanced or innovative stages. In principle, the companies were required how they evaluate themselves (network identity) and the intensity of their relationship (knowledge network) to other counterparts or business partners.

3.1 Knowledge Network Positioning

In this context, Quinn's (1996) in Khalil (2000) notion of organization's intellect will be integrated to illustrate the success the effectiveness deploying knowledge strategy within the organizations. This is because having the right amount of strategic resources does not imply organizations having the capability to effective manage knowledge.

Hence for the purposes of this study, the following guidelines (see Table 1) will be used to chart the organization's performance:

Table 1. Knowledge network positioning

Knowledge Strategy	Intellectual Capability
Core Stages	Know-what
Advanced Stages	Know-how + know why
Innovative Stages	Know-why + care why

- *Core Stages*: Organization having cognitive knowledge or know-what: this is the basic mastery of the discipline. It is essential but insufficient for commercial success. At this point the organization would have acquired the basic requirements to barrier of entry into the cluster.
- *Advanced stages*: Organizations have acquired capability in:
  a. **advanced skill or know-how:** this is the translation of book learning into effective execution and
  b. **System understanding or know-why:** this is the knowledge of the cause-and-effect relationship underlying a discipline and
- *Innovative Stages*: Organizations have acquired capability in:
  a. **System understanding or know-why**: this is the knowledge of the cause-and-effect relationship underlying a discipline and
  b. **Self-motivated creativity or care-why**: this is the will, motivation and adaptability needed for success. Creative and motivated groups outperform groups with greater financial or physical resources.

The findings of the study would primarily be using a social network analysis using mapping or "sociogram" introduced by Sharma (2005) in order to measure the relationships and flows between the people, groups, and other knowledge processing entities of the GLCs.

4 FINDINGS AND DISCUSSIONS

4.1 Network Positioning

From the survey, only 26 companies responded and fully-complied with the questionnaire. The GLCs were unique position in that they are homogeneous as a state-owned enterprise or corporation yet heterogeneous in their business dealings and interactions. The first level of analysis involved the segregation of the companies into three categories as core (denoted as nodes "C" advanced (denoted as nodes "B") and innovative (denoted as nodes "A). Hence in this study, the term node is used interchangeably with company. An aggregate summative scale was derived from the main sections of the questionnaire: organization structure, business strategy and knowledge management.

Each company (node) was randomly identified as company 1-26 in keeping with anonymity of the respondents. In the Table 4.1, a total of 9 companies were in the innovative position (A); 15 companies within the advanced position (B) and two belonging to the core position (C). Each position reflects the company's capability to effectively manage knowledge. The majority of the GLCs (57%) had acquired the advanced skill (know-how) and knows why in leveraging on organizational learning and high absorptive capacity in knowledge diffusion. Among the skills acquired were the abilities in information extraction, real-time collaboration, semantic web and expertise management location. From the survey, it was also discovered that 35 percent (node A) of the GLCs were discovered as being innovative companies illustrating high competencies in developing specific function knowledge, incorporating best practices programs and advocating communities of practice across its business relations. While 8 per cent (node C) of the companies were in core stages whereby firms have cognitive knowledge or know-what of knowledge management. Although this is essential in KM but it is insufficient for successful take-up of knowledge strategy initiatives within the organization.

The essential principle in the sociogram network model by Sharma (2005) illustrates the relations or linkages among and between the nodes. The respondents (organizations) were asked to evaluate their companies knowledge assets and resources and a connection was established when the value scores to these entities were similar. From Figure 1, the GLCs network appear to be "boundaryless" whereby it can be seen from the crosscutting of links (path) across the various nodes (company). The path in Figure 1 indicates that at least 7 companies (i.e.: nodes 10, 16, 2, 11, 9, 21 and 26) have established organizational ties across two separate network positioning of advanced and innovative. Companies 10 and 16 are in the node A (innovative) position whereas companies 2,11,9, 21 and 26 represents node B (advanced) positions. It is important to note that the path illustrates a "bridging capital" (Sack, 1998) which is created when there are organization ties in the sharing of knowledge and resources between the two different networks. However for the purposes of this paper the path is referred to as the "knowledge bridge" as the acronym befits the exchange or flow of knowledge assets between the nodes.

4.2. Significance of Organizational Relations

According to the social network analysis theory, the extent of relationship among organizations is measured according to three types of centrality (Sharma, 2005):

- Degree centrality (the number of direct connections a node has)
- Betweenness centrality (a node with high betweenness has great influence over what flows in the network indicating important links and single point of failure) and
- Closeness centrality (the measures of closeness of a node which are having the shortest paths to all others).

In the sociogram, the companies were also linked in relation to their network positions in order to measure their centrality (i.e.: influence or control) albeit direct or indirect in knowledge diffusion and sharing of resources.

From Table 3, node 20 has a higher degree of centrality in comparison to the other nodes which indicates that the company has resource attributes that are valued and sought after by its partners. In Figure 1, it can be seen that node 20 has an approximately 6 direct links or connections in contrast to the other nodes. Thus in the network, company 20 has high competencies in forging potential partnership or organization ties. Node 11 (see Figure 1) is identified as having the betweeness centrality. This company (node 11) may not have equivalence in the control of resource such as company 20 but it plays a significant role as a mediator in influencing the decision-making process in the network perhaps by decree of larger shareholding or years of experience across the GLCs industries. Hence without the link or ties with node 11, it can be deduced that other companies within the network would not be willing to engage in work collaborations. In this context, the element of trust and integrity plays a significant factor in organizational relations and is an added value to the network. In Table 3, node 2 holds the closeness centrality in which it is the shortest link to any other company. The shortest links or distances in order to transmit information accurately and in a timely way, while long distances transmit slowly and can distort the information (Sharma, 2005). As such the strength of the interorganizational ties of the network relies on company 2. Thus node 2 befits the communication profile of a business partner which is efficient in the know-how (example: advanced technology) and the know-why (example: expertise).

4.2.1 Network Clustering

In order to measure the level of competitive advantage in the GLCs interorganizational rela-

Figure 1. Sociogram of the GLCs

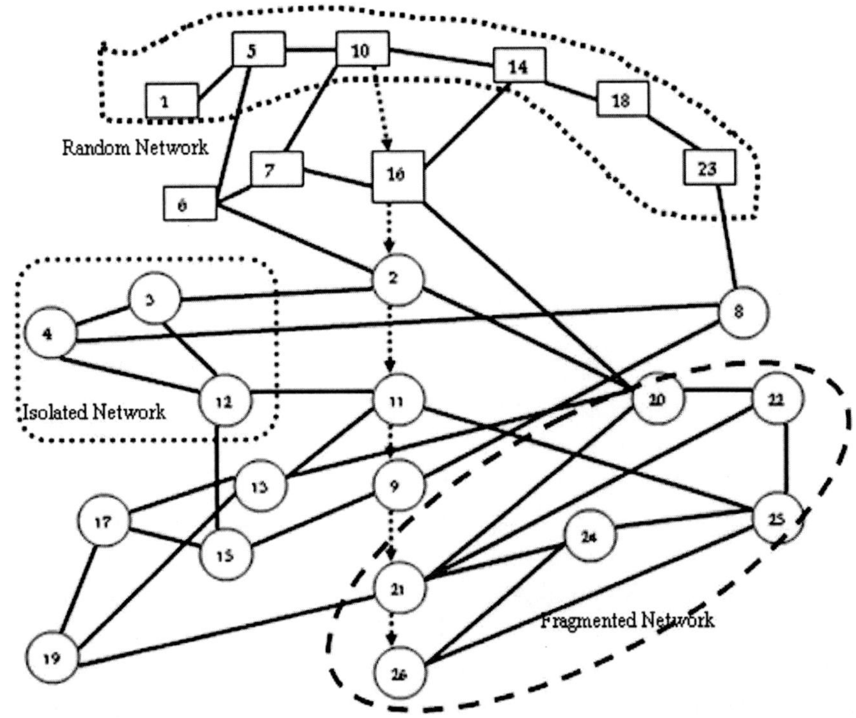

tions, were categorized into 3 distinct networks positioning (see Table 2). A number of studies (Watts & Stogartz, 1998; Sharma, 2005 & Toyoma, 2005) identified three types of clustering that developed across the network that were identified in the study as:

- **Random Network:** Links is identified on a random basis within the network. In Figure 10, a random network was formulated across the companies among the innovative companies (namely: nodes 1, 5, 10, 14, 18, 23 and 8). From Figure 2, the links (inner series), are not duplicates and unidirectional. The inner series represents the determinate values assigned by the organizations in forging knowledge collaboration in relation to other counterparts. This illustrates that the pattern of knowledge diffusion or level of interface among the nodes (outer series) are not frequent on the basis that each company have already acquired high competencies and creative in tapping opportunities among themselves.

Table 2. GLCs network positioning

Node	N	List of company (N-nodes)
C (core)	2	12, 15
B (advanced)	15	2, 3, 4, 8, 9, 11, 13, 17, 19, 20, 21, 22, 24, 25, 26
A (innovative)	9	1, 5, 6, 7, 10, 14, 16, 18, 23

Table 3. GLCs centrality

Centrality Type	Node (N)
Degree	20
Betweeness	11
Closeness	2

Nonetheless the disadvantage is that the random pattern can lead to the notion of "cliques" which may not be sufficient for long term sustenance and deters the discovery of new and unique knowledge.

- **Isolated Network:** An isolated network occurs when the nodes are not integrated well into the network and therefore represents both untapped skills and a high likelihood of turnover (Sharma, 2005). In the study, it was discovered that nodes 3, 4 and 12 (see Figure 2) were distant from the path in the network and also isolated from the other nodes in its constituents (nodes B). It can be argued that although the companies may have acquired advanced competencies to be accepted into the network, the nodes can be said to have an almost discrete organization perspective on network boundaries. As such these companies opted for a competitive rather than collaboration when interfacing with other organizations. Therefore any use of coloration would primarily be a tactical arrangement which is short-term in nature (McCarthy et.al, 2003). More importantly in the drawbacks of the isolated network is in the probability of not utilizing the knowledge of expertise that would add value to the network itself.
- **Fragmented Network:** In Figure 2, the fragmented network is illustrated by the accumulation of nodes 20, 21, 22, 24, 25 and 26. With fragmented networks Toyoma (2005) likened the situation when the average paths connecting the nodes are ubiquitous yet more frequent in contrast with random networks.

The mapping from Figure 1 confirms the latter finding in the study that node 2 (see Table 3) holds the position in the context of closeness centrality. In other words, company 2 plays a vital role in the communicating or knowledge diffusion by displaying the shortest links or distances across other nodes in the network. The nodes composition in Figure 2 highlights the "cliques" phenomenon among the nodes by illustrating a polarization trend whereby the nodes relate to themselves based on network positioning but not rather than across nodes/ organizations.

Figure 2. The symmetrical interorganizational relationships

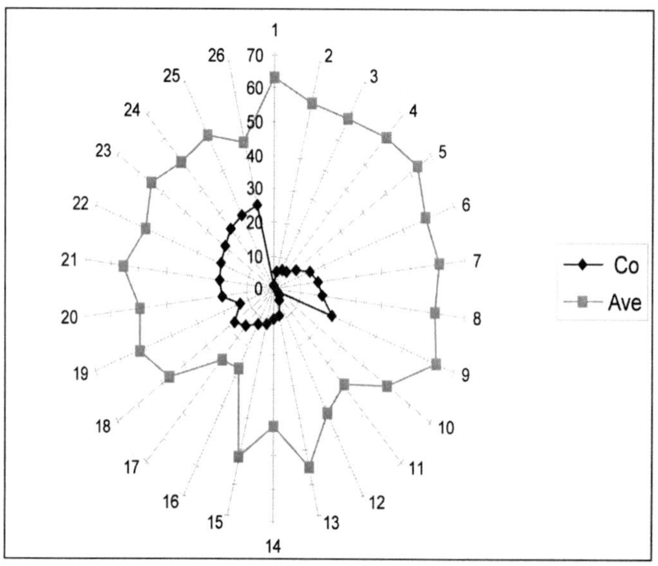

5 CONCLUSION

The findings of study were able to substantiate the earlier propositions that measured the scope (centrality) and evolution (knowledge bridge) of the network and illustrated the evolution (network positioning) in the GLCs relationships and performance in KM. It can be argued that in principle, the emergence of cliques or polarization between the two networks implies that there is a need for changes in knowledge diffusion among the companies and to proliferate KM to identify new and unique value propositions (example: best practices methodologies and mentoring techniques) through a more frequent and dynamic interface or collaboration across the different networks. The polarization affects suggests that there are opportunities to further improve interorganizational relations among the GLCs which will only add new value propositions to the knowledge network. However challenges lie in detangling or disengagements from the clique phenomenon. Although the cliques behavior among the network clusters may strengthened competitive advantage in relation to centrality of knowledge resources but it would not viable for long-term sustenance as earlier studies have shown that innovation and value propositions are typically develop beyond corporate boundaries and its existing competencies. This evidence highlights the necessity for the GLCs to explore beyond their core competencies and engage in new cross-industrial knowledge within emerging markets or networks.

Therefore the study proposed that the companies adopt a more embedded view on engaging interfaces and invest in cross-industrial partnerships across network boundaries. In reference to the patterns of links connecting the nodes within the knowledge network, it can be concluded that issues of trust and competencies rather than communication were among the main concerns of the GLCs. The GLCs have strength in technological and capital resources yet they are pressured to behave and engage in a more accountable manner especially in the circumstances of economic flux. As such, while that there are risks involved when engaging in organizational ties with a new partner, the GLCs in Malaysia could engage in a series of micro-relations engagements such as creating a community of practice through which it can access the required competencies and develop trust while learning about the potentials of new ideas or unique knowledge that would only enhance interorganizational relationships and ultimately performance in the long run.

REFERENCES

Aidemark & Sterner. (2003). A Framework for Strategic Balancing of Knowledge Management Initiatives. *International Conference on System Sciences 36th Proceeding.* Hawaii

Baum, J. A. C. (1996). Organizational Ecology. In Clegg, S., Hardy, C., & Nord, W. (Eds.), *Handbook of Organization Studies*. London: Sage.

Chatterjee (2002). *Knowledge Management-Theories, Issues and Challenges, IIT.*Kanpur, India.

Hansen & Oetinger. (2001). Introducing T-Shaped Managers: KM's Next Generation [March.]. *Harvard Business Review*, 106–116.

Housel & Bell. (2001). *Measuring and Managing Knowledge* (International Edition). New York: McGraw-Hill.

Kadushin.C (2004). Basic *Network Concepts. The Culture and Commerce of Publishing*. New York.

Kahal, S. (2001). *Business in Asia Pacific: Text and Cases*. Oxford, UK: Oxford University Press.

Khalil, T. (2000). *Management of Technology: The Key to Competitiveness and Wealth Creation* (International Editions). New York: McGraw-Hill.

Kotler.P, Jain. D, Maesincee.S (2002). Marketing Moves: A New Approach to Profits, Growth and Renewal. *Harvard Business School Press.*

Kuhlen (2003). *Change of Paradigm in KM-Framework for the Collaborative Production and Exchange of Knowledge*. University of Konstanz, Germany.

Levinthal, D. (1991). Random walks and organizational mortality. *Administrative Science Quarterly*, *36*, 397–420. doi:10.2307/2393202

Malhotra, Y. (2000). KM for E-Business Performance: Advancing Information Strategy to "Internet Time". *The Executive Journal, CRC Press*, *16*(4), 5–16.

McCarthy, A., Garavan, T., & Toole, T. (2003). HRD: working at the boundaries and interfaces of organizations. *Journal of European Industrial Training*, *27*(2), 58–72. doi:10.1108/03090590310468886

Muller, A., & Valikangas, L. (2002). Extending the boundary of corporate innovation. *Strategy and Leadership*, *30*(3), 4–9. doi:10.1108/10878570210427891

Nonaka, I., & Takeuchi, H. (1995). *The Knowledge-Creating Company*. New York: Oxford University Press.

Olkkonen, R. (2001). The Network Approach to International Sport Sponsorship Arrangement. *Journal of Business and Industrial Marketing*, *16*(4), 309–329. doi:10.1108/EUM0000000005503

Raak, A., & Paulus, A. (2001). A Sociological Systems Theory of Interoganizational Network Development in Health and Social Care. *Systems Research and Behavioral Science Syst. Res.18, 207-224.*New York: John Wiley & Sons Ltd.

Samara, A. (2007). Framework for Discovering KM Forces: The Fifth Element. *Journal of KM Practice*, *8*(1).

Sharma.A (2005). *Social Networks.* University of Texas, Austin.

Shelda, D. (2006). *Knowledge Management*. New York: John Wiley & Sons Australia, Ltd.

Shipilov, A., Rowley, T., & Aharonson, B. When do Networks Matter? (2006). A Study of Tie Formation and Decay. In Baum J.A.C, Dobrev, S. and Arjen van Witteloostuijn (Eds.), *Advances in Strategic Management* (Vol.23. June. Skyrme. D (1998). *Commercializing Organizational Knowledge.* ArkGroup Asia.Com.

Shipilov. A. V (2005). Should you bank on your network? Relational and positional embeddedness in the making of financial capital. *Strategic Organization, 3*(3).

Suliman Al-Hawamdeh & Thomas L. Hart. (2002). *Information and Knowledge Society.*New York: McGraw Hill Education (Asia).

Sveiby.K.E (2001).What is KM? Tapscott.D (2005). *Part II: Strategy in the Age of Collaboration, New Paradigm Learning Corporation (NPLC).* Canada.

Toyoma.K (2005). The Science of Social Networks. *Microsoft Research India.*Indian Institute of Science.

Weber. M (2002). Standardization in Knowledge Management – Towards a Common. *KM Framework in Europe, KM European Forum.*

Wigg.K.M (1999). What future KM users may expect. *Journal of KM, 3*(2),, West Yorkshire, England: MCB University Press

Chapter 5
Breaking Out from Lock-In: Regional Innovation Strategies in the German Ruhrgebiet

Gert-Jan Hospers
University of Twente, The Netherlands & Radboud University, The Netherlands

ABSTRACT

This chapter discusses strategies aimed at regional-economic structural change in the German Ruhrgebiet. The Ruhrgebiet used to be the largest industrial area in Western-Europe. After the crisis in the coal and steel industry the region pursued re-industrialisation policies in the 1960s and 1970s. These attempts were largely unsuccesful. Therefore, since the 1980s the involved actors gradually adopted regional innovation strategies. Thus, they were able to break out from the region's lock-in situation. The re-orientation of the Ruhrgebiet towards innovation based on the industries' expertise and past (e.g. environmental technology, energy and industrial tourism) is more successful than its earlier re-industrialisation attempts. Also for other old industrial areas in Europe this kind of place-based renewal might be the future.

INTRODUCTION

In many of Europe's old industrial regions, the process of structural change has resulted in substantial decline and rising unemployment. Ironically, those very areas that were the forerunners of the Industrial Revolution were forced to given way to new growth poles in the European economy. In particular regions specialised in 'smokestack' industries such as coal, iron and steel, heavy engineering, textiles and shipbuilding have been hard-hit by industrial decline (Hassink & Shin, 2005; Birch et al., 2008). The widespread effects of this de-industrialization process can be clearly seen over the years by the steady rise of European regions that received support from the EU's Structural Funds. Jointly, this regional mosaic of traditional industries – including such areas as the West-Midlands, Wales, Wallonia, the Ruhrgebiet, North-Pas-de-Calais, Lorraine and the Basque Country – has been termed Europe's 'Rustbelt' (Cooke, 1995). With a mix of strategies, European, national and local authorities have tried to rejuvenate these traditional regions since many

DOI: 10.4018/978-1-61692-846-9.ch005

Copyright © 2011, IGI Global. Copying or distributing in print or electronic forms without written permission of IGI Global is prohibited.

years. In the literature, the Ruhrgebiet, Wales and North-Pas-de Calais are often highlighted as areas where local parties have pursued such restructuring policies with varying degrees of success.

In the coming years the issue of how to deal with regional industrial decline will be high on the European policy agenda again. The fact is that the enlargement of the European Union with member states from Eastern-Europe has resulted in a significant expansion of Europe's Rustbelt. Especially in Poland, Hungary, Slovakia, Romania and Bulgaria many regional economies are still highly dependent upon smokestack industries (Lintz et al, 2007). Although the shakeout in inefficient heavy manufacturing has taken place, the restructuring of these traditional industrial regions is anything but completed. To note an example: the so-called 'GOP' in Silesia, being Poland's major economic area with industrial conurbations such as Kattowice, is still dominated by mining and steel production. These industries still employ a considerable part of the region's work force. Coal and steel have not only shaped the region's socio-economic structure, but also its landscape: because of the pollution problems in the GOP, public authorities have denoted it as an 'ecological disaster area' that needs restructuring to be able to comply to West-European standards.

Obviously, regional industrial decline once more is a topical European problem asking for appropriate and timely policy responses. In this context, the long-lasting experiences of older industrial regions in Western-Europe with restructuring might provide inspiration for how to deal with the problems in the rest of the continent. The chapter at hand analyses such an inspiration source, the case of the German Ruhrgebiet. This region is perhaps Europe's most well-known example of an old industrial area in which structural change has been occurring since about fifty years now (Birch et al., 2008). The article is organised as follows. First, to place the case in context, we give a rough sketch of the Ruhrgebiet and its economy. Next, we study the re-industrialisation strategies that have been pursued in the Ruhr as well as the associated problems caused by regional lock-in. Then, the focus is on the bottom-up strategies aimed at regional innovation. After that, we assess the restructuring efforts of the Ruhrgebiet and have a look at the area's future. Finally, we trace some implications the Ruhr-case might have for the other regions that are part of Europe's Rustbelt.

THE RUHRGEBIET IN SHORT

Traditionally, the German Ruhrgebiet, situated in the Federal State of North-Rhine Westphalia, has been the largest industrial centre in Europe. With its 5.3 million people on an area of 4,400 cubic kilometres the region is also one of Europe's most densely populated conurbations (RVR, 2007a). The Ruhrgebiet is neither a historical or political entity but rather a functional area that grew on the basis of its industrial geography. Seen in geological terms, the Ruhrgebiet belongs to the north-west European coal belt running from Silesia via the Ruhr, Belgium and North-France to England. Although coal has been its common geographic denominator, the area does not possess a single uniform landscape. The Rhine marks the western border of the Ruhrgebiet, while the region is structured by the three tributaries of this river: the Ruhr in the south, the Lippe in the north and the Emscher in between. These rivers have given their name to three of the four zones comprising the towns and cities in the Ruhrgebiet: the Ruhrzone (Werden, Hattingen, Witten, Hagen and Schwerte), the Emscherzone (Oberhausen, Bottrop, Gelsenkirchen and Herne) and the Lippezone (Wesel, Dorsten, Marl, Lünen and Hamm). In between the Emscherzone and the Lippezone we find the old industrial heart of the Ruhrgebiet. This is the Hellwegzone, named after the Hellweg, a centuries-old transport and trade route dating back to Hanseatic times. Here, the mediaeval towns Duisburg, Essen, Bochum, Dortmund and Unna developed in large industrial cities when the Ruhr was industrialised. Administratively, all the towns and cities and their parishes are linked in the Association of Local Authorities in the Ruhrgebiet, the Regionalverband Ruhrgebiet (RVR). The headquarters of the RVR is in Essen.

The Ruhrgebiet is hardly more than 160 years old. Although coal has been dug in the Ruhr valley since the 14th century for heating, forging and casting uses, the mining industry was still relatively unimportant. The arrival of the Industrial Revolution was the prerequisite for the coal-mining boom the Ruhrgebiet experienced in the 19th century. Local entrepreneurs like Dinnendahl and Harkort put in operation steam engines that pumped off ground water and enabled people to sink shafts (Pohl & Ponthöfer, 2002). The coal mined in the Ruhr Valley formed the basis for the iron and steel production. The industrialisation of the Ruhr also marked the start of the region's extensive transport connections, such as roads, railways and canals. Supported by progressive Prussian economic policies, after 1850 an enormous industrial boom in the Ruhrgebiet could set in. People from all over Germany came to the area to find work; between 1850 and 1925 the population rose from around 400,000 to 3,800.000 (RVR, 2006). After the two World Wars, in which many industrial sites were destructed, it was predominantly the Ruhrgebiet that contributed to Germany's economic miracle in the 1950s. In 1956 the mining industry reached its peak: at that year 124,600 tons of coal was produced, offering work to nearly half a million of people. In the 1958, however, the coal-mining crisis started, which put a brake on employment in the Ruhr. Cheap imports from countries such as the USA and competition from substitutes for coal (crude oil, natural gas) resulted in closings of many mines and drastic job losses. This situation deteriorated with the world-wide crisis in 1974: due to increasing imports from low-wage countries, decreasing demand for cars and ships and overcapacity also many steel firms were forced to shut down. Since then, the coal and steel industry in the Ruhr came in a heavy crisis, asking for strategies accommodating structural change. It is only since the last two-and-a-half decades that the Ruhrgebiet get chances to regain its competitiveness within Europe.

Coal and steel set their stamp on the demographic development of the Ruhrgebiet as well. Population figures in region rose to a height of 5.7 million people in 1961. With the onset of the crises the population began to decline to about 5.3 million (RVR, 2007a). The Ruhr suffers, like many German regions, from demographic shrinkage: the proportion of old people continues to increase, whereas that of young people is showing a steady decline. Within Germany the Ruhrgebiet does differ, however, in terms of the composition and identity of its population. The regional social structure is a combination of a traditional local working-class population, a growing middle class (jointly: 89,5%) as well as (descendants of) immigrants (10,5%) (RVR, 2007a). During its history the Ruhr got two large waves of immigration. From 1890 until 1918 many Polish enjoying German rights came to work in the Ruhr industry, followed in the 1960s by people from Mediterranean origin. In particular Turkish, Yugoslavians and Italians arrived in the region to accept the low-paid and unskilled industry jobs that were unpopular among the Germans themselves. As a result of the crisis many of the miners and steel workers lost their job. Although new jobs have been created, the greatest social problem in the Ruhrgebiet continues to be its high levels of unemployment. Due to their common heritage, one might expect a strong regional identity in the cities making up the Ruhrgebiet. To the contrary: as the higher authorities in the past did not want to concentrate power in the Ruhr, there is no city that can claim a natural central position within the region. Consequently, the heart of the citizens does not lie with the area as a whole but with their city (Lagendijk & Van Houtum, 2001). This strong 'pride of place' – often symbolized by the local football club – has led to a climate of rivalry between the Ruhr cities. This intercity

Table 1. Sectoral employment rates in % of those employed as well as unemployment rates in the Ruhrgebiet and Germany (1961-2009) (source: Goch, 2002; RVR, 2009)

	Primary sector		Secondary sector		Tertiary sector		Unemployment rate	
	Ruhr	Germany	Ruhr	Germany	Ruhr	Germany	Ruhr	Germany
1961	2.4	13.6	61.3	46.6	36.3	38.8	unknown	0.5
1970	1.5	9.1	58.4	49.4	40.0	41.5	0.6	0.5
1980	1.4	5.3	51.7	45.3	47.0	49.4	5.3	3.5
1990	1.2	3.6	44.4	40.6	54.4	55.8	10.8	6.6
2000*	1.2	2.5	33.3	33.5	65.4	64.0	12.2	8.1
2009*	1.1	2.2	27.8	25.5	71.1	72.3	11.0	7.7

*The sector employment rates and unemployment rates for Germany in 2000 and 2009 are for Germany *after* the re-unification, thus including economic data for the federal states that used to be part of Eastern-Germany.

competition has hindered the development of a common identity and a common strategy for the Ruhr. Rather, the cities are inclined to strive for the local rather than the regional interest. In the last decade, however, local parties such as the KVR increasingly have tried to put a stop to this 'parochialism' and to replace it by an identity of a common Ruhr legacy instead – but until now without much success.

THE ECONOMY OF THE RUHRGEBIET

Today's economic structure of the Ruhrgebiet has been the result of the region's distinctive economy history at the one hand and global economy-wide developments at the other hand. The coal and steel crisis in the 1960s and 1970s has affected the whole Ruhr economy and can be still felt today. Over the last forty years more than 500,000 jobs have been lost in these basic industries, being one of the explanations why the share of employees in the Ruhrgebiet that is working in the secondary sector has decreased from 61.3% in 1961 to 27.8% in 2009 (see Table 1) (Goch, 2002; RVR, 2009). Still, however, there are people employed in coal and steel production, concentrated on a few firms such as Thyssen-Krupp and RAG. Nevertheless, the jobs in this sector continue to sink. The process of closing down collieries in the Ruhrgebiet will go on in the coming years, even if only six mines are in operation now. Moreover, due to high-technology investments less and less workers are needed to control the gigantic equipment in the mines and steel factories. Consequently, most of the remaining jobs in the future are to be found in high-qualified work (such as engineering) rather than production work. This trend of upgrading the economic base can be seen in the whole Ruhr economy. Generally, the secondary sector of the 'new' Ruhrgebiet consists of modern branches now, often represented by large, worldwide known concerns. Examples of manufacturing activities in which the region has achieved (inter)national comparative advantage in recent years include sophisticated energy supply, solar technology, medical technology, industrial design and last but not least environmental technology.

In line with the European-wide expansion of services, it is the tertiary sector that is dominating the Ruhr economy now, both in terms of output and employment. The proportion of people working in services in the Ruhrgebiet has risen from 36.3% in 1961 to 71.1% in 2009, a share slightly lower than the average of Germany as a whole (72,3%) (Goch, 2002; RVR, 2009). Within the inherent heterogeneous category of services especially business-oriented services, household services (like retail and tourism) and banking & insurance services have been growing in the Ruhrgebiet over the years. The strongest branch in household services is the retail industry, as can be observed by anyone visiting one of the many mega-shopping centres (e.g. Centr-O in Oberhausen, the Hansazentrum in Bottrop and the Drehscheibe in Bochum). Retail is followed by health and veterinary services, transport and communication services, wholesale trade, legal and business consultants and real estate agencies. Recent growth areas within the regional service sector are computer and internet services, multimedia, advertising, telecommunications, engineering consultancy and – perhaps a little surprising to some people – tourism (e.g. industrial heritage). Although in absolute terms the development of the tertiary sector in the Ruhrgebiet looks quite favourable, compared with the average of Germany, another view emerges: in comparison with other large agglomerations in Germany the Ruhrgebiet still shows less capacity to generate employment in the business-oriented services. In this respect, it has been suggested that in the last 20-30 years there has been a strong correlation between the number of business services and the development of jobs in the manufacturing sector in the German economy. This 'parallel thesis' suggests that tertiarisation in the Ruhrgebiet cannot be separated from developments simultaneously taking place in the region's manufacturing sector (Illeris, 1996; Bade and Niebuhr, 1998). Indeed, the growth of information and communication services in Dortmund and logistic services in Duisburg, for instance, both can be linked to the city's originally strong electronic respectively transport industries.

The growth of the service sector in the Ruhrgebiet has by no means been enough to compensate for the loss of the manufacturing industries since the 1970s. Despite the successes achieved in the process of structural change, the level of unemployment in the Ruhrgebiet is still relatively high (RVR, 2009). At the moment, the Ruhr labour market lags behind the job development in Germany: in 2009 about 11% of people in the region are out of work compared with 7.7% in the nation as a whole (see table 1 for comparisons in earlier years). Although these differences in job rates are still considerable, compared with the situation in the period 1980-1995 the Ruhr labour market has improved. In this period, also called 'the negative uncoupling', unemployment in the Ruhr reached record levels in 1987 and 1988: in both years the unemployment rate was 15.1%, which strongly contrasted with the national average of 8.1%. The relatively favourable development of the Ruhr labour market later on in the nineties has been primarily boosted by a wave of new firms in the region (Goch, 2002). The share of self-employed, for example, rose from 8.4% to 13.4% of the working population between 1995 and 2000 (Projekt Ruhr, 2002). These figures were above the national average for business start-ups. Most new ventures started in this period are knowledge-based and operate in the field of high-technologies such as information technology and solar technology. Often, they are spin-offs of technology centres or universities located in the Ruhrgebiet. To be sure, these new firms are often started by young entrepreneurs that have just graduated from university. The start-ups, however, offer few possibilities for older employees to get back to work since they do not have the qualifications needed. Nonetheless, the new companies have eased the difficult situation of the regional labour market a bit and may be a source of employment for current and future generations in the Ruhrgebiet.

RE-INDUSTRIALISATION AND LOCK-IN

The strategies pursued in the Ruhrgebiet to bring about structural change since the 1960s fall into two categories: re-industrialisation and regional innovation – with the last strategy being the most successful one. Until 1984 most efforts to counter industrial decline in the Ruhr were aimed at preserving the traditional regional-economic structure. The catchword of the local parties was 'defence is the best attack'; consequently the region was largely re-industrialised in the 1960s and 1970s (Pohl & Ponthöfer, 2002). Confronted with an increasingly shrinking world-market for coal and steel large concerns such as Thyssen, Mannesmann and Krupp, for example, tried to keep competitive by additional investments that should raise their scale and productivity. Furthermore, within the regional basic industries cooperation increased since the 1960s, a trend that led to several mergers between former competitors and closer linkages with users and suppliers. In addition, starting in 1970s the large concerns bought up firms in associated branches such as processing and industrial technologies, but mostly outside the region. This strategy of 'external diversification' transformed the coal and steel concerns that were merely operating on a local scale into large multinational, mixed concerns. At the same time, however, this policy did not result in many new activities within the Ruhr itself. Despite all the efforts to keep competitive, many mines and plants were forced to close down. The closings, took place gradually and were socially controlled. In mining, for instance, the workers that were fired got large sums of financial compensation; alternatively, they were simply retired at an age of 49 years. Simultaneously, however, a new generation of young miners and steel workers was educated with subsidies by the local government. In short, the structural change approach in the Ruhrgebiet in that time was merely defensive: the local actors still believed that the region's future would lie in coal and steel.

This is not to say that offensive policies were entirely absent in the Ruhr during the 1960s and 1970s. At the outset of both crises, especially local authorities did their utmost to look beyond the heavy industries' short-term interests and think about long-term regional growth (Grabher, 1993). For the first time in its history, the region got institutes for higher education: in 1965 the Ruhr University was set up, followed by the establishment of other universities and polytechnics throughout the region. Next, to diversify the economic structure the local government tried to attract inward investments, predominantly in industries then unknown to the Ruhr such as micro-electronics, automotive and chemicals. Apart from the arrival of car-manufacturer Opel to the Ruhr (that built a factory in Bochum) this policy largely failed though (Hospers, 2004). Most of the time the government's restructuring initiatives were frustrated by the local industries. The reaction of Gustav Krupp to the establishment of higher education in the Ruhr, for instance, was illustrative for the mindset of most of his colleagues: 'What we need in the Ruhr, are muscles, not brains'. To note another example: large concerns refused to give their sites for the attraction of inward investment or stated such unattractive conditions that potential investors finally gave up. In turn, the local authorities were giving in rapidly. With regard to this refusal to sell land, the government did not mind very much, as, after all, the traditional industries paid high income taxes. Likewise, industry lobbies for intervention by the local government often succeeded, for the relationships between managers and politicians in the Ruhr community were close. Soon, therefore, the original offensive attempts were overwhelmed by a self-sustaining coalition of local parties that hampered structural change in the Ruhr rather than supporting it.

It was only in the mid 1980s that the local actors realised that solely relying on the past was not an appropriate regional restructuring policy. After more than twenty years this defensive re-industrialisation strategy was gradually replaced by a more offensive approach. Why did it took so long before the Ruhr parties changed their mind – although it was clear for the outside world that coal and steel were anything but drivers for growth in Western-Europe? The clue might lie in a phenomenon characteristic for old industrial regions known as 'lock-in', referring to the possibility that regions become locked-in into rigid trajectories (Grabher, 1993; Hassink & Shin, 2005). During the re-industrialisation phase of the Ruhrgebiet three types of lock-in were playing a role: economic, institutional and cognitive lock-in. First, the Ruhr's monostructure resulted in 'economic lock-in', paralyzing entrepreneurship, innovation and flexibility. Most firms in the Ruhrgebiet were directly linked with the few large companies dominating the regional economy. These close linkages reduced the need for firms to look outside the region and to innovate. Entrepreneurial spirits were neither activated among individual workers, as they were cared for by their employers from the cradle to the grave. This 'rigid specialization trap' led to a kind of ossification in the Ruhr that prevented firms to react quickly and effectively to exogenous impulses, i.e. the declining demand for coal and steel (Grabher, 1993). Second, the Ruhrgebiet suffered from 'institutional lock-in'. The dense regional institutional tissue led to a self-sustaining coalition of local business men, politicians, labour unions and workers, whose only shared interest was the preservation of the existing structure. In this setting new initiatives did not readily emerge, thus putting a brake on structural change. This is what Granovetter calls 'the weakness of strong ties' (Granovetter, 1973). Third, the Ruhrgebiet had to cope with cognitive lock-in', both inside and outside the region. The local parties had an over-optimistic view of the region's potential: they did not want to believe that the crises were of a structural rather than cyclical nature. The outside world, however, had a one-dimensional, negative view of the Ruhrgebiet as an unattractive, polluted industrial region. As a matter of fact, this was wrong: at that time the area was one of the greenest regions within Germany (RVR, 2006). If anything, the cognitive distance between the region's identity and image may offer an additional explanation why the Ruhrgebiet failed in attracting inward investment that could contribute to its renewal.

BREAKING OUT FROM LOCK-IN

Since the mid-1980s structural change policy in the Ruhrgebiet has been gradually changed from a re-industrialisation strategy towards a regional innovation approach. More than before, emphasis was laid now on the development of new, future-oriented branches of economic activity around the old industries in the region. For one thing, this change in orientation came from the large coal and steel concerns themselves: despite all their earlier efforts they had to recognise that staying in their traditional business was a dead-end strategy. Ultimately, they realized that industrial decline was not a cyclical, but a structural trend. The increasingly competitive world market simply forced them to diversify internally as well. Although willy-nilly, firms as RAG, Thyssen and Krupp diversified beyond coal and steel and invested in related growth branches like plant engineering, environmental technology and control services (Pohl & Ponthöfer, 2002). For another thing, local authorities played an important part in bringing about a 'break-out' from the Ruhrgebiet's lock-in situation. In 1984 the State of North-Rhine-Westphalia changed its industrial policy in a technology policy and developed a programme aimed at 'sunrise technologies' with a focus on environmental technology (Ache, 1994). Simultaneously, most municipalities within the Ruhr stopped their policy efforts to attract inward investment. It was decided to focus on innovation instead and to set up local technology transfer centres providing advice and services to starters. Within a decade the Ruhrgebiet already counted 29 of such centres, of which some have become very successful (RVR, 2006). Examples are the Technology Centre Dortmund and the Duisburg Electronic Park that both house many high-tech businesses employing hundreds of people.

An interesting example of regional innovation in the Ruhrgebiet is the region's diversification strategy in the field of environmental technology (Schönert, 1995; RVR, 2007b). Interestingly, this branch of activity has its roots in the local coal and steel industry that constantly has searched for innovative ways to keep pollution levels as low as possible. Thanks to the strict environmental rules and the high demand for clean technologies among local firms the Ruhr could build up much expertise in how to counter environmental damage. At the moment, the Ruhr has grown into the centre of environmental technology research in Germany. The cluster has created new employment in the region as well: about 100,000 people are working in this branch and their number is still growing. Local firms, universities, research institutes (e.g. the Soil Protection Centre and the Environmental and Packaging R&D Centre) and environmental agencies closely cooperate, the result being many innovative applications (RVR, 2006). Although the market for these innovations used to be mainly locally concentrated, the Ruhr environmental cluster increasingly sells its output outside the region. Other branches that grew out of the region's industrial past and in which the Ruhrgebiet has developed a comparative advantage now are energy supplies and waste disposal. Due to the massive amounts of energy resources needed and waste produced by the coal and steel factories R&D in the field of renewable resources, recycling and waste combustion was stimulated in the Ruhrgebiet at a relatively early period. Just as is the case with environmental technology, these branches are future-oriented, while they paradoxically emerged from the region's tradition in heavy manufacturing.

The regional innovation strategy the local parties in the Ruhr pursued since the mid eighties was not only novel in terms of its focus on regional renewal. Also the way the approach towards structural change was organized differed from the past. To stimulate the region's endogenous potential, a bottom-up approach rather than top-down strategy was chosen for. The initiative IBA (Emscher Park International Building Exhibition), which lasted from 1989 to 1999, is a case in point for this shift towards decentralising responsibilities in matters of structural change in the Ruhr (Cooke, 1995; Sieverts, 2005). This public-private project was aimed at the economic, ecological and social reconstruction of a densely populated area of 800 square kilometres near the river Emscher that had suffered much from industrial exploitation. Besides coping with structural change in this district, the aim of the project was a modernization of institutions and procedures of regional policy. Consequently, the project's official subtitle was: 'Workshop for the Future of Old Industrial Regions'. Within the context of the IBA the State of North-Rhine Westphalia called upon towns, companies, architects, citizens and interest groups in the Ruhr to make single project proposals that would fit in one of the following central guideline projects: (1) the renovation of the Emscher Landscape Park, (2) the ecological improvement of the Emscher river, (3) the new utilisation of industrial buildings, (4) the development of new working locations and (5) the development of new housing forms and municipal districts. The philosophy of the IBA to give the region itself a say in the Ruhr's revitalisation ('renewal from within') met approval in the local community: in ten years time 123 cooperative projects were implemented, varying from the set up of technology centres, the renovation of apartments and the restoration of industrial monuments for tourist purposes. Especially the last project has obtained world-wide attention, not in the last place because the Ruhr proudly presents its renovated industrial heritage as a unique tourist and leisure attraction now. With campaigns like 'The Ruhrgebiet... is hard to beat', offers like the 'RuhrTopCard' and a region-wide Route of Industrial Heritage, the area has tried to raise the number of tourists and day-trippers to the area and to give the Ruhrgebiet a more positive image in the outside world.

Inspired by the experiences of the IBA, public and private actors in the Ruhr have launched several projects on the way to renew the region. Representative examples of such projects are the cases of 'E-City Dortmund' and 'Solar City

Gelsenkirchen' (Pohl & Ponthöfer, 2002; RVR, 2007b). The E-City Dortmund-initiative (dortmund-project) was set up in 2000 as an answer to the decision of Thyssen-Krupp to close the local steelworks. To compensate for the resulting job losses, the city of Dortmund, Thyssen-Krupp, consulting firm McKinsey and some other local organisations formulated a vision for the city based on the motto: strengthen what is strong, invest rather than subsidise and initiate pilot projects in public-private partnership. Because manufacturing, information technology and logistics have played a leading role in Dortmund since long, the local parties have done their utmost to create jobs in the interface of these sectors (such as e-logistics and robotics) by investing in incubator centres and start-up promotion. Part of the new technology infrastructure is created on Thyssen's former steel industry site. With such measures, Dortmund may transform 'from steeltown to e-city'. Also the city of Gelsenkirchen explicitly draws upon its economic legacy. In the industrial heydays, energy supply for steel production was one of the main drivers of the local economy; Gelsenkirchen was called 'the city of the thousand fires' (following the many small fires in the steelworks). This background as an 'energy city' was recognized in the 1995 decision of local authorities and companies to construct a Science Park in the field of energy technology on the site of a run-down factory. The niche the park found was solar technology, a modern field in which the expertise of the traditional sectors could be combined with state-of-the-start high-technology such as photovoltaics. The park has been able to attract Europe's biggest solar cell factory (Shell Solar International), solar housing estates and a number of other solar-related activities. Consequently, Gelsenkirchen tries to raise a profile as a hot spot in solar and other modern renewable technologies. With a wink at the past the location brands itself as 'the city of thousands suns' or simply 'Solar City Gelsenkirchen'.

TAKING STOCK OF THE RUHR'S STRATEGIES

Obviously, it is hard to evaluate the overall impact of the regional innovation approach in the Ruhr since the mid-1980s. The added value or difference made by the particular efforts of the local actors is hard to isolate from factors beyond their control, including international, domestic and regional forces and trends. What would have happened anyway, without a deliberate transition of the regional change strategy? Besides, with regard to some initiatives it is too soon to draw conclusions on their effectiveness. Nevertheless, some preliminary observations can be made about the impact the changing policy orientation has had on the Ruhrgebiet. On a general level, we think that regional innovation has been a more adequate response to regional restructuring than the re-industrialisation efforts in the 1960s and 1970s. What is more, the regional innovation strategies in the Ruhr can be viewed as a creative policy experiment, both in its substantive content and in its procedural set-up.

For a start, the experiences in the Ruhr suggest that the lock-in situation can be broken by the involved actors indeed – and in particular by the local government – by drawing on the existing regional-economic structure. To be sure, it was the market that forced the large firms in the Ruhrgebiet already to look beyond their traditional core business since the 1980s. With its policy change in 1984, however, the State of North-Rhine Westphalia certainly has catalysed and focused this process of regional diversification. It finally realised that modern high-tech technologies per se without a base in the industrial past (unrelated diversification or re-industrialization) simply were a bridge too far for the region. Instead, the re-orientation towards technologies related to the industries' expertise (related diversification or regional innovation) has born fruit over the years. Among such technologies as energy technology and waste disposal especially environmental

Table 2. The environmental sector vis-à-vis an old industrial region (source: translated and adapted after Schönert, 1995 and RVR, 2006)

Characteristics environmental sector	Characteristics old industrial region
- emerging sector with high returns	- at end of product life-cycle
- new and high-quality employment	- large unemployment figures
- high degree of technology-intensity	- relatively low levels of R&D
- international competitive sector	- fierce international competition
- usually diversified company structure	- dominance of few large firms
- endogenous potential for the future	- searching for new markets
- improvement of ecological situation	- considerable ecological problems

technology is a promising 'new combination' for the traditional firms in the Ruhr. Table 2 lists the general advantages the environmental sector has for old industrial regions in decline (Schönert, 1995; RVR, 2006). On the hand, today a growing demand for clean technologies can be observed. This structural trend comes from the greater attention in current post-industrial society for issues that have to do with 'the quality of life' such as sustainability. On the other hand, old industrial regions themselves can benefit from a supply-side effect from investments in environmental technology: such investments can contribute to an improvement of the region's ecology that often has been damaged due to industrial (over) exploitation. Thus, in the current inter-territorial competition a region may become more attractive for residents, firms and visitors and enlarge its development chances. In our view, therefore, the Ruhr has chosen the right track by focusing on the environmental technology cluster and related activities (e.g. solar technology) as future growth branches replacing the heavy industries of the past.

Next, the IBA project and the subsequent bottom-up initiatives in the Ruhrgebiet can be seen as an institutional innovation that has facilitated regional structural change. In a sense, with the IBA the State of North-Rhine Westphalia ran a risk by adopting a change strategy that was explicitly aimed at 'renewal from within'. As a matter of fact, the same parties whose conservative attitudes and actions had led to lock-in were asked now to generate innovative ideas within the state's central guideline projects. In assessing the IBA, some authors contend that the traditional consensus culture among local business men, politicians and the civil society indeed has resulted in less innovative projects than could have been the case. Nonetheless, in ten years time the IBA has dealt with 123 projects with as concrete results among other things the cleaning of 300 square kilometres green space, the construction of 17 technology centres, 6,000 flats and the restoration of many industrial monuments (RVR, 2007b). Set aside these results, the idea to promote 'industrial tourism' in former mines, cooking plants and steelworks certainly has been an innovative idea that the former local industry coalition has brought forward. Also the recent dortmund-project shows that the local business community has abandoned its earlier defensive thinking: Thyssen-Krupp closed its Dortmund steel works while trying to secure jobs for the city by participating in a public-private partnership. By tapping creative potential on the spot, the IBA has raised the awareness of the importance of joint local action in countering industrial decline. As such, the project has been really able to realise its original sub-goal to be an experimental and creative 'workshop for the future of industrial areas'. It is this very role of local empowerment and responsiveness for which the IBA and the bottom-up actions that followed have to be praised primarily. The fact is that the quantitative effects of the decentralised change strategy in the Ruhr are probably far less significant than the qualitative effects. For example, the technology centres and industrial tourism sites built with IBA-funds have not been able to replace the enormous number of jobs lost through industrial decline. In view of the structural problems on the Ruhr labour market, the employment effects of the regional innovation strategies should not be exaggerated (Hospers, 2004). At the same time, the IBA has boosted local confidence and the region's image. The latter is confirmed by the growth of tourism in the Ruhrgebiet: between 1990 and 2008 the number of tourists that visited the region has doubled (RVR, 2009). Undoubtedly, the designation of the Ruhrgebiet as European Capital of Culture 2010 (RUHR, 2010) will improve the region's image as a destination for industrial tourism.

PERSPECTIVES FOR THE RUHRGEBIET

The assessment of the regional innovation strategies in the Ruhr cannot be but ambivalent. While both the diversification and bottom-up strategy have turned out to be adequate responses towards regional restructuring, the economic future of the region still is surrounded by uncertainty. On the one hand, to be successful, restructuring strategies in the Ruhrgebiet should draw on the region's particular economic and institutional past. Earlier attempts to rejuvenate the region failed to recognise this important aspect of contingency. Hence, it will be the history of the Ruhrgebiet that shows the way to new economic futures. On the other hand, it is questionable whether such an approach will suffice when looking to what has been realised so far in the region. We saw that environmental technology, energy technology and industrial tourism have been chosen as new branches combining locally built assets to take advantage of global trends in a creative way. But will these activities separately create enough additional jobs to rejuvenate an entire industrial region in decline? Among the future growth branches, the Ruhr's environmental cluster perhaps has the best prospects. The Ruhrgebiet has grown into Germany's centre in this technology now and there are reasons to believe that demand for this technology will be stable or grow in our post-industrial society in which sustainability has become a society-wide concern. But in terms of employment the sector still falls short to be a structural solution for the problems on the Ruhr labour market. The same is true for the sustainable energy branch and the industrial heritage sector.

Instead of developing these activities merely separately, we think it is time now for the public and private actors in the Ruhrgebiet to look where these new branches can complement and possibly reinforce each other. At the moment, for example, industrial monuments in the region are utilised and marketed merely for tourist purposes, while they also might be useful places for starters in new technologies. What is more, the future re-use of the Ruhr's industrial legacy is not restricted to tourism and technology. The striking words of urban planning guru Jane Jacobs may serve as a motto in this context: 'New ideas must use old buildings' (Jacobs, 1961, p. 189). Experiences in Northern-America and the United Kingdom demonstrate how this credo can find practical application. In cities such as New York, Vancouver, Manchester and Newcastle a variety of modern service activities have been housed in industrial heritage. Industrial monuments turn out to be useful places for the establishment of call centres, conference rooms, consultancy services, studios and trendy shops (Hospers, 2002; Mennel et al., 2007). These examples show how the recycling of industrial heritage may fit in the framework of other localised regeneration strategies. Such an extended re-use of industrial relicts for economic purposes is just one of the possible ways how the Ruhrgebiet could combine and coordinate its individual renewal strategies in a more innovative way – and thus could contribute to a more integral approach towards structural change.

Obviously, the scaling up, coordination and branding of the separate renewal initiatives in the Ruhrgebiet asks for joint strategic action encompassing public and private stakeholders in the entire region. There are serious doubts, however, whether the Ruhr will be able to show sufficient organising capacity and leadership 'from within' in order to formulate and implement an integral regional vision for the future. Again, history plays a role here: the drive of higher authorities in the past to prevent the concentration of power in the Ruhr has resulted in a climate of parochialism between cities within the area. This striving for local interest keeps the cities divided and is the reason why many commentators see the emergence of a highly-debated Ruhrstadt as hardly feasible (Lagendijk & Van Houtum, 2001; RVR, 2007a). Although confronted with similar problems, the municipalities making up the Ruhr still prefer to think in solutions benefiting their town or city rather than the region as a whole. It is striking, for example, how the Ruhr cities individually present themselves as the 'places to be', while they could realise synergy by working together. As examples, just think of the aforementioned projects of E-City Dortmund and Solar City Gelsenkirchen. This inter-city competition in the Ruhr has reduced the potential to strengthen organisational capacity and will be an obstacle to future collective action. The main challenge the cities in the Ruhrgebiet face for the future is to join forces. Hopefully, the RUHR 2010-project – in which the cities in the Ruhrgebiet cooperate in the field of industrial culture – is leading the way forward.

SOME CONCLUSIONS AND LESSONS

The enlargement of the European Union has enlarged Europe's Rustbelt as well. Especially in Poland, Hungary, Slovakia, Bulgaria and Romania a number of traditional industrial regions can be found that need to be restructured to be able to revitalize. Although a specific case, the process of structural change in the German Ruhrgebiet might offer some general lessons about how to counter industrial decline in a regional context. First, the Ruhr-case reveals that regional restructuring is a long-winded process lasting decades rather than years. The importance of this insight, so pessimistic but yet so true, can hardly be overemphasized: in dealing with regional restructuring 'quick wins' are hardly possible owing to the persistence of an area's past. Second, the ups and downs in the Ruhrgebiet point to the following lesson: rejuvenation strategies should ideally depart from the existing economic and institutional structure of the region in question. Usually, we think, regional innovation approaches building upon the region's assets will make more sense than simple, rootless re-industrialisation strategies. Third, if older industrial regions are to face a favourable future, they may benefit the most from related diversification. Diversifying the economy can break the industrial monostructure and simultaneously enlarge the area's absorptive capacity for new economic developments. Finally, the case of the Ruhr demonstrates how important it is to involve a range of local stakeholders to secure the support and understanding of regional restructuring strategies. Unlike top-down measures coming from outside, bottom-up policies can tap from the creative potential on the spot. Even in traditional industrial areas locked-in into rigid trajectories – like the Ruhr in the 1960s and 1970s – this probably will work.

In conclusion, we think that the experiences in the Ruhrgebiet are a useful learning device for other 'smokestack areas'. The case of the Ruhr may offer some ideas and traps for industrial regions in Eastern-Europe joining Europe's Rustbelt. After years of experimenting and with all the remaining problems the Ruhrgebiet at least shows that structural change is possible and may result in such interesting 'new combinations' as environmental technology and industrial tourism. Obviously, it is tempting to use such promising restructuring experiences as basic ingredients for the formulation of a 'best practice' to recommend for application in the rest of Europe. Doing so would be a contradiction in terms though. If anything, one overall lesson of the case of the Ruhrgebiet is the importance of what the economist Friedrich von Hayek aptly has called 'the particular circumstances of time and space' in economic life (Von Hayek, 1948). Likewise, appropriate strategies for regional renewal in an enlarged EU should ideally emerge from a careful consideration of what is suitable, acceptable and feasible within the particular local context. Although there is no magic recipe for rejuvenating European regions hit by industrial decline, the Ruhr-case is still useful: in every industrial region the precise nature and rate of structural change is not only determined by the particularities of time and place, but also contingent on policy responses by the local community. It is from this notion, we believe, that there is still hope for a sustainable future for Europe's old industrial regions.

REFERENCES

Ache, P. (1994). *Wirtschaft im Ruhrgebiet*. Working Paper 1994/123. University of Dortmund: Institute for Spatial Planning.

Bade, F. J., & Niebuhr, A. (1998). *Zur Stabilität des räumlichen Strukturwandels*. Working Paper 1998/158. University of Dortmund: Institute for Spatial Planning.

Birch, K., Mackinnon, D., & Cumbers, A. (2008). Old industrial regions in Europe: a comparative assessment of economic performance. [iFirst article]. *Regional Studies*, *42*, 1–19.

Cooke, P. (1995). *The Rise of the Rustbelt*. London: UCL Press.

Goch, S. (2002). Betterment without airs: social, cultural and political consequences of the deindustrialization in the Ruhr Area. *International Review of Social History*, *3*, 91–99.

Grabher, G. (1993). *The Embedded Firm: on the Socio-Economics of Industrial Networks*. London: Routledge.

Granovetter, M. (1973). The strength of weak ties. *American Journal of Sociology*, *78*, 360–380. doi:10.1086/225469

Hassink, R., & Shin, D. H. (2005). Guest editorial 'The restructuring of old industrial areas in Europe and Asia'. *Environment & Planning A*, *37*, 571–580. doi:10.1068/a36273

Hospers, G. J. (2002). Industrial heritage tourism and regional restructuring in the European Union. *European Planning Studies*, *10*, 397–404. doi:10.1080/09654310220121112

Hospers, G. J. (2004). *Regional-Economic Change in Europe: a Neo-Schumpeterian Vision*. Münster, London: LIT.

Illeris, S. (1996). *The Service Economy: a Geographical Perspective*. Chichester: Wiley.

Jacobs, J. (1961). *The Death and Life of Great American Cities*. New York: Random House.

Lagendijk, A., & van Houtum, H. (2001). Contextualising regional identity and imagination in the construction of polycentric urban regions: the cases of the Ruhr Area and Basque Country. *Urban Studies (Edinburgh, Scotland)*, *38*, 747–768. doi:10.1080/00420980120035321

Lintz, G., Müller, B., & Schmude, K. (2007). The future of industrial cities and regions in central and eastern Europe. *Geoforum*, *38*, 512–519. doi:10.1016/j.geoforum.2006.11.011

Mennel, T., Steffens, J., & Klemek, C. (2007). *Block by Block: Jane Jacobs and the Future of New York*. New York: The Municipal Art Society of New York.

Pohl, W., & Ponthöfer, L. (2002). *Innovationsraum Ruhrgebiet*. Berlin: Cornelsen.

Projekt Ruhr. (2002). *The Ruhrgebiet*. Essen: Projekt Ruhr.

RVR. (2006). *The Ruhrgebiet: Facts and Figures*. Essen: RVR.

RVR. (2007a). *Kleiner Zahlenspiegel der Metropole Ruhr 2007*. Essen: RVR.

RVR. (2007b). *Innovationsbericht Ruhr 2006: Neue Ansätze einer Innovationsorientierten Regionalpolitik*. Essen: RVR.

RVR. (2009). *RVR Online Zahlen & Statistik*. Essen: RVR.

Schönert, M. (1995). *Umweltwirtschaft als Hoffnung für altindustrielle Regionen? Chancen und Risiken für die Regionalentwicklung*. Working Paper 1995/141. University of Dortmund: Institute for Spatial Planning.

Sieverts, T. (2005). *Zwischenstadt: Zwischen Ort und Welt, Raum und Zeit, Stadt und Land* (3rd ed.). Basel: Birkhäuser.

Von Hayek, F. (1948). *Individualism and Economic Order*. Chicago: University of Chicago Press.

Chapter 6
Alliances may Explain the Significant Growth of Biotechnology Companies in the Critical Period 1996-2001?

Tomas Gabriel Bas
University Adolfo Ibañez, Chile

ABSTRACT

The biotechnology is a very complex sector and its growth depends on many variables like intellectual properties, venture capital, clusters, public policies, alliances, amongst others. In biotechnology, the complexity surrounding alliances can be observed. The market globalization, the exorbitant costs of R&D and the rapid changes in technology, are arguably in the midst of the principal reasons that push companies to establish alliances. Biotechnology companies use this instrument to develop external features in the search for resources and missing expertise. This chapter sets out to identify whether such alliances in biotechnology companies are an advantage in themselves, sufficient for the acquisition of new capabilities and whether they help the growth of these companies. For this approach, a private database of companies (900 companies including years 1996-2001) in the two most advanced countries in this sector: United States and United Kingdom will be used.

INTRODUCTION

The chemical industry, considered as the first to have had a scientific base, constitutes a very heterogeneous sector. Molecular biology is complemented by the chemical industry in its steps, and the integration of two sciences have formed the young biotechnology sector, which is indivisible. In addition, the development of companies from 1920s to now results in the emergence of large multinational companies that work today in biotechnology. The biotechnology companies were restructured while being interested in the always-promising sectors of pharmacy and agrochemicals. These two sectors form oligopoly structures in an environment of very high competition. In these industries, the R&D department is the primary component of importance, and the costs to complete an innovating product are often very high. For example, it will necessarily take from ten to

DOI: 10.4018/978-1-61692-846-9.ch006

Copyright © 2011, IGI Global. Copying or distributing in print or electronic forms without written permission of IGI Global is prohibited.

twelve years and 500 to 800 million US dollars, on average, for a new drug to be made available on the market, included in these figures are the opportunity costs of development (Sachwald, 1994; Drew, 1999; DiMasi, 2001).

This article has a double profile. The first one implies the ambiguity in reason of a complex causal structural industrial cooperation (asymmetric differences). The second profile involves a small number of studies made on the relationship between alliances, advantages and performance (taken in the broad sense: profitability, but also growth, stability, etc) to explain by them the growth of biotechnology companies occurred in the critical period 1996-2001.

First, the methodology and hypothesis used in this work will be described. Secondly, a literature review regarding the theoretical aspects and forms of alliances will be presented. Thirdly, the way in which these alliances contribute to the strategic aims of the partners and the way biotechnology firms cooperate will be described. Finally, the article shows the results obtained through this research and conclusions are drawn suggesting that alliances by itself neither explain nor justify the pace of growth of emerging biotechnology companies. Within this scenario, the question that we will try to answer is: Why biotechnology managers choose collaborations instead of other possibilities? In addition, we want to know, what are the advantages and the disadvantages that the managers can encounter while engaging in the alliances (does this practice really increase the company's knowledge, open the door to intellectual property, make easier access to venture capital and ensure growth?).

We created a database of 900 biotechnology companies from more than 3500 biotechnology companies in the world with and without alliances in the United States and United Kingdom. We selected these countries because they are the leaders in this sector, not only by the number of companies, but also by their quality. We will analyze if the biotechnology companies with alliances develop an advantage in the acquisition of new capabilities, in relation to social capital, intellectual property, venture capital and finally, in which sector they are more active (human health, agriculture, environment).

METHODOLOGY

We used secondary information to create an original and exclusive database of biotechnology companies from the United States and the United Kingdom for the period 1996-2001. SPSS statistical software was used to analyze the data for regressions and correlations. Because several growth promoters are involved in the success or failure of biotechnology companies, we used different dependent and independent variables. The dependent variable was the fast growth of the companies, which was measured by the increase of 50% or more in the number of employees during the period 1996 to 2001. We use the number of employees because we work with private and public companies. We cannot use the income data from private companies, because for these companies that information is not publicly available. The independent variables were treated in a metric approach, for example the age of the companies, or dichotomy form by yes/no (supply/absence of alliances). The variables were: the age of the company (variable metric calculated over a number of years since the foundation), area of exploitation, such a human health or agro/bio (dichotomy yes/no), patents (yes/no), venture capital (yes/no) and finally, alliances (yes/no).

These factors influence the behaviour of the companies at a point in time, such as now, and they show very different performances in an environment of similar characteristics. The literature shows that the companies do not have the same characteristics, nor the same competences or routines; the companies generate a variety of different capabilities, as if each one of them were equipped with limited rationality (Nelson & Winter, 1982).

The companies do not have identical resources and they do not react in the same way either to same internal or external stimuli, which results in certain companies growing much more quickly than others (Azoulay & Weinstein, 2000; Peteraf & Bergen, 2003; Rothaermel & Deeds, 2004). This complexity surrounding the various factors allowed us to construct our research assumptions.

THE HYPOTHESIS

Technological and commercial alliances between specialized biotechnology companies (SBC) and large companies (generally multinationals) are not sufficient to support the fast growth of the SBC. We assume that the alliances by itself do not explain the fast growth or the acquisition of capabilities in biotechnology companies. Circumstances place the large companies and the SBC in a kind of constant interdependence in the search for expertise, resources and knowledge. This situation shows that cooperation is a route almost forced upon the SBC. Alliances can provide financial resources and complementary capabilities to the specialized companies in biotechnology (Hagedoorn & Schakenraad, 1994). These resources can be crystallized in activities like R&D, marketing, manufacture, distribution, protection and defence of the intellectual property as well as approval systems by national organizations (like the FDA in the United States). However, the SBC needs more than a simple alliance for growth. They need a "virtuous circle" with the interaction of all factors such as venture capital, intellectual property and knowledge.

THE THEORETICAL ASPECTS

For several years, alliances between independent companies have not ceased to multiply, in order to start R&D programs, production scheduling or for development and marketing of technological products (Lewis, 1990; Hagedoorn & Schakenraad, 1994; Combe, 1996; Gulati, 1998; Rothaermel & Deeds, 2004). The explanations, which we find in the literature, that justify the existence of such alliances, are numerous and sometimes ambiguous. Several authors simply claim that economic globalization is partly responsible for the growth of cooperation between companies (Lewis, 1990; Hagedoorn, 1993; Policet & Noel, 1994; Yoshino & Rangan, 1995; Dussauge, & Garrette, 1995; 1999; Doz, 1996; Gulati, 1998; 1999). Also, this technological environment is in constant transformation and the speed of change is very fast (Pisano & Mang, 1993).

The term "alliance" or "cooperation", as Yoshino and Rangan (1995) and Ingham Mothe (2000) explain, generates very heterogeneous realities. This concept is sometimes unclear, because it is employed to express all the contractual forms, such as joint ventures, consortia, licenses, distribution and R&D agreements, etc. The definition of alliance or cooperation is relatively vague, but also rich and evolutionary (Bouayad, 1996). In fact, to define alliance is a complex task. This is due to the ambiguity of the terms and the generic use made of these concepts that cover innumerable possibilities ("coalition", "partnership", "agreement", "cooperation between firms", etc). Economic and management scholars have contributed partly to this confusion, from the great quantity of definitions and classifications that they have produced (Combe, 1995). In the same way, the concepts of cooperation or alliance are often ambivalent (Dussauge & Garrette, 1999). The expression "cooperation" relates to the relations established in a durable way to divide limited resources, without putting the autonomy of the partners in jeopardy. In addition, the duration of alliance, according to several authors, is more often of a strategic nature (Aliouat, 1996). Alliances are contractual, formal cooperation agreements, as opposed to the multiple informal forms of cooperation. The definitions show the diversity of concepts. Combe (1995) defines alliance as an association of a formal or informal character, between two or several concurrent companies (or potential competitors) or complementary, with or without financial participation. The allied companies try to rather develop, produce or market goods by sharing their competences than resorting to the commercial contract whose range is limited to the short term or total integration, marked by the disappearance of an entity. This concept coincides with that proposed by Jolly (1995). The author adds that the acquisitions of unilateral grant of licenses, research under contract or total fusions are operations that escape to the alliance. For Gulati (1999) strategic alliances are voluntary cooperation agreements that imply exchanges, division or co-development and which can include contributions such as financing, technology or specific goods.

The growth of high technology companies, based on knowledge, plays a very important role in the development of the agreement. However, this role can be played in various ways. Either the companies choose internal development independently, or they prefer external growth hierarchically or cooperatively. In this chapter, we will focus on external growth, particularly the option of alliance in high technology industry[1].

The alliance strategy can be useful, as Niosi (1994, 2003) explains, for limiting and controlling, the growing costs of internal development (R&D) of high technology. In addition, alliance can help the partners to rise in the market hierarchy (Hagedoorn, 1995; Aliouat, 1996; Nerkar & Roberts, 2004) or development and or access to foreign market (Harrison et al., 2001; Rothaermel & Deeds, 2004). However, Jacquemin and Remiche (1988) illustrate the difficulty in classifying technological alliances, because they are intermediate forms of organization between the hierarchies and the markets.

Knowledge sharing has a crucial role in alliances (Badaracco, 1991). In such situations, each partner contributes with a part of his resources and competencies including technology, equipment, financing, know-how, etc. In response to this contribution, the partners will seek to benefit from the results, which can be shared, or not, in the intellectual property or any other forms of advantage considered in the agreement (Niosi, 1994). The knowledge is created through the partners in the alliance (Badaracco, 1991, Inkpen, 1996, Doz, 1996, Doz & Hamel, 1998, Ingham & Mothe, 2000) as a result of the collective learning generated by the partners during the cooperation agreement. More the agreement is longer, dynamic and interactive, more the possibilities of learning and accumulation of knowledge will increase for the partners. The experience of partner is critical for the success of the alliance (Zahra & George, 2002).

THE ALLIANCE MANAGEMENT

Before approaching the modes of alliance management, it is interesting to note a series of recommendations being used for the development and negotiation of Gaudin (1988) describe like a "reciprocal charter of rights and obligations" in cooperation. This critical charter can be used as the base for better negotiating the agreement. The charter should include program objectives; agreement duration; the implement choice; clarification of mutual interest; instruments of direction, execution and control of the agreement; rules or principles of exploitation and how to protect results; and finally, the liquidation of the cooperation agreement.

Taking into account these elements and stipulating the measurements and objectives most simply and clearly, the alliance agreement should obviously be facilitated (Killing, 1988). Thus, the strategies of a manager will depend on the partner objectives, the time of the alliance, the type of technology to be exploited and the market to be used. It would also be necessary to consider the characteristics of each manager, because there are not two managers who will react in an identical way when facing the same constraint (Nelson, 1994; Williams, 1994; Carroll, 1994). In addition, it is important to consider environmental uncertainty in management. This contingency can be caused by a request of technology, by the preferences of the customer, the actions of the competitors, the governmental policies, and the suppliers' capabilities, but also by the potential partners.

The experience and how the partners are organized to work with each other is one of the keys for successful alliance, (Loranger & Roos, 1992; Zahra & George, 2002). A skillful alliance manager is able to face the problems and overcome the difficulties more easily. The difficulties we refer to are the social environment including language and culture, the other partners, and legal framework (Yoshino & Rangan, 1995). However, it is necessary that the objectives of the allies are as complementary and explicit as possible. It is essential also to establish a joint data carrier and that the process is as interactive as possible (Lorange & Roos, 1992). Thus, the alliance success will depend on the experience, quality, and the manager's talent (Woiceshyn & Hartel, 1996; Zahra & George, 2002).

The alliance management is not an easy task due to the multiplicity of constraints. Managers must compose, within the existing legislation, the structures of the market in which the alliance evolves or with the existing capacities of communication (AECEC, 1992). The size of the partners (asymmetric influence) is another characteristic that should be considered because the input met by the multinational corporation in alliances should not be necessarily the same that of an SBC. In addition, the modes of alliance management will change according to whether they are national or international, high-tech or mature technology, formal or informal, as much as their goals, sometimes, which present divergent characteristics (Niosi, 1994). In fact, the modes of alliance management should normally be specific to the culture of the companies that begin in the cooperation. Moreover, in any agreement, it will be necessary also to consider the management of the intellectual property that will result from this cooperation agreement (Lerner & Merges, 1997). Another important point is the degree of opportunism shown by one or both partners expressed during the cooperation agreement (Axelrod, 1984).

The alliance is a learning organization and represents a source of know-how advantage that depends on the basic intangible knowledge of its members (Moingeon & Ramanantsoa, 1995; Nonaka et al., 1996). Thus, learning is the only way to survive within cooperation, but also to obtain a differential growth. The organizational learning takes place in a process of social interaction through which an accumulation of knowledge occurs (organizational or not), which can contain knowledge or know-how (Ingham & Mothe, 2000). Knowledge is an abstract concept, which is built consciously or unconsciously by the interpretation of a group of information acquired through experience and meditation in the practice, which could give to its owner a mental and/or physical skill in its art (Albino et al., 1999). These authors argue that, from a structural point of view, knowledge is build by the information produced by the persons within the organizations. In their turn, the organizations are used as support or, they offer the context so that the process of knowledge founding takes place thanks to the interaction of the community. The creation of knowledge must thus be a construction in constant interaction (Nonaka et al., 1996). This opinion is divided by Inkpen (1996) who argues that the generation of new knowledge is based on the capabilities of transformation deployed by the managers. They are responsible for setting up the conditions of change necessary for the absorption or the assimilation of all new knowledge. This assimilation represents a process of learning accumulation of their individual skills, as well as the routines belonging to each organization are made profitable by the cooperative project (Albino et al., 1999).

THE ALLIANCES OF BIOTECHNOLOGY COMPANIES

Biotechnology is a technological process founded on knowledge and can be defined as the application of science and engineering to the methods, processes and techniques (referred to as recombination of the DNA to, and/or within biological systems. This definition includes recombination of the DNA, cellular fusion and its related techniques, as well as the advanced techniques like the bioprocess engineering. All these techniques applied directly or indirectly to microorganisms, humans, animal or vegetable cells or parts of these, can facilitate the development and production of new molecules and cells, new organisms and procedures to improve in an innovative way, which already exist. In this context, university researchers including molecular biologists, medical doctors, as well as genetic and chemical engineers dominate the general concepts surrounding molecular biology. However, these same scientists have neither the skills of management to organize companies, nor financing skill to market their products and this is why they are often constrained to search the partners for a cooperation agreement (Powell et al., 1996).

The approach between the specialized biotechnology companies and the large companies begin usually with a research contract, information exchange or simply consultation. Then, if the circumstances are favourable, the companies will be able to think of constituting partnerships that are much more formal. These alliances are often cemented, partly through the support of governments. Being given the importance attached to biotechnology, the governments of the industrialized countries encourage universities and their researchers to patent and market the know-how of their products (OECD, 1987).

Alliances in biotechnology are different from alliances in other technologies. For example, biotechnology is unable to produce prototypes; it must, contrary to the other sectors, face the uncertainty of living organisms, with the ethics and the danger of environmental contamination. The product development in biotechnology is generally longer, more complex, and with higher uncertainty, and in the case of products regarding human health, pre-clinical and clinical phases need to be taken into account. We can consider quasi-immediate outputs from others high-tech products like software, contrary to the biotechnology, whose return on the investments requires several years (Woiceshyn & Hartel, 1996). The legal aspect of biotechnology processes is expensive, complex and slow, contrary to technology information for example (Powell et al., 1996). Thus, uncertainty related to the market is omnipresent.

The objective of strategic alliances between the large companies (generally pharmaceutical) and the specialized biotechnology companies (SBC), is to obtain an advantage for the acquisition of new capabilities, which are very expensive or impossible to obtain inside, or which would take too much time to reproduce inside the company. There is a kind of information exchange (Botkin & Matthews, 1995), which is possible only if each company agrees to lose a certain degree of autonomy. Thus, the SBC will lose a few advantages, especially at the decisional level, in order to gain other advantages. These advantages can be represented in the SBC by the access to financing, laboratory equipments, large international markets, or by the admission to certain markets whose cultural barriers are difficult to cross, as can be the case of Japan or China (Barley et al., 1992).

WHY THE BIOTECHNOLOGY COMPANIES SEEK TO COOPERATE?

The high costs of biotechnology R&D, the long duration of clinical trials, as well as the expenditure of marketing and product commercialization, exceed the financial possibilities of the small-specialized biotechnology companies. To fill this gap, these companies must sign agreements of cooperation with large companies or with peers. These alliances not only provide new sources of financing for the R&D, but they also make possible the participation of various team members to diversify and reduce uncertainties inherent to the development of new products.

In general, alliances between small/large; small/small or large/large companies materialize to combine complementary competences and resources. In this direction, the gaps of the small companies are related to industrial R&D, manufacturing, marketing, obtaining patents and accelerate the governmental approval of products (Baum et al., 2000; Statistics Canada, 2000; Bas & Niosi, 2007). The pharmaceutical companies research new innovating products to fill their pipeline. The sectors where the large companies take part in agreements with smaller companies are often the pharmaceutical, agro/bio, and chemical sector (Shan et al., 1994).

Generally, the SBC will establish alliances with:

- *Multinationals Pharmaceuticals Companies*: By the experience, the potential to increase the probability to patent and exportation, the financial capacity, the distribution networks, and the skill in R&D.
- *Universities*: By the intellectual capital and the equipment of laboratories.
- *Governmental laboratories*: By the high capabilities in R&D and financials contacts.

Forrest and Martin (1992) analyze also the principal reasons that lead the companies to establish cooperation agreements in biotechnology. According to these authors, the allies seek mainly:

- The possibility of fast technology exploitation;
- The medium-term incomes generation;
- The risk and uncertainties partition in the new products development;
- The access to the financing, otherwise very difficult;
- Credibility;
- Expertise in R&D, which the partner has.

The large companies often looking for agreement with SBC, when certain competences are difficult to obtain in the large corporations, and the goal is to increase the number of products in their pipeline (Forrest & Martin, 1992). In addition, many competences of the large companies, in particular in biology, genetics, and proteomic do not exist in the small ones. The large companies thus prefer to join the small ones, rather than with other great multinationals (Martin & Forrest, 1994) in order to preserve their options on new technology.

For a few years (just to the crash of high technologies in 2001), biotechnology has been one of the high-tech sectors having one of the fastest growths in the world. According to a study published in 2000 by Industry Canada, this expansion was four times more quickly than the average of the economy in general. The fast growth of the biotechnology companies has been very high, but on the other hand, it is limited to a very restricted number of companies (Niosi & Bas, 2001; Bas & Niosi, 2007). Small companies accounting for 80% of total compose the biotechnology sector and they are usually very young (less of ten years) (Industry Canada, 2000). Moreover, the companies of this size are very innovating (Stephan & Everhart, 1998; Audretsch, 2001) and generally well equipped with intellectual resources. The theory of the resources of the company and the theory of the competences postulate that the internal capabilities have a direct effect on the growth of the companies (Penrose, 1959; Hamel & Prahalad, 1993; Foss, 1997; 2000).

Table 1. The state of the biotechnology in the world in 2002 (the incomes are in millions dollars)

	USA	UK	Canada	Germany	France	Japan
Incomes from Biotechnology	31,749	3,323	2,893	1,190	859	N/A
Number of Public Companies	339	48	89	20	8	2
Number of Private Companies	1,118	347	302	N/A	N/A	N/A
Total Number of Companies	1,457	395	391	365	240	271

Source: Statistics Canada (2003)

Table 2. The specialized biotechnology companies (SBC) in the USA, Canada and U-K in 2000-2001

	USA	Canada	UK
SBC in Human Health (%)	56%	70%	50%
SBC Total Revenues (mill. US$)	31,000 mill. US$	2,800 mill. US$	3,300 mill. US$
SBC Total R&D Expenditures	15,700 mill. US$	1,000 mill. US$	2,000 mill. US$
Venture Capital Invested in SBC	1,330 mill. US$	488 mill. US$	564 mill. US$

Sources: KPMG; BIO.org; Ernst &Young; British Information Services.

RESULTS

We evaluate the dimension and the dynamics of this technology in the years of more vitality and growth (1996-2001) and we have a well description of the stakes with the biotechnology companies that are confronted before 2001 and after this year. Table 1 and Table 2 present a brief summary of biotechnology situation immediately after 2001 for the principal countries working in this sector. The first table shows how the United States has fully mastered on all levels. First, their prevalence appears by the number of biotechnology companies, bench near 1,500 companies, follow-up far by the United Kingdom with near 400 companies similar to Canada. Germany arrives in fourth place with 365 companies, but that is due to the youth and very small size of its companies. With regard to the public companies (quoted on the stock exchange), once again the United States is at the lead, with 339 companies; Canada occupies this time the second place with 89 companies and the United Kingdom is third with 48 companies. In addition the United States biotechnology companies dominate largely over the incomes plan, with 31,749 million US$, followed by far by the United Kingdom with 3,323 million US$ and Canada by 2,893 million US$. In addition, the reduced number of companies of other countries, the income achieve by there is more limited. This table clearly shows the uncontested leadership of the United States, follow-up by the United Kingdom and Canada. These three countries dominate jointly, with Germany that appears on the horizon like the fourth future world force.

Table 2 shows a particularity, because 70% of the Canadian SBC belongs to the human health sector, whereas this number is 56% in USA and 50% in UK. These statistics illustrate the human health like the more represented segment compares to another sector like agro/bio or environment. As explained previously, cooperation agreement between SBC and large pharmaceutical companies can play an important role for benefiting the hu-

Table 3. United States correlation

		VD Fast Growth	Patents	Human Health	Alliances	Venture Capital	Age
VD Fast Growth	Pearson Sig. (2-tails) N	1 . 137					
Patents	Pearson Sig. (2-tails) N	.168 .050 137	1 . 137				
Human Health	Pearson Sig. (2-tails) N	.330** .000 137	.242** .004 137	1 . 137			
Alliances	Pearson Sig. (2-tails) N	.109 .203 137	.363** .000 137	.423** .000 137	1 . 137		
Venture Capital	Pearson Sig. (2-tails) N	.445** .000 137	.330** .000 137	.242* .004 137	.344** .000 137	1 . 137	
Age	Pearson Sig. (2-tails) N	-.298** .000 .137	-.164 .056 137	-.269** .001 137	-.087 .309 137	-.125 .145 137	1 . 137

**The correlation is significant at 0.01 and we have a multi co linearity with the variables.
*The correlation is significant at 0.05

man health in detriment of another biotechnology sectors. Also, the lack of ethical problems or public approval, as it is the case of agro/bio in Europe, plays an outstanding role. Another interesting element to emphasize is the numbers of revenues generated by the SBC, those that are dominated by the USA with 31,000 millions US$ followed by UK with 3,300 millions US$ and Canada with 2,800 millions US$. The expenses in R&D show the interest of these countries to this particular technology. The model shows the USA like a comfortable leader in R&D investment with 15,700 million US$, followed far behind by UK with 2,000 million US$ and Canada with 1,000 million US$. Another key factor when we considered the development of the biotechnology companies is the Venture Capital, whose investment is dominated once again by the USA with 1,330 million US$, followed by UK with 564 million US$ and Canada with 488 million US$.

UNITED STATES CORRELATION AND LOGISTICAL REGRESSION

Starting from data obtained from the random sample, we proceeded to make a linear correlation (Pearson) (Table 3). The correlations show multi co-linearity between particular variables like venture capital, patents, human health sector and alliances. This fact is simple to explain, because companies that establish alliances, obtained patent from their inventions, operate in human health sector, and obtained venture capital, are those that presented a fast growth. However, they are strongly correlated and we are in presence of multi co-linearity. This phenomenon appears in the correlations between the variables venture capital and patents, with a correlation of 0.363; the variables alliances and patents have a correlation of 0.363; alliances and human health present a correlation of 0.423 and finally, between the variables alliances and venture capital the correlation is 0.344. Considering this very strong co linearity, we must choose a variable that best explains both the fast

Table 4. Logistical regression of US companies with the independent variable: Alliance

Case in study = 137.
Dependent Variable: Fast Growth (Yes/No)
Independent Variable: Alliance (Yes/No)

Table 5. Omnibus test of model

	Chi-square	df	Sig
Step 1	1.641	1	.200
	1.641	1	.200
	1.641	1	.200

Table 6. Model summary

Step	-2 Log likehood	Cox & Snell R-square	Nagelkerke R-square
1	188.274	.012	.016

Table 7. Classification table

Observed		VD Fast Growth		Percentage Correct
		1(No)	2(Yes)	
Step 1 VD Growth	1(No)	34	34	50.0
	2(Yes)	27	42	60.9
Overall Percentage				55.5

Table 8. Variables in the equation

	B	S.E.	Wald	Df	Sig	Exp. (b)
Alliance	.442	.346	1.631	1	.202	1.556
Constant	-.672	.565	1.417	1	.234	.511

growth of the companies and their stagnation. The venture capital showed the best estimate to explain the fast growth of the companies, with a correlation of 0.445, whereas the alliances shown the meagre estimation with 0.109 i.e. Americans companies which had access to the venture capital more often presented a stronger growth in comparison with those whose access to this type of financing was restricted or null. In addition, as presented in Table 3, most of the companies can see return in a kind of "growth virtuous circle". The companies belonging to this circle are: in human health sector, have patents in America, obtained venture capital, and have established long-term alliances with industrial partners.

Tables 4, 5, 6, 7, and 8 show the impact and the importance to be involved in others areas rather than human health, because other sectors

Table 9. United Kingdom correlation

		VD Fast Growth	Patents	Human Health	Alliances	Venture Capital	Age
VD Fast Growth	Pearson Sig. (2-tails) N	1 . 49					
Patents	Pearson Sig. (2-tails) N	.398** .005 49	1 . 49				
Human Health	Pearson Sig. (2-tails) N	.361* .011 49	.617** .000 49	1 . 49			
Alliances	Pearson Sig. (2-tails) N	.285* .047 49	.529** .000 49	.564** .000 49	1 . 49		
Venture Capital	Pearson Sig. (2-tails) N	.306* .033 49	.531** .000 49	.413** .003 49	.600** .000 49	1 . 49	
Age	Pearson Sig. (2-tails) N	-.146 .318 49	-.281 .051 49	-.298* .038 49	-.223 .123 49	-.262 .069 49	1 . 49

**The correlation is significant at 0.01 and we have a multi co linearity with the variables.

*The correlation is significant at 0.05

were in stagnation or at best, had a very slow growth. Thus, our logistic regression cannot explain large growth and stagnation by only one variable, alliance (60.9%). To obtain better results is necessary to employ other independent variables (venture capital, patents and human health) in interrelation.

UNITED KINGDOM CORRELATION AND LOGISTICAL REGRESSION

Starting from data obtained from United Kingdom random sample, we proceeded to make a linear correlation (Pearson), which is presented in Table 9. It shows the independent variables being able to influence the dependent variable (VD) (fast growth of the UK biotechnology companies). We used the same approach as in the case of the United States by correlating the independent variables, i.e. patents, human health, venture capital and alliances, between them and with the dependent variable (called VD growth) in order to control the co-linearity.

We can observe in Table 9 that the companies in the United Kingdom have very strong co-linearity between ranges of variables. This situation is independent of the phenomenon of multi co-linearity and easily observable in the correlations between alliances and venture capital with 0.600; human health and venture capital, 0.413; human health and alliances with a correlation of 0.564; patents and venture capital 0.531; patents and alliances 0.529 and finally, patents and human health with a correlation of 0.617. As in the case of the matrix correlation of the United States, we confronted the same decision as for the variable explaining the differential growth of the UK companies. The correlation shows that the independent variable "Patent" explains best the fast growth of the companies because it is most strongly dependent with a correlation of 0.398. The British companies with American patents had a stronger growth than those, which do not have. The correlation between the independent variable

Table 10. Logistical regression of UK companies with the independent variable: Alliance

Cases in Study = 49.
Dependent Variable: Fast Growth (Yes/No)
Independent Variable: Alliance (Yes/No)

Table 11. Omnibus test of model

	Chi Square	df	Sig.
Step 1	4.081	1	.043
Block	4.081	1	.043
Model	4.081	1	.043

Table 12. Model summary

Step	-2 Log likehood	Cox & Snell R-square	Nagelkerke R-square
1	63.827	.080	.107

Table 13. Classification table

Observed		Predicted VD Fast Growth		Percentage Correct
		1(No)	2(Yes)	
Step 1 VD Fast Growth	1 (No)	12	13	48.0
	2 (Yes)	5	19	79.2
Overall percentage				63.3

Table 14. Variables in the equation

	B	S.E.	Wald	DF	Sig	Exp. (b)
Alliance	1.255	.643	3.814	1	.051	3.508
Constant	-2.130	1.124	3.594	1	.058	.119

alliances with the dependent variable Growth showed a score of 0.285. However, more important still, we can find that the alliances by itself cannot explain the growth of the biotechnology companies. To explain that we need the virtuous circle with the other variables (venture capital, alliances, human health) to take an active part in the process. These data confirm the assumptions predicting the importance to work in interaction with a virtuous circle and their factors.

We chose to regress the dependent variable with alliance – the variable that yielded the best estimate. 79.2% of companies that growth established a cooperation agreement (Tables 10, 11, 12, 13, & 14). The other independent variables – venture capital or patents – yielded somewhat similar decisive results.

CONCLUSION

The extensive work done from 1980 in the management of biotechnology companies, particularly on the alliances, has tried to present these like the central axis of the growth of biotechnology companies. Although the cooperation agreements are extremely important in the search of missing competencies and resources (access to capital, distribution chains, asymmetrical reduction, greater and better protection of the intellectual property, etc.), we could also say that the alliances by itself do not explained and nor justified the fast growth of many biotechnology companies during the critical period 1996-2001. The alliances are very important for the acquisition of market, capabilities, knowledge and resources, but this is only one variable in a portrait that we could describe as "typical" in biotechnology, which is observable in the "virtuous circle" growth of the companies. The companies that make experimental research in human health, usually protect their inventions by patents, can attract venture capital or any other form of financing and will growth more quickly and better than others will. This strategy may open the doors to potential alliances. A good patent (for example obtained in the United States) also grants prestige for the company. This can draw the attention of the large pharmaceutical companies in search of competences of the SBC and seem for establishes alliances. The power of all these factors under the umbrella of the "virtuous circle" will be positioning favorably the SBC to a fast growth.

REFERENCES

Affaires Extérieures et Commerce Extérieur Canada (AECEC). (1992). *Les alliances stratégiques: passeport pour l'Europe*. Ottawa: Prospectus Publications Limited.

Albino. A.; Garavelli, A. & Schiuma, G. (1999). Knowledge Transfer and Inter-firm Relationships in Industrial Districts: The Role of the Leader Firm. *Technovation*, 9, 53-63.

Aliouat, B. (1996). Les stratégies de coopération industrielle. Collection Gestion. Série: Politique générale, finance et marketing. Ed. Economica. Paris.

Audretsch, D. (2001). The Role of Small Firms in U.S. Biotechnology Clusters. *Small Business Economics*, *17*, 3–15. doi:10.1023/A:1011140014334

Axelrod, R. (1984). *The Evolution of Cooperation*. New York: Basic Books.

Azoulay, N., & Weinstein, O. (2000). Les compétences de la firme. *Revue d'Économie Industrielle*, *93*, 117–154. doi:10.3406/rei.2000.3007

Badaracco, J. (1991). *The Knowledge Link. How Firms Compete through Strategic Alliances*. Boston: Harvard Business School Press.

Barley, S., Freeman, J., & Hybels, R. (1992). Strategic Alliances in Commercial Biotechnology. In Nohria, N., & Eccles, R. (Eds.), *Networks and Organizations: Structure, Form, and Action*. Boston: Harvard Business School Press.

Bas, T. G., & Niosi, J. (2007). The Issue of Asymmetrical Growth in Specialized Biotechnology Firms in the United States and the United Kingdom. *International Journal of Biotechnology*, *9*(1), 87–100. doi:10.1504/IJBT.2007.012170

Baum, J., Calabrese, T., & Silverman, B. (2000). Don't Go it Alone: Alliance Network Composition and Startup' Performance in Canadian Biotechnology. *Strategic Management Journal*, *21*, 267–294. doi:10.1002/(SICI)1097-0266(200003)21:3<267::AID-SMJ89>3.0.CO;2-8

BIS (British Information Services). (2003). *Biotechnology in the UK – The Government View*. Retrieved from www.usembassy.org.uk

Botkin, J. W., & Matthews, J. B. (1992). *Winning Combinations. The Coming Wave of Entrepreneurial Partnerships between Large and Small Companies*. New York: Wiley.

Bouayad, A. (1996). *Les alliances stratégiques*. Paris: Dunod.

Carroll, G. (1994). A Sociological View on Why Firms Differ. In Rumelt, R., Schendel, D., & Teece, D. (Eds.), *Fundamental Issues in Strategy*. Boston: Harvard Business School Press.

Combe, E. (1995). *Alliances entre firmes et course technologique*. Paris: Économique.

Combe, E. (1996). Alliances en R&D, course à l'innovation et gain stratégique; éléments théoriques et application au segment des DRAM. *Revue d'économie industrielle*. 78, 4e trimestre.

Danneels, E. (2002). The Dynamics of Product Innovation and Firm Competences. *Strategic Management Journal*, *23*, 1095–1121. doi:10.1002/smj.275

DiMasi, J. (2001). *Backgrounder: A Methodology for Counting Costs for Pharmaceuticals R&D. Centre for the Study of Drug Development*. Boston: Tufts University.

Doz, Y. (1996). The Evolution of Cooperation in Strategic Alliances: Initial Conditions or Learning Processes? *Strategic Management Journal*, *17*, 55–83. doi:10.1002/smj.4250171006

Drews, J. (1999). *Quest of Tomorrow's Medicines*. New York: Springer-Verlag.

Dussauge, P., & Garrette, B. (1995). *Les alliances entre firmes concurrentes: Évolutions, conséquences pour les partenaires et impact sur la concurrence. Les cahiers de recherche. Paris: Chambre de commerce et d'industrie de Paris. Dussauge, P. & Garrette, B. (1999): Cooperative Strategy. Competing Successfully through Strategic Alliances*. Chichester, UK: Wiley.

Ernst & Young. (2003). *Resilience: Americas Biotechnology Report 2003*. Retrieved from www.ey.com.

Forrest, J., & Martin, M. (1992). Strategic Alliances Between Large and Small Research Intensive Organizations. Experiences in the Biotechnology Industry. *R & D Management*, *22*(1), 42–53. doi:10.1111/j.1467-9310.1992.tb00787.x

Foss, N. (1997). *Resources, Firms and Strategies*. Oxford, UK: Oxford University Press.Foss, N. (2000). Equilibrium vs Evolution in the Resource-based Perspective: The Conflicting Legacies of Demsetz and Penrose. In Foss, N. & Robertson, P. (Eds.), *Resources, Technology and Strategy. Explorations in the Resource-based Perspective*. Routledge Advances in Management and Business Studies.

Gaudin, J.-H. (1988). Structures internes des entreprises conjointes et intérêts privés. In Jacquemin, A. & Remiche, B (Eds), Coopération entre entreprises. *Entreprises conjointes stratégies industrielles et pouvoirs publics*. Paris: De Boeck université.

Gulati, R. (1998). Alliances and Networks. *Strategic Management Journal*, *19*, 293–317. doi:10.1002/(SICI)1097-0266(199804)19:4<293::AID-SMJ982>3.0.CO;2-M

Gulati, R. (1999). Network Location and Learning: The Influence of Network Resources and Firm Capabilities on Alliance Formation. *Strategic Management Journal*, *20*, 397–420. doi:10.1002/(SICI)1097-0266(199905)20:5<397::AID-SMJ35>3.0.CO;2-K

Hagedoorn, J. (1993). Understanding the Rationale of Strategic Technology Partnering: Interorganizational Modes of Cooperation and Sectorial Differences. *Strategic Management Journal*, *14*, 371–385. doi:10.1002/smj.4250140505

Hagedoorn, J. (1995). A Note on International Market Leaders and Networks of Strategic Technology. *Strategic Management Journal, 16*, 241–250. doi:10.1002/smj.4250160307

Hagedoorn, J., & Schakenraad, J. (1994). The Effect of Strategic Technology Alliances on Company Performance. *Strategic Management Journal, 15*, 291–309. doi:10.1002/smj.4250150404

Hamel, G., & Prahalad, C. (1993). Strategy as Stretch and Leverage [March-April.]. *Harvard Business Review*, 75–84.

Harrison, J., Hitt, M., Hoskisson, R., & Ireland, R. (2001). Resource Complementarity in Business Combinations: Expending the Logic to Organizational Alliances. *Journal of Management, 27*, 679–690. doi:10.1177/014920630102700605

Industrie Canada. (2000). Profil économique du secteur de la biotechnologie canadienne. Équipe de recherche et d'analyse. *Direction générale des sciences de la vie*. Ottawa.

Ingham, M. & Mothe, C. (2000). La Coopération en recherche et développement. Les déterminants de l'apprentissage organisationnel. *Revue française de gestion*, 107, 71-82, Janvier-février.

Inkpen, A. (1996). Creating Knowledge Trough Collaboration. *California Management Review, 39*(1).

Jacquemin, A., & Remiche, B. (1988). *Coopération entre entreprises. Entreprises conjointes, stratégies industrielles et pouvoirs publics. Association Internationale de Droit Économique*. De Boeck-Wesmael.

Jolly, D. (1995). Manager un partenariat technologique. *L'Expansion Management Review, 77*, 72–82.

Killing, P. (1988). Understanding Alliances: The Role of Task and Organizational Complexity. In Contractor, F., & Lorange, P. (Eds.), *Cooperative Strategies in International Business*. Lexington, MA: Lexington Books.

KPMG Corporate Finance inc. (2002). Overview of Selected Issues of the Biotechnology Industry in Four Countries. Centre québecois de valorisation des biotechnologies, Ministère de l'Industrie et du Commerce, Gouvernement du Québec, Investissement Québec, BioQuébec.Retrieved from www.kpmg.com.

Lerner, J., & Merges, R. (1997). The Control of Strategic Alliances: An Empirical Analysis of Biotechnology Collaborations. National Bureau of Economic Research. NBER Working Paper Series No 6014. Cambridge, MA.

Lewis, J. (1990). *Partnerships for Profit*. London: Structuring and Managing Strategic Alliances. Collier Macmillan Publishers.

Lorange, P., & Roos, J. (1992). *Strategic Alliances, Formation, Implementation, and Evolution*. USA: Blackwell publishers.

Martin, M., & Forrest, J. (1994). *Strategic Alliances Between Large and Small Research Intensive Organizations: Experiences in the Biotechnology Industry. Discussion Papers No 132*. Halifax: Dalhousie University, Center for International Business Studies.

Moinjeon, B., & Ramanantsoa, B. (1995). Comment rendre l'entreprise apprenante. *L'Expansion Management Review., 78*, 96–103.

Nelson, R. (1994). Why do Firms Differ and How Does It Matter. In Rumelt, R., Schendel, D., & Teece, D. (Eds.), *Fundamental Issues in Strategy*. Boston: Harvard Business School Press.

Nelson, R., & Winter, S. (1982). *An Evolutionary Theory of Economic Change Harvard*. Cambridge, MA: University Press.

Nerkar, A., & Roberts, P. (2004). Technological and Product – Markett Experience and the Success of New Product. Introduction in the Pharmaceutical Industry. *Strategic Management Journal, 25*, 779–799. doi:10.1002/smj.417

Niosi, J. (1994). *New Technology Policy and Social Innovation in the Firm*. London: Printer Publishers Inc.

Niosi, J. (2003). Alliances are not Enough Explaining Rapid Growth in Biotechnology. *Research Policy*, *32*, 737–750. doi:10.1016/S0048-7333(02)00083-5

Niosi, J., & Bas, T. G. (2001). The Competencies of Regions – Canada's Clusters in Biotechnology. *Small Business Economics*, *17*, 31–42. doi:10.1023/A:1011114220694

Nonaka, I., Takeuchi, H., & Umemoto, K. (1996). A Theory of Organizational Knowledge Creation. *International Journal of Technology Management*, *11*(7/8), 832–845.

(1987). *OECD*. Paris: Biotechnology and Patent Protection.

Penrose, E. (1959). *The Theory of the Growth of the Firms*. White Plains, NY: M.E. Sharpe, Inc.

Peteraf, M., & Bergen, M. (2003). Scanning Dynamic Competitive Landscapes: A Market-Based and Resource-based Framework. *Strategic Management Journal*, *24*, 1027–1041. doi:10.1002/smj.325

Pisano, G., & Mang, P. (1993). Collaborative Product Development and the Market for Know-how: Strategies and Structures in the Biotechnology Industry. *Research on Technological Innovation. Management and Policy*, *5*, 109–136.

Policet, A., & Noël, A. (1994). *Les alliances: une stratégie de développement pour les nouvelles PME technologiques? Cahier de recherche. Centre d'études en administration internationale*. CETAI.

Powell, W., Koput, K., & Smith-Doerr, L. (1996). Interorganizational Collaboration and the Locus of Innovation: Networks of Learning in Biotechnology. *Administrative Science Quarterly*, *41*, 1–16. doi:10.2307/2393988

Rothaermel, F., & Deeds, D. (2004). Exploration and Exploitation Alliances in Biotechnology: A System of New Product Development. *Strategic Management Journal*, *25*, 201–221. doi:10.1002/smj.376

Sachwald, F. (1994). *Les défis de la mondialisation, innovation et concurrence. Institut français des relations internationales*. Paris: Masson.

Shan, W., Walker, G., & Kogut, B. (1994). Interfirm Cooperation and Startup Innovation in the Biotechnology Industry. *Strategic Management Journal*, *15*, 387–394. doi:10.1002/smj.4250150505

(2003). *Statistics Canada*. Ottawa: Biotechnology Use and Development.

Statistique Canada. (2000): Comment expliquer la croissance rapide parmi les entreprises canadiennes de biotechnologie? *Bulletin de l'analyse en innovation*. 88-003-XIF. Ottawa.

Stephan, P., & Everhart, S. (1998). The Changing Rewards to Science: The Case of Biotechnology. *Small Business Economics*, *10*, 137–151. doi:10.1023/A:1007929424290

Williams, J. (1994). Why do companies Differ and How does it Matter. In Rumelt, R., Schendel, D., & Teece, D. (Eds.), *Fundamental Issues in Strategy*. Boston: Harvard Business School Press.

Woiceshyn, J., & Hartel, D. (1996). Strategies and Performance of Canadian Biotechnology Firms: An Empirical Investigation. *Technovation*, *16*, 231–243. doi:10.1016/0166-4972(95)00066-6

Yoshino, M., & Rangan, U. S. (1995). *Strategic Alliances*. Boston: Harvard Business School Press.

Zahra, S., & George, G. (2002). Absorptive Capacity: A Review, Reconceptualization, and Extension. *Academy of Management Review*, *27*, 185–203. doi:10.2307/4134351

ENDNOTE

1 Through the word "high technology", which is rather vague in its definition, we want to refer to biotechnology basically, but also to advanced materials and electronics. For more details to see OECD, codes CAB (http://www.oecd.org).

Chapter 7
Optimal Policy for Biopharmaceutical Drugs Innovation and Access in India

Rakhi Rashmi[1]
University of Nice, France

ABSTRACT

In theory, patents work by providing the inventor an incentive to invent in the first place and then to disclose. Disclosure to the public is rewarded by giving the inventor a monopoly. As product patent and higher patent protection has been advocated by Art 27.1 of the TRIPs agreement on the basis that for greater innovation through transfer of technology is a necessity in developing countries like India as it provides capital to fund expensive innovations, who are otherwise not be able to fund expensive innovations on its own. On the other hand, at the same time drugs are also related with the health of the people and to take care of the health of the people is the utmost priority of any Government and there are issues like accessibility with regard to strong patent protection to biopharma products and data exclusivity. Also as per Art 7 of the TRIPs transfer of technology has to occur to the developing countries in order to promote technological innovations, which is conducive to social and economic welfare. Therefore, striking the right balance between incentive and public access creates a tension is essential. This study suggests optimal policy (Patent and other regulations) to have a balance between biopharma drugs innovation and their access in India while complying with the provisions of the TRIPs agreement by broadly categorising variables such as (1) patent policy such as the scope of biotech patents and the extent of the right in terms of breadth and length; and (2) regulatory environment such as the taxation incentive, Investment policy, Government initiative for the development of this sector etc.

DOI: 10.4018/978-1-61692-846-9.ch007

Copyright © 2011, IGI Global. Copying or distributing in print or electronic forms without written permission of IGI Global is prohibited.

INTRODUCTION

Useful inventions in the field of biotechnology have contributed significantly in recent years for the benefit of the humanity as many different technologies in chemistry and biology are being combined to develop new therapeutics (Bioventure on Global Health, (n.d.).). For example, advances in the recombinant DNA technology, study of the cell growth, gene therapy proteomics, and bioinformatics contribute to the development of proteins can provide cures for many chronicle and hereditary disease as Alzheimer disease (Bioventure on Global Health, (n.d.). These inventions are important for a country like India where there is widespread of these diseases. According to World Health Organization Special Programme for Research and Training in Tropical Diseases (WHO/TDR) neglected diseases are leishmaniasis, schistosomiasis, onchocerciasis, lymphatic filariasis, Chagas disease, malaria, leprosy, African trypanosomiasis, tuberculosis and dengue. It is further categorized into most neglected and neglected diseases.

Neglected diseases are considered to be the highest number of patients (Commission on Intelligent Property Rights, 2003). Three main neglected diseases, HIV, tuberculosis and malaria accounts for 40% of deaths caused by infectious diseases (Bioventure for Global Health, 2006).

India is burdened with a larger HIV/AIDS epidemic than any other country in the world. It accounts for almost 13 percent of the 40 million people living with HIV/AIDS globally and over 69 percent of the 7.4 million people living with HIV/AIDS in the Asia and Pacific region in 2003 (NACO report, 2003). At the same time investment for these drugs innovations is negligible (Figure 1), therefore availability through technology transfer from the multinational innovator companies are desired.

Higher patent standards and data exclusivity provisions have been advocated on the ground that the same will contribute to greater transfer of technology which will lead to greater innovation in India (R. Rashmi, 2007). In the absence of such provisions they are reluctant in introducing new drugs in India. In biotechnology sector, discovery of entirely new drug takes years and costs million of dollars, where as the copy of the same can be manufactured in very little time and in fraction of the money spend in the discovery of new drugs. In biotech innovation only 22 percent of drugs that enter clinical trials eventually receive FDA approval (Masi, Hansen, & Grabowski, 2003). Also, it costs about $400 million, on an average, in out-of-pocket expenses to develop a new drug (Masi, Hansen, & Grabowski, 2003). Also in the absence of such, modern medicine which reaches barely 30% of India's own population and there is no coverage for over 600 million people (Pharma, 2001). Thus, in order to recoup the high and rising costs of biotech R&D, inventors need to capture enough of the economic returns to make their investment worthwhile through stronger patent protection. As patents grant an exclusive right to exploit a specific product or process for a set period of time, which protects new products from competitors, and enable exclusive right to market. Harding's metaphor of tragedy of commons has been used as a powerful justification for privatizing common property (Hardin, 1968) and the stronger patent protection is seen as crucial for the commercial success for the biopharmaceutical companies as they sustain the large and risky R&D expenditure needed for the product innovation (Lall, 2003). It also enables them to recoup the significant investments they have made in developing and discovering the new products and processes and bringing them to the market. Further, patent protection enables companies to generate sufficient income to support future research and develop new products. Patents, therefore, are the lynchpins of the biopharmaceutical industry (Parker, 2001). Thus, from the private interest point of view, patents are important as a reward to the innovator to stimulate private investment

Figure 1. Market for neglected diseases drugs

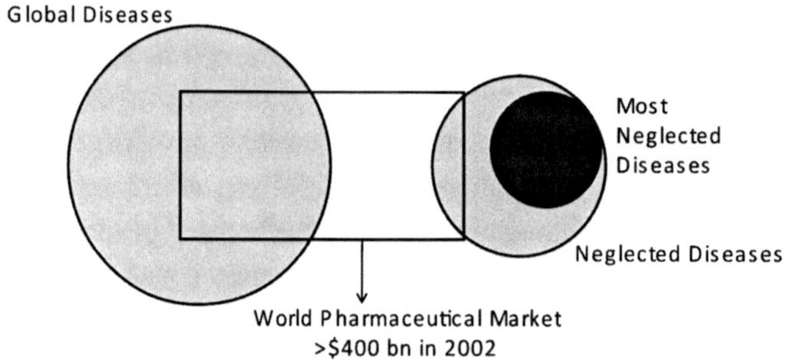

for research and development, which leads to economic growth.

Also the drugs innovations are the part of the industry which works on the profit, but at the same time the drugs are also related with the health of the people and to take care of the health of the people is the utmost priority of any government. At the same time strong patent protection does a poor job of increasing incentives to conduct R&D on products that mostly benefit poor population. In fact the WHA 59.24 resolution reiterates that IP can hamper medical research activities for the development of new products where potential paying market is small or uncertain and may lead to high prices and barriers to access to medicines (WHO IGWG, 2006). (see Figure 1)

Also if we analyse to data provided by business standard then we will find that the per capita expenditure on drugs in India are critically low as India share only 1.2% of the world's drug's production whereas her share of the world population is 16% (Table 1).

If we talk about the importance of biotechnology, it is unquestionable as biotechnology research has much to promise for the human society. Beyond the obvious benefits of biotechnology tools, such as genomics promise to increase the pace of target molecule validation, which will in turn shorten the time required to develop a number of drugs for the same disease (UNCTAD report, 2004). Also, there are many practical applications that are beneficial to human society directly. Rare

Table 1. Country's position in drug production

Country	% of world's drug production	% of world's population
USA	28.2%	4.7%
Germany	7.7%	1.5%
France	7.1%	1.1%
UK	3.4%	1.1%
Brazil	1.7%	2.8%
India	1.2%	16.1%

Source: Business Standard, February 19, 1997

blood types might be created from specialized use of biotechnology. Paralysis from spinal cord trauma may be reversible by using stem cells to replenish damaged and severed nerve cells in the spinal column. Also there are chances that badly needed organs will seemingly be by the application of biotechnology Damaged skin from fires and accidents can be replaced with skin tissue that is grown in the lab from the patient's own stem cells; perhaps even severely scarred tissue from fires can be healed by injecting the damaged skin with potent stem cells.

Such necessary technology needs to be developed by putting more and more investment into R&D and transferred to developing countries to reduce the spread and impact of disease which improves socio economic standing of the improvised populations (Evenson & Westphal, 1995). Now question comes up that if the potential is so great of biotechnology research, then what is all the opposition and controversy for? Actually patenting of DNA sequence is related with the blueprint of life. So, several moral and ethical issues are attached with the patenting of the same.

Also, as discussed before, there are issues like accessibility which have been raised by the health advocates with regard to strong patent protection to biopharma products. They have showed their concern about the impact of the higher patent standards and data exclusivity on access to drugs. As price increase shall be a regular feature and not an accident of strengthening of patent protection in developing countries (Correa, 1998). Prices are crucial because the drug industry differ from other industries where consumer resistance to monopoly prices is limited (Keyala, 2001-02).

Thus, there are challenges related to research and development, creation of investment capital, technology transfer, patentability and intellectual property, affordability in pricing, regulatory issues and public health, morality and ethics. Central to these are two key factors: innovation and accessibility to the products of biotechnology. Policies that foster a balance between innovation and facilitating technology diffusion need to be put in place. Innovation in Indian biopharma industry needs support. As per the report of Commission on Intellectual property, Innovation and public health (CIPIH), where task was to assess the impact of current incentive mechanism for R&D in stimulating needs driven innovation and formulate concrete proposals for action by national and international stakeholders who have found that there is great divide between the possibilities offered by recent and ongoing scientific advances and the difficulty of translating them into products and deliver them in ways that can benefit poor people. Innovation is useless in the absence of favourable conditions for poor people in low income countries to access new preventive and treatment products. The demanding TRIPS provisions for stronger patent protection are not to be read in isolation. They have to be interpreted in the light of other provisions found in the preambular Article 7 of the text. The text attempts to balance the rights and privileges of the patent holder with his obligations and responsibilities to the society. Article 7 of the TRIPs takes into account the need to promote effective and adequate protection of IPRs but at the same time stresses the need to ensure that measures and procedures to enforce IPRs which do not themselves become barriers to the legitimate trade. In theory, patents work by providing the inventor an incentive to invent in the first place and then to disclose. Disclosure to the public is rewarded by giving the inventor a monopoly. Striking the right balance between incentive and public access creates a tension that must be carefully balanced.

Proper framing, adoption and implementation of strong patent policy to promote biopharma technology in general through foreign technology transfer, stimulating push and pull factors as an important tool to reduce the cost of R&D and create viable market for promoting innovation for drugs of neglected diseases and also checking abuse of higher patents such as high prices by strengthening flexibility provisions of TRIPs along with price regulations, increasing generic competition, tax incentive etc the Government can maintain balance between innovation and access to drugs. Pharmaceutical health policy is crucial part of health policy however the goals of these policies often directly conflict with each other. Therefore Government face dilemma of balancing a dual role of encouraging industrial production and concurrently attempting to curtail drug prices (Kamath, 2001/02).

Therefore this study attempts to suggest optimal policy (Patent and other regulations) to have a balance between biopharma drugs innovation and their access in India while complying with the provisions of the TRIPs agreement. The paper discusses optimal scope of biotech patent in maintaining a balance between free flow of technology which will create innovative strength later and access to drugs, it advances the notion that a patent is not the only legal right that influences innovation and access but rather a collection of other independent variables too that have impact on the incentive to invest in innovation. Altogether these variables can be broadly categorized as:

1. patent policy such as the scope of biotech patents and the extent of the right in terms of breadth and length; and
2. regulatory environment such as the taxation incentive, Investment policy, Government initiative for the development of this sector etc.

OBJECTIVE

The main objective of this paper is to suggest optimal policy for biopharma drugs innovation and their access by taking into account essential part of the controversy with regard to biotechnology patent on the compliance with the provisions of TRIPs along with Indian drug regulations, fiscal and trade legislations to achieve the same.

METHODOLOGY

The paper also suggests optimal policy and mechanism to accelerate the innovation or biopharma drugs for the neglected diseases by taking into account the problems envisaged in metaphor tragedy of commons of Garret Hardin (Hardin, 1968) which has been used as powerful justification for privatizing commons property and also by considering analytically the suggestions of the metaphor tragedy of anticommons by Michael A Heller and Rebeca Eisenberg (Heller & Eisenberg, 1998) which says that more intellectual property rights may lead paradoxically to fewer useful products for improving human health. The paper suggests some mechanisms to address the mismatch between need for health technologies to specifically meet developing country's health need and the business section tradition of lack of willingness to meet that need.

The paper is divided into two segments. In the first segment optimal patent policy has been designed in order to have a balance between access and innovation. In the second segment, the paper makes a suggestive analysis for optimal regulations which are relevant for the technology transfer and foreign investment in Indian biopharma sector other than patents. The first part of the paper scrutinizes the amended provisions of Indian Patent Act 1970 (all three amendments) and suggests optimal patent length and patent breadth. As for a balanced patent policy, patent subject matter, patenting requirements and patent breadth are three basic tools for policy makers to consider. In order to design a balanced policy which could be used to enhance both innovation and access (Encaoua & Martinez, 2006) taking into account inherent flexibilities of the TRIPs provisions have been taken into account as the agreement leaves considerable room for legislating national law (Correa, 1998). So, the study takes into account the India needs and circumstance, also the flexibilities such as compulsory licensing provision, Parallel importation, Bolar provisions etc which are inherent in the TRIPs, for the promotion of the societal interest and provides adequate safeguards and deterrent to prevent and control abuses of patents in an effective, efficient and predictable manner to have balance between both the interest groups.

The second part of the chapter takes into accounts drug regulations and fiscal policies of Government of India and analyse taking into account India's need and circumstance as the study is of the view that the optimal scope of patent policy needs to be balanced along with the coherent interaction with other regulatory or economic policies in order to attract investment flow in the country (Love, 2003).

CASE STUDY

Part I: Optimal Patent Policy

Optimal Patent Policy for India to Balance Drug Innovation and Access

Scope, is an important measure of the degree of patent protection (Aoki, Kubo & Yamane, 2006). The scope of patent is defined by its breadth, its length and by the degree of novelty (Van Dijk, 1996). The length of the patent protection characterizes the duration of monopoly power; the breadth of patent defines range of products that are encompassed by the claims of the patent. Therefore protects the patent holder against potential imitators (Merge & Nelson, 1990). On the basis of the above discussion, this segment is divided into two sub segments: optimal length and breadth of the patent for India. Degree of novelty is included in patent breadth segment. Each will be discussed in the light of the issues in controversies and then take into account interest of both the groups for the suggestion of best suitable option.

i. Optimal Breadth for Biopharmaceutical Patenting in India

A patent holder has the right to prevent others from making, selling or using the invention protected by the patent, so the *breadth of a patent* is the extent of protection granted to patent holders against imitators and follow-on inventors (Encaoua & Martinez, 2006). In principle, patent breadth is determined by the claims accorded by the patent examiners to the patentee, defining the boundaries between what is protected and what is not, and by the courts interpretation of these claims during litigation procedures (Organization for Economic Cooperation and Development report, 2004).

As mentioned previously, a biotechnology patent could in practice contain product claims encompassing either in the form of substance of cloned genes, recombinant proteins, monoclonal antibodies, plasmids, vectors etc or composition of matter such as multivalent vaccines, pharmaceutical mixtures etc (Otto, 2004). The scope of biotechnology patenting consists of several factors and the acceptability and non acceptability of the same determines the broadness or narrowness of biotech patents (Bostyn, 2004/05). These factors are: animal and plant patenting, microorganism patenting, patentability of genes and DNA sequence etc (Bostyn, 2004/05).

So, patent claims can have literal as well as broader interpretations. On the scope of biotech patentability currently, two schools of thought exist with different interpretations. The first school states that there is no scientific basis to support the patenting of genes and genomes, which are discoveries at best, while the second school states that characterization of genes is not straightforward, so it does constitute an inventive step and therefore should be patentable (Bostyn, 2004/05). Different countries have different practices and following to different schools. In order to find the optimal breadth of patent for India the study takes into account each components discussed before which constitute the scope of biotech patents in the light of the needs and circumstances of India by making comparison of the practices in US, EU. Thus, this segment of the paper first analyses the Indian patent policy (Indian Patent Act 1970 with all amendments) in the light of requirements of TRIPs agreement and also in order to understand the current scope of the biotechnology patent. Then taking the needs and circumstances of India and compare the same with the provisions of TRIPs agreement, US and EU laws to know the maximum limit to maneuver the same in order to maintain the international patenting standard which should ensure fair and reasonable remuneration to the patent holder, and facilitate prompt transfer and dissemination of technology while maintaining public interest.

Most liberal approach for biotech patents is adopted by Australia as the landmark judgment of Australian patent office (APO) held in Rank hovis Mc Dougall case (Rank hovis McDougall case, 1978) that living organisms are patentable implying that they are inventible. Other than human beings there is no prohibition in Australian patent Act against patenting of life forms. This liberal approach has been followed by USA if we refer to the landmark judgment of Diamond Vs Chakrabarty (United State Supreme Court, 1980) which declared inventions related to genetic engineering and living organisms as inventive. The court granted patent to a genetically modified bacterium and held that inventor's claim is not hitherto unknown natural phenomenon but to a non naturally occurring manufacture or composition of matter or product of human ingenuity having a distinctive name, character and use." Looking at the trend of the judicial decisions in US, it can be easily inferred that patent can be granted for developing manufacturing processes of the biological substance which is discovered in the nature and isolated from its source (Diamon v. Diehr, 1980). Thus after the Supreme court decision in Chakrabarty case it became apparent that non naturally occurring organisms that have been manmade or man altered satisfy the novelty requirement. The same decision paved the way for patentability of transgenic animal when it qualifies under the standard of novelty. Genes are also qualified to be patented if they are isolated and purified form but not if they are simply in the form in which the scientist discovered the sequence (Erramouspe, 1996). So, in US the isolated and purified gene sequences are awarded patent protection in accordance with the novelty standard as per USC 102[2]. On the issue of utility requirement, it is generally assumed that scientist create transgenic animals with a specific use in mind. So, transgenic animals and human gene sequences are clearly useful and utility requirement does not pose a significant barrier to the patent protection. If the patent applicant of gene sequence is capable of showing that his sequence can be utilized for different types of markers, probes and primers for various genetic research then that is enough to qualify the utility requirement under 35 USC 101[3]. About the non obviousness requirement, the basic tenets remained the same as for ascertaining the obviousness of an invention, the invention must be viewed in the light of other inventions in the prior art the invention will not be granted a patent. Moreover, in re Duel case (Deuel, 1995), the Federal circuit seemed to relax the obviousness standard by stating that general motivation to search for some gene that exists does not make obvious a specifically defined gene that is subsequently obtained as a result of that search. More is needed and it is not found here. This decision allowed patents to be granted for DNA molecules even if the method for finding the DNA was obvious to satisfy the non obviousness requirement. Looking at all the landmark decisions of US on the cardinal principle of patenting, it can be said that the scope of biotech patents in US is quite broad which include the patenting of animals and plants as in Diamond vs. Chakrabarty, the Supreme court held that everything under the sun is patentable if they satisfy the cardinal principle of patenting.

Now, if we take the example of Europe then we will find that although as per Article 4[4] of the Biotech Directive (98/44/EC) animals and plant variety is not permitted for patenting but Article 4.2[5] of the same directive states that inventions which concerns plants or animals shall be patentable if the technical feasibility of the invention is not confined to a particular plant or animal variety. In 1998 the European patent office released a directive and clarified matters related to patentability of biotechnological inventions. As per Article 5 of EC directive (98/44/EC)[6], DNA sequences are patentable subject matter as these are considered as synthetic molecules isolated from the organisms and characterized and produced as recombinant molecules containing the information in the natural genes once if qualifies the cardinal principle of novelty, inventiveness and industrial applicability. In case of Biogen vs. Medeva,[7] the court of appeal ruled that even one miniscule part of the genetic code is responsible for the production of relevant protein and then that circumstance the process of choosing a particular sequence among many is an inventive step of invention. The European Directive on Biotechnological Inventions (No. 96/9/EC of March 11, 1996)(Patentable Subject Matter, 2008) establishes that "biological material" and substances isolated from nature (such as new antibiotics) will be considered patentable (Patentable Subject Matter, 2008). European Directive (98/44/EC) provides broader definition of "Biological material" to include "any material containing genetic information and capable of reproducing itself or being reproduced in a biological system" and bring that under the scope of patentable subject matter.

The above discussion was on the practices of cardinal principal of patentability in the policies of USA and Europe, however, the debate has never really been the ability of biotechnology to satisfy the requirements but more contentious debate hovered around the social and ethical aspect of biotech patents regardless of their novelty, utility and non obviousness.

Although, from the above discussion, it is evident that by making a comparison with US patent laws and European patent law is quite conservative in terms of biotechnology patenting. This basically stems from the fact that UK along with Europe prefer to stick to the notion of not wanting to grant patents which would invoke mixed reactions from different sections of the society. The European Union's patent system has some starkly different rules and regulations regarding patents. Central to these differences is the European Patent Convention's (hereinafter "EPC") statutory moral utility doctrine. On account of the differing structures and level of use of the moral utility doctrine, the US and EU responded differently to biotechnological innovations in general (Chambers, 2002). In fact, the US system adapted more readily and actually promoted biotechnological innovation, whereas the EU system seemed for a time to stifle it. European law tries to strike a balance via the Council Directive 98/44 on the Legal Protection of biotechnological Inventions (hereinafter "Directive 98/44/EC"), which is an attempt to reign in the EPC's strong codified moral utility doctrine. As article 6 of the EPC states that inventions must not violate ordre public or morality (Chambers, 2002). If they do, then citizens are given standing to show that it violates ordre public or morality and can block the issuance of the patent. In many EU member states, unlike the US there are laws prohibiting certain biotech inventions from being patented (38). Under EPC 52 (NACO report, 2003) gene therapy patents are prohibited "because they are not susceptible to industrial application (Chambers, 2008)." Unlike the US system, the Article 53(a) of the EPC prohibits patents for inventions that are contrary to the morals of society. This concept is called ordre public gives "automatic standing to concerned citizens, empowering them to challenge individual patents on the ground that issuance would be morally offensive and allowing the use of the judicial process to shape the law regulating bio tech patents (Chambers, 2008). This type of standing, public order is not available in the US. The issue of patentability of the animals being against the morality or public order was dealt in Harvard oncomouse case in which patent was granted to a genetically modified mouse carrying a specific called an activated oncogene. Although In EU in Florey/relaxin case[8], it was held that RNA for deriving the corresponding cDNA encoding a human protein was not immoral and neither is the patenting of a DNA fragment unethical. However, the grant of the patent application was questioned on the basis of public order and morality under Article 53 (a)[9] in which it was held that transgenic animal is patentable but denied the patent to rodent species. In principle under EC directive 98/44/EC transgenic plants and animals are patentable in Europe but not plant and animal varieties.

Now after this discussion if we see the provisions of the TRIPs agreement of WTO, required to be complied by India seems to go with EU practices. As the TRIPS provides that patents shall be available for any inventions, whether products or processes, in all fields of technology provided they are new, involve an inventive step (non-obvious) and are capable of industrial application (useful). It is important to note at the onset that TRIPS in this regard is similar to the requirements of the EPC. Therefore, it seems that the International community is currently following the EPC in this regard. In addition, article 27 of TRIPS has a statutory moral utility doctrine, again similar to the EPC over against the US patent code. Though US case law shows a general trend towards minimizing the US common law moral utility doctrine, cases minimizing the moral utility doctrine (such as Chakrabarty) were decided before the passage of the TRIPS.

Therefore, it seems that what patent protection TRIPS provides, resembles the EPC over against the US patent law system. Finally, TRIPS like the EPC provides patents for "technological innovations." In other words, US business methods under the EPC and under TRIPS do not seem patentable, unless they are patents for technological innovations.

Now in the wake of the above discussion, it is important to find out the scope of biotech patenting in India. In order to bring the Indian law in compliance with the TRIPs provisions of Art 27(3)[10] under which the members are under obligation to provide patents for microorganism a new clause to section 3 was added by Patent Amendment Act 2002. This provision excluded from patentability, plants, animals in whole or any part thereof other than microorganisms but includes seeds varieties of seeds varieties species and essentially biological processes for the production or propagation of plants and animals. So, inventions related to animals are not patentable in India as a whole. Also, disallows patenting of living subject matter but draws an exception to microorganism patenting.

Thus, India recognized the patenting of microorganism by Patent Amendment Act. Although the landmark judgment of Dimminiaco AG vs. Controller of patents (2002) had paved the way for microorganism patenting way back in 2001 in which it was held by Kolkata high court that the biotech matter as patentable even if the end product of a process is a living virus/ microorganism / living entity. Further the court affirmed that the patent claimed was useful as it protected poultry against contagious disease and the end product resulted in a new article. The court came out with 'vendibility test' as the most effective test to determine whether the process of manufacture ought to be patentable or not. As per the court, a vendible product is one that can be commercially transacted. Thus, a more flexible attitude had been shown. A year after the decision of this case microorganism got the validity to be patented by the Patent Amendment Act 2002. Thus we can make out that although Patent amendment Act 2005 has accepted living entities of artificial origin such as microorganism and vaccines which fulfils the cardinal principle of patentability and not harmful to human, animal or plant health or unethical. Presently there is controversy with regard to the definition of microorganisms as there is absence of a working definition of microorganism in the TRIPs text. In the absence of the same, there are difficulties to clearly differentiate the microbes claimed from the naturally occurring microbes and show that the biological material is an outcome of invention and not a mere discovery.

In the wake of the same, a key question comes up – whether India should adopt a very narrow and limited definition of microorganism to exclude everything other than microscopic organisms including algae, bacteria, fungi, protozoa and viruses or whether it should expand the scope as in the Europe where all biological material containing genetic information and capable of reproducing itself or being reproduced in a biological system?

On the issue of pharmaceutical and microorganism patenting in India, a Technical Expert Group (TEG) was formed under the chairmanship of Dr RA Mashelkar and the group is popularly known as Mashelker Committee which has submitted its two reports, one in 2006 and second in 2009. In both of its reports the committee has concluded that excluding microorganism from patentability would contravene TRIPs. Therefore it can be said that it is in the interest of India to opt for the higher life form patenting by choosing expanded definition of microorganism by biological material and include DNA fragments, genes and proteins to keep pace with the advances of biotechnology and broader definition will give companies reasons to expand their research in these areas and ensure that research will continue. The same cardinal principle of novelty, inventive step and utility should be satisfied in order to patent a microorganism. For novelty criterion to be satisfied, the microorganism should be isolated from the nature for the first time.

At the same time to have proper check in order to maintain the morality and ethics, the need to be a properly and efficiently framed legislation based on moral utility doctrine similar to Article 53 of EC Directive 98/ 44. In this regard it is apt to mention that Government of India has formulated ethics policies to harmonize with Ethical Guidelines for Biomedical Research on Human Subjects developed by the Indian Council of Medical Research in 2000. The Ethics committee set up by the Government of India covers the areas of basic research, genetics, genomics, and education and legal aspects. Also it is important to mention about the Government of India's recent policy on stem cells which is directly linked with morality and ethics. In India, the new stem cell policy which came into force last year has put in place a mechanism for greater private participation and compliance. Review and monitoring of stem cell research has been decentralised at the institutional level. Stem cell research and its funding have been classified into three levels — permissive, restricted and prohibited.

Permissive researches are disease-specific and are being addressed locally at institutional levels. Restricted and prohibited level proposals are to be referred to the central committee and this addresses broader issues like cloning, cell-based engineering and works on reproduction. According to the Policy, stem cell research should be promoted in the country in the view of its potential for clinical use. Research based on stem cells derived from adults, bone marrow or foetal cord blood may be undertaken after obtaining appropriate informed consent and with adequate safety measures. For embryonic stem cell research, embryos should not be generated for the sole purpose of obtaining stem cells. Only surplus, spare or supernumerary embryos can be used after obtaining informed consent of both spouses. Such collection of embryos should be done only from registered Assisted Reproductive Technique (ART) clinics. To safeguard national interests, it is also perceived that all human genetic research, stem cell research and stem cell research involving international collaboration must be undertaken after formal clearance of the national government. Consent form for use in collection of tissue to be used in human stem cell research has also been prepared by the National Bioethics Committee. This policy of India shows that Government of India is liberal towards the biotech innovation at the same time by keeping proper check on the practice so that they should be done maintaining morality and ethics.

Overall, from the above discussion we can make out that in recent years Indian patent legislation has broadened the scope of patentability for biotech products but still not as broad in scope as Australia and US. The scope of biotech patent should be designed in the similar line as EU by keeping the scope broader by extending the same to more biotechnological applications, once they satisfy the cardinal principal of patentability within the framework of TRIPs. Along with broader breadth of patent, there is need to formulate strict morality provisions in the same legislation. In this regard a very active role of judiciary is desirable also efficient and judicious members in the courts are highly recommended.

Now, we will discuss another controversial aspect of India's Patent Act which is related with the breadth of the patent such as exclusion from patentability for derivatives of known substances, unless it can be shown that they are significantly more efficacious than the original substance (India Patents Act, §3(d))[11].

On August 6, 2007, on a writ was petition filed by Novartis AG and its subsidiary, Novartis India, challenging the legal validity of Section 3(d)[12] of the Indian Patents Act (IPA). Challenge was done by Novartis AG and its subsidiary on the ground that Section 3 (d) denies its rights under Article 27 of the TRIPS Agreement, which obligates WTO member states to provide patent protection to all fields of technology without discrimination, and therefore violates the obligations under the TRIPS Agreement. Second, it argued that in the absence of a definition or guideline, phrases like 'enhancement of the known efficacy' or 'differ significantly in properties with regard to efficacy' give uncontrolled as well as unguided powers to the controller of patents. The same would result in an arbitrary exercise of power, and violates the right to equality under Article 14 of the Constitution of India.

On the issue of non compliance with TRIPs the court refused to examine whether Section 3(d) violates the obligations under the TRIPS Agreement[13] and rejected Glivec patent application on the ground that a patent cannot be granted for "the mere discovery of a new form of a known substance which does not result in the enhancement of the known efficacy of that substance." While making the decision the court equated the meaning of efficacy with a therapeutic effect on the body, the court also accepted the argument of the respondents that "...petitioner is not a novice to the pharmacology field but it being pharmaceutical giant in the whole of the world, cannot plead that they don't know what is meant by enhancement of a known efficacy and they cannot show the derivatives differ significantly in properties with regard to efficacy". the court further held that "... in sum and substance what the amended section with the explanation prescribes is the test to decide whether the discovery is an invention or not is that the patent applicant should show the discovery has resulted in the enhancement of the known efficacy of that substance and if the discovery is nothing other than the derivative of a known substance, then, it must be shown that the properties in the derivatives differ significantly with regard to efficacy". Hence it was held that a patent applicant has to show enhanced therapeutic effect in order to obtain a patent for a new form of a known substance or for its derivatives.

So, what the patent applicant is expected to show is how effective the new discovery made would be in healing a disease/having a good effect on the body". Patenting of drugs, the protection to various forms of same substance (salts, esters, ethers, polymorphs, metabolites, pure form, particle size, isomers, mixture, etc.) is often seen as 'ever-greening' (extending incremental protection to a subsisting patent) and hence such protection is objected. As on denial of patent grant to derivatives of known substances unless there's proof that they differ in terms of 'efficacy' can be a denial to a demonstrably new and better version of an existing drug. Due to this many genuinely better drugs are actually improvements on existing formulations.

Ever-greening refers to an extension of a patent monopoly, achieved by executing trivial and insignificant changes to an already existing patented product, whereas 'incremental innovations' are sequential developments that build on the original patented product and may be of tremendous value in a country like India. Incremental innovation has value not just in improving therapeutic efficacy, but also in providing significant benefits in terms of drug delivery, patient safety and compliance. Therefore, such incremental development needs to be encouraged by the Indian patent regime. Incremental innovation should be patented but not frivolous patents so that patent officials can differentiate between frivolous patents and incremental patents. As far as the implication of the judgment is concerned, it suggests that the meaning of the term 'efficacy' should be judged in light of the therapeutic effect of the new form on the body. The judgment accepts patent for incremental innovation only where the cardinal principal of patent is satisfied as a combination of two drugs may offer substantial improvement in therapeutic effect and may be held patentable.

Thus, it can be interpreted that instead of placing Swiss claims outside the purview of patent protection, the judgment allowed such claims in limited circumstances. As, the application of therapeutic effect as a standard for efficacy does not rule out the possibility of incremental innovation but in a limited way which needs to be broadened. For instance, a combination of two drugs may offer substantial improvement in therapeutic effect and may be held patentable. This should be further broadened and it is recommended that patents on salts, esters etc. should indeed be granted if such products meet the internationally accepted conditions of novelty, involving an inventive step, and capable of industrial application (TRIPS Article 27(1)) in order to have better flow of foreign investment in India. To deny the grant of patent to an invention worthy of protection on the ground that it is likely to be misused is the flaw in the policy. If the patent rights are abused such as charging excessive prices then there are safeguards or flexibilities to curb the same in the regulation itself. Just on apprehension laws should not be framed. Section 3(d) reflects India's attitude to minimize the impact of product by granting it to only those substances that are truly innovative but at the same time it has not defined clearly which needs to be defined.

Allowing broader patents on biotechnology will encourage investment in studies for valuable search to combat diseases. This is good for the progress of science and ensures balance between the interests of the consumer and the innovator.

Broader patentability will act as incentive because biotechnology is a high risk and a high cost form of research which ultimately improves public health and Indian patent standard will in the line of international patent standard. So, it will gain confidence of the investors and newer drugs will be introduced in India early and not after several years of its launch in developed countries. In the absence of such, foreign investors will not be interested in investing in India. For example Novartis preferred to invest in China rather than India. Therefore in order to compete with China India needs to keep her policy framework pro investment too.

ii. Optimal Length for Biotechnology Patent in India

The length of the patent protection characterizes the duration of monopoly power. By the patent (Amendment) Act 2002, India implemented the TRIPs required 20 years patent term as an obligation to WTO as its member to comply with Article 33. Prior to 1999, Indian law provided for relatively short patent terms for pharmaceutical processes term was only five to 7 years. So, the length of the pharma patents has been increased after the 2002 amendment to Indian Patent Act 1970 which is fixed.

Maximum duration of protection is decided on a case to case basis which varies in different sectors. So, it is not possible to know with any certainty what, exactly, is the optimal duration of protection for any particular invention. For the pharmaceutical products, the requirement for Government regulatory approval to market pharmaceuticals limits the effective life of the patent monopoly. Thus longer patent protection is desirable also because biopharmaceutical industry is highly technical and investment intensive. So, the duration of patent should be maximum in order to promote investment for the innovation. In this context we will discuss issue in controversy which is with regard to data exclusivity and Article 39.3 of the TRIPs text. Although data exclusivity or protection is part of drug regulation but it is discussed in this section as it can be termed as the substitute of patent life. There is demand of data exclusivity by the MNCs for having law to protect the clinical trial and other data used to obtain marketing approval of new pharmaceutical products in order to comply with TRIPs provision. Data Exclusivity is a period during which Government health authorities respect confidentially of data submitted to them for a fixed period of time. During that period, regulatory authorities may not disclose or rely on test and clinical trial data of one company to grant marketing approval for generic drugs (Data Protection,(n.d.).).

The multinational pharmaceutical firms claim that TRIPs provision requires "data exclusivity," therefore India should implement data exclusivity legislation. However, generic pharmaceutical companies have been advocating 'new chemical entity' to be defined on the lines of novelty under the patent regime, while the pharma MNC have been advocating a definition to mean a new pharma product which has been introduced for the first time in a country irrespective of the fact whether it is patented or not (Data Protection, (n.d.).). Another controversial matter is that what constitutes 'unfair commercial use' under Article 39.3[14]. In US and EU any use of data submitted by the originator, for granting approval to a subsequent applicant without the authorization of the originator of the data must be considered as unfair commercial use. So there is an obligation for granting the originator of data a period of exclusivity during which national authorities would not be permitted during the exclusivity period to rely on data they have received in order to assess subsequent applications for the registration of similar products.

As per the scholars of developing countries Article 39.3 does not require the recognition of exclusive rights, but protection in the frame work of unfair competition rules (Data Protection, 2008).

Data Exclusivity and General Biopharma Drugs Innovation (other than Drugs for Neglected Diseases)

Bringing solution to the different and opposing views and interests of public health and strong IP protection is important. The predicament before the developing countries like India is how to strike a fine balance between public health and the demand of increased IP protection. Before attempting to analyze these issues it is important to understand the meaning of 'data exclusivity, its genesis, impact on drugs and the reasons for conferment of this privilege. As per Article 39.3, if in respect of pharmaceutical or agricultural chemical products which utilize a 'new chemical entity', the regulatory authority of a country requires submission of 'undisclosed clinical and test data' for granting marketing approval it is under obligation to protect the data against 'unfair commercial use' and also to protect the data against 'disclosure' where it is necessary to protect the public or unless steps are taken to ensure that the data are protected against 'unfair commercial use'.

The disclosure is permitted only in case to protect the public and secondly, where the data is protected from unfair commercial use. Thus, we can make out from the content that individual members can take steps as they deem fit for protecting data against 'unfair commercial use'. The purpose of Article 39.3 can be inferred from paragraph 1 of the same article that is to '*ensure effective protection against unfair competition'*. Article 39.1 provides-

In the course of ensuring effective protection against unfair competition as provided in Article 10bis of the Paris Convention (1967), Members shall protect undisclosed information in accordance with paragraph 2 below with data submitted to governments or governmental agencies in accordance with paragraph below.

'Unfair commercial use' is not defined in the agreement thus there are different interpretations. The term 'unfair competition' has been defined in article 10bis of Paris Convention but that does not identify the actual practice of 'unfair commercial use.

Article 6 of the WIPO model on protection against unfair competition provisions lays down the acts that would amount to unfair competition in respect of secret information, as it says:

Article 6(1) -Any act or practice, in the course of industrial or commercial activities, that results in the disclosure, acquisition or use by others of secret information without the consent of the person lawfully in control of that information (hereinafter referred to as "the rightful holder") and in a manner contrary to honest commercial practices shall constitute an act of unfair competition.

Two points emerge from this general principle, firstly the disclosure, acquisition or use of the secret information must be without the consent of person lawfully in control of that information and secondly it must be in a manner contrary to the 'honest commercial practices'. Thus, the protection is to be granted against "unfair commercial use" of confidential data. Thus, Article 39.3 does permit a national regulatory authority to rely on data in its possession to assess subsequent applications, relating to the same drug, since this would not imply any "unfair commercial use (TRIPS and Public Health, 2001)."

Although the TRIPs text only speaks about the data protection against unfair commercial practices but at the same time such protection is desirable for country like India in order to attract continuous foreign investment to step forward towards the path of novel drug industry. A the Satwant Reddy Committee constituted by Government of India has also recognized that not providing data exclusivity for pharmaceuticals could adversely impact FDI and discourage the launch of new products in India [15].

So, it can be said that it is a good move on the part of the Government by recognizing the importance of protection of the confidentiality of data against its unauthorized use in order to protect the proprietary interests of scientists and to maintain the economic incentives for further pharmaceutical research and development by more inflow of foreign investment. The Government must seek a middle path and provide five years of data exclusivity either starting from the date on which the company markets its product or ending with the expiry of patents whichever is earlier provided that company must have filed a patent in India. Five year of exclusivity seems to be ideal to attract investment. Such legislation is important in order to give strong competition to the competitors like China which gives 6 years of data exclusivity by creating an environment which will attract foreign investors. According to National Institute of Health, lack of data exclusivity in India is the primary reason of India remaining at 9th position whereas China ranked 2nd in funding given by the NIH outside the US (Nair, 2004). Also as per Pfizer India, the lack of data protection is part of the reason that "people are talking about India but investing in China" (KPMG, 2006).

As investors in the field of bio medical sciences or biotechnology look for the country where their intellectual property (IP) enjoys the best protection, and where the environment is most conducive to the creation of new IP. The implementation of DE will support and accelerate the development and growth of biotech industry as it will be key to pharmaceutical companies' decision on the location of clinical trials. India presently has the brain power and economic advantages to carry out clinical trials in local hospitals other public research institutes. Having experience in conducting clinical trials is already a competitive advantage for India to increasing foreign direct investment in pharmaceutical sector. In many instances, a patent will cover the discovery at hand. However, more and more compounds which are not patent protected are being developed and thus in these instances data exclusivity is the only available intellectual property protection. DE would actually allow for more products to be available in the pipeline for generic companies. After DE period expires, the data becomes public domain and can be referenced by generics companies. The same approach of stronger patent protection has been suggested only for general biopharma drugs which does not include neglected diseases drugs taking the viewpoint of the concept of tragedy of commons suggested by Garret Hardin (Hardin, 1968) as at present circumstances India needs to learn technology innovation in this sector for which interactive learning is good way. Such interactive learning can come through the collaboration with MNCs.

So, in order to attract foreign investment for biopharma drug innovation five years of data exclusivity is needed as this will gain the confidence of the foreign investors to invest in India and will give edge over its competitor China. On the other hand it is apprehended that implementation of data exclusivity will bring devastating effect for the India biopharmaceutical industry which basically has a generic base. On this issue it should be kept in mind that India is different from the other developing countries in various aspects. India has a well-developed Pharmaceutical industry, which has already put stable feet in US generic market. Also, India has good outsourcing opportunities for clinical trials, R&D and technical services. India has highly trained scientists. Thus, we can make out data exclusivity for attracting innovations and for the development of industry Indian manufacturers generally work on off patented molecules and will keep on working on off patented molecule as around $ 72 billion worth drugs will go off patented between 2006 and 2010 (Hati, 2006). As per the Biospectrum ABLE survey June 2008, exports account for 58% in the biopharma sector and as per the estimate of Frost and Sullivan, the biogenerics markets in Europe and US alone has the potential to generate sales of $16.4 billion by 2011 at an average growth rate of 69.8%. Also, India's generic manufacturers are less interested in India's market or in other loosely regulated markets where consumers are relatively poor than they are in the wealthy markets of US and Europe. According to a joint study released by BioPlan Associates and the Society for Industrial Microbiology, Europe and the U.S. are expected to be the largest markets for biogenerics once the regulatory hurdle is overcome.

In April, the EMEA gave approval to Sandoz to market Omnitrope, a generic version of Pfizer's human growth hormone. Sandoz also asked the FDA to approve Omnitrope in the U.S. market. In addition to the E.U., Australia approved Omnitrope sales in 2007.

The FDA is still on the way to developing a regulatory pathway for biogenerics in the U.S. where the barriers to the development of biogenerics may be more political than scientific, as many of the world's major biotech players are anxious to defend their expiring blockbuster patents. Therefore Indian generic manufactures will continue manufacturing and marketing of generics into the vast and profitable US market, there will not be any adverse effect on Indian Generic industry (see Table 2). In this way it can be concluded that exclusivity of data for suggested period of time can work as pull factor in enhancing the general biopharmaceutical innovations.

For the innovation of biopharma drugs for the neglected diseases the chapter is of the view that the case of neglected diseases drug innovation and access should be treated differently taking into account the argument put forward by Heller and Eisenberg while using the metaphor tragedy of anticommons that privatization or more intellectual property rights may lead paradoxically to fewer useful products for improving health. This argument has been taken up to create a balance towards public health. Due to the high transaction costs of bargaining, heterogeneous interests among owners and cognitive biases of researchers. Privatisation must be more carefully deployed to serve the public good of biomedical research. As per the views of Heller and Eisenberg (Heller & Eisenberg, 1998) policy makers should seek to ensure coherent boundaries of upstream patents and to minimize restrictive licensing practices that interfere with the downstream development otherwise more upstream rights may lead paradoxically to fewer useful products for improving

Table 2. Patent expiries of biotech drugs from 2001 to 2007

Brand Name	Generic Name	Marketer	Patent Expiry
Cerezyme/ Ceredase	Alglucerase	Genzyme	2001
Humulin	Human Insulin	Eli Lilly	2001
Intron A	Interferon alpha-2b	Schering-Plough	2002
Avonex	Interferon beta-1a	Blogen	2003
Humatrope	Somatropin	Eli Lilly	2003
Genotropin	Somatropin	Genentech/Pharmacia	2003
Betaseron	Interferon beta-1b	Chiron	2003
Nutropin	Somatropin	Genentech	2003
Epogen/Procrit	Epoetin alfa	Amgen, Johnson & Johnson, and Sankyo	2004
Recormon	Epoetin beta	Roche	2005
Novolin	Human Insulin	Novo Nordisk	2005
Activase	Alteplase	Genetech, Boehringer Ingel-helm, Mitsubishi, and Kyowa Hakko Kogyo	2005
Protropin	Somatrem	Genetech	2005
Neupogen	Filgrastim	Amgen and Roche	2006
Engerix-B	Hepatitis B vaccine	G.S.K.	2006
Filgrastim	GM-C S F	Amgen	2007

Source: Datamonitor 2006

human health. Taking into account this argument in context of biopharma drug innovation for neglected diseases the research recommends open source for the development of drugs along with advance market commitment for the marketing of the same and towards the access of such drugs the optimal use of flexibility provisions of TRIPs agreement.

Also, for having a balance towards public health, Government of India has included section 107 A (a)[16] of the India Patent Act 2005, so called Bolar provision to allow generic manufacturers to start producing a patented drug in limited quantities during the period of the patent for collecting data submitted to a drug approval authority. This exception enables generics to enter the market soon after the patent expires (Bashir, 2005). Bolar provision will enable smooth flow of technical skills and knowledge which in turn will give boost to the domestic companies to invest in research. At the same time, the empirical data is not clear whether data exclusivity definitely has adverse impact on the prices of drugs. The multinational pharmaceutical companies practice the policy of differential pricing that is to say that the prices of the drugs vary from country to country depending on the purchasing capacities of the consumers.

There can be different reasons for the adoption of manufacturers and also to avoid compulsory licensing by the Government. As after the decision of Roche vs. Cipla (2008) judgment was decided in favour of Cipla as there was huge price difference between Roche's patented drugs and Cipla generic drugs. As after the decision, companies such as Pfizer is developing a three tier price model for its anti cancer drug stuent for Indian market (Reddy, 2008). Otherwise the general practice of the MNCs are that they adopt same price in their imports into the western market. Although parallel imports in many countries in the west are prevented by legislation but still there are cases of parallel importation. Parallel importation is a practice of importing patented drugs by the distributor meant for one jurisdiction into his home jurisdiction means if a drug company launches its product at different prices in different regions then it can be the possibility that online pharmacy start buying from the cheaper market and import them for higher prices in the western market. Section 107 A (b) of Indian Patent Act 2005 exempts from infringement on importation of patented products by any person from a person who is duly authorized under the law to produce or sell or distribute the product.

Other than this the flexibilities, which are inherent in TRIPs like compulsory licensing under Article under Article 31[17] or the parallel importing from other countries, can always be used as a tool for making the drugs more affordable in cases of emergencies and on the grounds permitted by the TRIPS agreement. Compulsory Licensing is "*an authorization given by a National authority to a person, without or against the consent of the title-holder, for the exploitation of a subject matter protected by a patent or other intellectual property rights(Correa, 1999)*": (detail in the next section).

Thus, 20 years of patent protection is fine for giving incentive for the innovation as at least the innovator will get 10 years of effective patent life to realize his money. This will be more efficient if it is combined with 5 years of Data exclusivity against misappropriation as data exclusivity will bring more confidence in the minds of the investors to invest in India. At the same time we should remember that not only the longer patent length will make patent length optimal as biopharma sector is related with enormous social value.

For the optimal patent design there is a need to have flexibility for the curtailment, of the monopoly during the duration of patent protection of 20 years in case of extreme social need or national and the biopharmaceutical sector is connected with the health of the people. For the same, efficient use of compulsory mechanism is permitted by the TRIPs agreement under Art 31. Under licensing mechanism, Government permit persons other than the patentee to use a patented invention prior to the expiration of the patent without the consent of the patentee, where the national interest so requires in case of national emergency or extreme emergency. Hence, by the compulsory licensing mechanism the maximum duration of patent protection can be reduced (Singham, 2000).

Effectiveness of Compulsory Licenses (CL)

In order to know the effect of compulsory licensing on India we will take into account the example of Thailand and Brazil who have used this compulsory licensing mechanism. Then we can make out the effectiveness of compulsory licenses in curbing the negativity of the broader patent protection such as high prices of the drugs.

Taking the case of Thailand first where the first set of Thai compulsory licenses was issued after the negotiations with patent holders for affordable prices failed. In late 2006 and early 2007, Thailand took active steps to promote access to medicines for its people, when it issued a compulsory license on two HIV drugs and one heart disease drug. In January 2007, compulsory license for Kaletra patented by Abott, and Plavix patented by Sanofi Aventis. These actions are seen as the most serious attempt to date to override patents (Kerry & Lee, 2007) Abott initially responded by withholding a number of new medications from the Thai market including the heat stable form of Kaletra. The company has since offered the medicine to Thailand and 39 other countries for US$ 1000/patient/year, although it continues to withhold other medications.

As a result of compulsory license, Merck immediately cut the price of first line antiretroviral Efavirenz from 1400 baht /bottle to 767 baht/bottle in Thailand and slashed the prices for of other developing countries too (Gerhardsen, 2007).

Abott laboratories cut the price of second line drug lopinavir / ritonavir a heat stable form of the same drug that does not require refrigeration, to $ 1,000 / month from $ 2, 200 / month for 45 lower and middle income countries.

Thai Government has also issued 4 compulsory licenses for 4 cancer drugs Docetexal, Letrozole, Erlotinib, Imatinib. After this, Thailand has finally issued three CL on Docetaxel, Letrozole and Erlotinib to allow universal access to essential medicine for all beneficiaries of the publicly financed national Health Security System. Novartis made a last minute offer to the Thai Government to provide Glivec free to all patients index the universal healthcare scheme as a trade off for not seeing its patent overridden.

In this way seeing the Thailand practice we can conclude that compulsory license is an effective mechanism to curb the high prices of the drugs as a result of stranger patent protection and a balance between innovation and access.

After Thailand, we take the example of Brazil. In case of Brazil threat of compulsory license pressured pharmaceuticals trail companies like Abott Merck and Roche (manufacturer of Lopinavir, indinavir, Nelfinavir and saquinavir respectively) to substantially reduce the price, thus enabling 100,000 people to receive free treatment (Gerhardsen, 2007). In 2006 Brazil Minister of health to violate Abott laboratories patent on Kaletra, an antireteroviral drug used as part of anti AIDs medication. Abott labs not only lowered the price by 46% but also distributed the drugs free of cost to 1, 63,000 HIV carriers which is about 27% of the total number of HIV patients. On the precedents around the world particularly in Brazil and Thailand, where drugs that are used for HIV therapy have been grated compulsory license and successfully used to bring down the prices of the drugs (See Annexure 1). It will be interesting to find whether the same thing can happen in India? In India, compulsory license can be issued under section 84 of the Indian Patent Act 2005, on certain specified grounds (such as excessive pricing, inadequate supply of product etc) after 3 years of the patent grant. Although India has yet to use compulsory license but it is apt to mention about Indian firm Natco pharmaceuticals plea for the country's countries first 'compulsory license" to patent office as it bids to make generic' copies of Pfizer's sutent and Roche's Tarceva cancer drug. One drug called erlotinib was patented in India last year by Swiss manufacturer Roche under the brand name Tarceva. The other is also a cancer drug called Sunitinib and is sold by US manufacturer Pfizer under the brand name Sutent. The case is still in the court but if Natco gets the verdict in its favour then that will be the first such case.

In case of Hoffman La Roche Ltd Vs. Cipla ltd (2008) the court denied the injunction and held that the cases of medicines specially the life saving drugs such as Erlotinib is granted then the court in effect be stifling with Article 21 so far as those would have or could have access to Erlotinib and court cannot be unmindful of the right of the general public to access life saving drugs by issuing injunction. As a result of which the foreign drug makers plan to offer medicines at different tiers of prices for Government supply, patient access programmes, and hospitals in rural areas. The companies such as Pfizer are developing a three tier price model for their anti cancer drug sutent (Roche vs NATCO, 2008). Thus, medicines are likely to be affordable. MNCs may adopt the same prices in their Indian markets as in their foreign markets. In this case, Roche filed a case against Indian generic company Cipla in early 2008 for seeking permanent injunction for restricting Cipla from manufacturing, offer for sale, selling and exporting the drug evolution. The prices of generic version of Cipla drug was 1600 tablet while the Roche price in the Indian market is Rs 4800/tablet (Hoffman La Roche Ltd vs. Cipla ltd, 2008).

So, looking at these examples it can be said that in India too compulsory license can be used as an important tool to bring down the price and maintain the balance between innovation and access. The decision of this case can be seen as the balancing judgment for the private patent right and public health and is seen as judiciary imposed compulsory license in India which was decided on the ground that the drug erlotinib was sold at one third of the price Roche patented anti cancer drug Tarceva and the same was imported by Roche into the Indian market not manufacturing locally so the long term supply of such important drug can be in question. Also if we also take the example of Novartis case (the famous Glivec case) then we will find that when the case was pending in the court Indian Health Minister asked Novartis to withdraw the case otherwise Government will take resort of compulsory licensing. At that time Novartis did not take back the complaint but after the adverse judgment of the Chennai High court, the decision of not going for appeal to the Supreme Court for long can be understood that the threat of compulsory license worked.

Seeing the abovementioned example of perceived threat of compulsory license worked in bringing down the drugs prices it can be said that even the threat of CL works in bringing down the price of the drugs even without actual imposition. Although the imposition of Compulsory license has certainly worked for making the drug affordable (See annexure 1 highlighted once are the threats of CL). We saw in the abovementioned examples successful usage of Compulsory license by developing country like Thailand and Brazil. The decision of Cipla case reflect that judiciary can play a vital role in balancing access and innovation by protecting unfettered monopolies on the vital medicines and also balancing between Indian obligation to global trade laws and constitutional and human rights obligations. Although the judgment can be taken as the landmark which confirms that statutory right in the form of patents over the invention which is useful for the public health can be curtailed when public interest demands. After this judgment the foreign drug makers plan to offer medicines at different tiers of prices for Government supply, patient access programmes, and hospitals in rural areas and not profit organizations. Thus, medicines are likely to be affordable. MNCs may adopt the same prices in their Indian markets as in their foreign markets. So, looking at these examples it can be said that in India too compulsory license can be used as important tool to bring down the price and maintain the balance between innovation and access.

On the other hand at the same time, there is another dimension of this issue as the drug companies worry that countries will abuse compulsory licensing provision, employing there in the absence of any public health crisis, simply because the Government wants to pay less for the drugs. But it is interesting to note that none of the interviewed generic companies were really interested in using section 92 A[18] as they did not think that the margins would have been great and most of them wished to focus on the regulated markets of developed countries of course now it presents a huge business opportunity for them as they must have recalculated potential business gains from these markets. Also, in order to curb misuse or mischievous use of compulsory licensing, broader scope of patentability as broad as possible within the limit of morality and ethics by deleting most of the restrictive provisions on patentability. This will give a real business impetus in investing in innovations. The abovementioned cases are compulsory licensing in the case of pharma drugs and not biopharma drugs. Till date there is no case of biologic compulsory licensees. The growth of this sector depends on the Government move to create regulatory pathways for biogenerics at the moment which is lacking at the moment.

Empirical evidences suggest that stronger intellectual property protection will facilitate the flow of FDI in the developing countries. A study conducted by Mansfield indicated that effective intellectual property rights protection could be an important factor in securing foreign direct investment and technology transfer especially in high technology industry like biopharmaceutical. As mentioned previously technology transfer is vital for developing countries like India for the economic as well as for improving her standard of health and environment. The arguments that IP rights increases costs is actually Government posed taxes and tariffs that raise the price of life saving drugs. For example, in India the combined taxes and tariffs on imported medicines are 55 percent (Wilson, 2008). To control the drugs prices, there is proper drug price control mechanism of Drug Price Control Order (DPCO). Moreover, there are always the weapon of compulsory licensing to undermine IP in case of national emergency or other circumstances of extreme urgency or in case of public non commercial use. As there is need to balance between innovation and access.

We will talk about these concepts at length on how these will help in neglected diseases drugs innovation. In this segment we will focus on the pull side of the profit equation. It has been discussed before that there are two main phases of drug innovation, upstream research phase and downstream development phase which is long clinical trial phase and are quite expensive whereas the market for the same is low. So firstly we will discuss the best suited mechanism to enhance R&D of the neglected diseases drugs and then we will proceed with the market aspect for the same. There are different mechanisms as resort to solve the crisis of the neglected diseases drugs innovation and their applicability and among them one is open source method about which the study is of the view that the same is one of the best suited for accelerating R&D for such drugs and for making the market for the same roper adoption of Advance Market Commitment (AMC) and implementation seems to be useful. As per the open source mechanism one is free to use a part of the commons but one should not use the intellectual property rights to fully privatize or enclose the commons. This can be a good kind of practice to minimize the problems related with the tragedy of commons and tragedy of anticommons and balanced solution towards the problem of neglected diseases drug innovation in a balanced way. Under the open source mechanism there can be no restrictions on the modifications of the invention provided unless some other conditions are attached to and licenses can limit or restrict, if not prohibit outright commercialisation of the invention (Srinivas, 2006).

Under the open source, different types of rights and obligations can be created as per the objectives and aims of the project which can be exercised by opting for a license that facilitate the open source principle and applications. Although the open source does not claim to resolve all the problems associated or arising out of patent system but it claims that they provide workable solutions based on alternative perspective on exclusive rights and collective rights and obligations (Srinivas, 2006). Such kind of system if powered by Advance market commitment which ascertains guaranteed market for these products then it will be effective tool to cure neglected diseases problem. Under the open source solution drug research should be initiated by the mutual cooperation of public and private players according to their strength and their expertise in the stages of drug innovation. Due to the advances in life sciences over the years the pharmaceutical industry has spawned wide changes which have impacted major aspects of drug research including organizational and managerial aspects. As per Prof Franco Malerba and Luigi orsenigo (Malerba & Orsenigo, 2001) drug innovative process are categorized in different stages; the production of new scientific knowledge, the development of this knowledge in applied knowledge and the use of the latter for the production and marketing of a new product. A network collaboration between the various actors specialized in these stages of innovative process in which they are relatively more efficient is desirable and in every stage (i.e. In pre clinical and production stages) it is possible to cooperate and share. In this way open source mechanism can be more effective in bringing solution to the problem of the neglected diseases drugs crisis.

Also, on the issue of the availability of the genetic material which has been raised by the developing country like India which are not available due to the product patent protection can also be not a problem at least for the neglected diseases drugs innovation if open source is adopted for the neglected diseases drugs innovation. As open source method generally avoid the use of patents to deny access to data or proliferation of patents with fragmented proprietary rights (Srinivas, 2006) as the main objective of this mechanism is to ensure that intellectual property rights do not result in blocking data access to other researchers and companies (Srinivas, 2006). The accessibility of the data bases should need some sort of formalities which should be simple and flexible. Such system will make data available to the genuine users and contributors to the technology without making it free for all. Such system can be quite effective in case of developing countries. In this regard concept of Limited Common Property (LCP) suggested by Carol Rose (Rose, 1998) is apt to mention which says that it is hybrid form of property of an institution or community which is neither totally private nor totally public in which a formed with common objective such as neglected diseases drugs innovation and can hold some part of the data as LCP and some part in public domain. The specific community working on neglected diseases can claim or enforce patent rights in a manner which will not restrict the rights of the other member to engage in research or development of follow on innovation. Thus in this way there will be balance between private interest and public interest. The approach of the president of John Hopkins University Prof William R Brody seems to be a useful one. According to same network can be build for open source to develop as well as to test drugs in which license for the new compound developed can be assigned to the network. The network will hold the license for the associated IP and create database for sharing information. The consortium will grant a royalty free license firms that agree make and distribute the drug at a cost that is much less than that of the new proprietary drugs.

Such practice will help in more innovation in this regard. Early stage funding for the same is recommended for more innovation through this mechanism along with AMC which guarantees market for such drugs. Also, in case of India such mechanism can play major role in drug research and development where there is qualified human resource in abundance as such research can take place successfully and more efficiently if the private sector, Government, University all join hands together in close cooperation.

So, overall it can be concluded that optimal scope of patent policy for India needs to be broader with higher life form patentability, incremental innovation and 5 years of data exclusivity in order to attract investment towards the innovation of biopharma drugs for the general drugs other than for the neglected diseases drug innovation is the need of the hour with a proper access proper implementation of compulsory license mechanism, other regulatory and fiscal policies to have a proper balance. Also there needs to be some clarity in the patent legislation like patentability of biomolecules like nucleic acid along with the properly defined efficacy standard in order to prove invention. For the neglected diseases drugs innovation specific measures can be taken which can be lesser use of patents such as proper adoption and implementation of open source mechanism with public private partnership for the upstream research with early stage and clinical development innovation funding and Advance market commitment for creating market for the product. Then there will be proper development as it has been suggested that pharmaceutical companies can co-opt open source model but it may not substitute for original drug R&D (Munos, 2006). Also if we see the health expenditure of Government of India then we will find that it is very low if we compared to other countries or even if we compare the same with the proportion of population (see Figure 2).

Figure 2. India and world

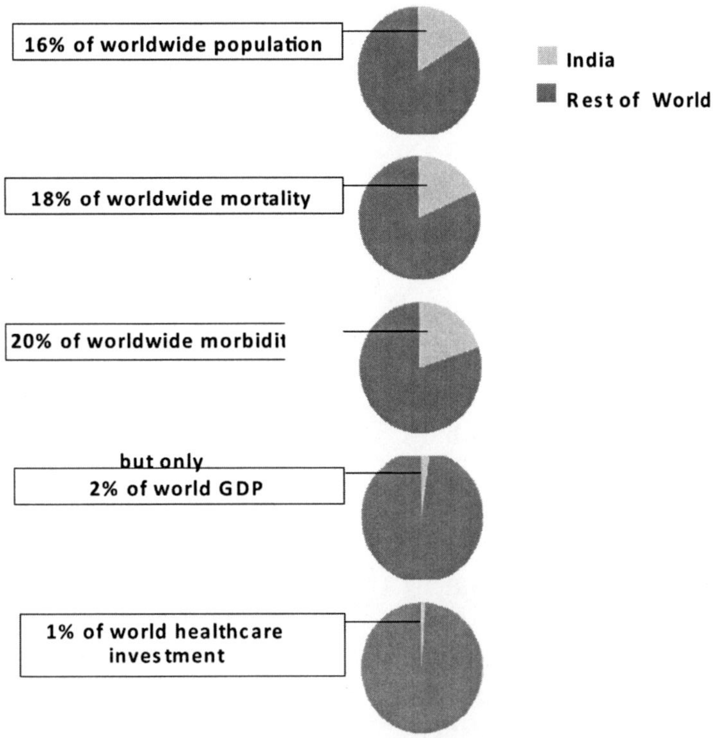

The health expenditure of the Government needs to be revised and upgraded heavily.

Part II: Optimal Regulatory Policy

In order to determine the right location of drug business for MNCs, the regulatory environment of a country plays an important role. A research based sector such as the biotech industry is completely dependent upon the different Governmental control through its regulations which governs all the levels of biopharma supply chain from the process of research, development, manufacturing and eventual marketing of new therapeutic products. These need to be framed in such different manner which attracts investment along with the properly framed regulation to procure access. Government incentives such as direct research funding, tax breaks on R&D expenditure or fast track approvals also forms push funding policies which is important for the neglected diseases drugs innovation. We will discuss Indian Government initiatives towards push funding for general as well as neglected diseases drugs innovation in this section.

As mentioned previously an efficient rule based system of IPR strikes a balance between creation and dissemination along with the coherent interaction with other regulatory or economic or fiscal policy (Love, 2003). According to the report of the Investment Commission 2006 (Investment Commission Report, 2006), the major impediments to investment have been identified as investment restrictions and/ or entry route barriers, absence of long-term policies, inflexible labour laws, bureaucratic delays, discretionary interpretation, vested interest, bias and subjective practices, high cost of entry, transactions and exit; ineffective dispute resolution, poor infrastructure. These impediments can be categorized in two groups, tariff barriers

Table 3. Recent legislation changes to clinical trial rules in India

Legislation amended	Date	Impact
Schedule Y of Drugs and Cosmetic Rules in India	2005	This will ensure the uniform quality of clinical research in India in accordance with International Good clinical practice guidelines for the generation of clinical data on drugs
Drugs and Cosmetics Act 1940	2005	Removal of phase lag in clinical trials in India under new legislation, it is now possible to run trials concurrently in India with overseas trials
National GLP Compliance Monitoring Authority	April 2002	Evaluating and approving facilities to Good Laboratory Practice Standards in accordance with OECD guidelines.

Source: www.uktradeinvest.gov.uk/uki/biotechnology accessed on 2nd November 2006

and non tariff barriers. Tariff Barriers in India arise due to higher rate of taxation, royalty, interest, gains from sale of capital assets located in India higher fees for technical services etc.

Non-tariff barriers in India include attitudes and bias toward foreign products, a rigid distribution system, and Government bureaucracy. Among all the factors, Government bureaucracy poses the most important non-tariff barrier to entry in India and which frustrate foreign investors. We will discuss this in the light of drug regulations of India. So, in this part of the paper first we will identify non tariff barriers and tariff barriers for the foreign investment in biopharmaceutical sector by looking at the Government of India drug and fiscal policy to eliminate the same. Now, firstly we will discuss about the tariff barriers and Governments initiative to eliminate the same.

At the moment, major impediments for the global companies are the complex and cumbersome processes involved in getting regulatory approvals in India which is quite time consuming too. The regulatory uncertainties about time of approval, involvement of multiple agencies for the approval of biotech products, and for processing import/ export licenses are other important stumbling blocks in planning a clinical trial. In addition, ethics committees generally don't function well because of red tapism. Whole lots of documents are required for exporting human samples which are much in comparison to US and Europe. Also, approval processes are frustratingly lengthy, such as those for breeding or importing genetically modified animals. Also, most institutions lack organized patient databases and as a result the patient recruitment rates are either under estimated or over estimated. Additionally, there is problem of inefficient investigators. The cost of imported equipment and biological reagents is also high in the country. The main problem which has been found in the study regarding India's federal regulatory system is bureaucratic setup and corruption. The bureaucratic scuttle has been particularly hard on biologics manufacturers in India, who must seek approval from multiple state, district, and federal agencies for routine activities such as the importation of recombinant molecules and cell cultures for research purposes[19]. India has recently begun to tackle its red tape to accelerate drug development.

In 2005, India's legislature enacted an amendment to the prevailing Drugs Control Act, modifying its rules on clinical trials (Table 3) (Basu, 2004). Previously, for example, Phase 3 trials in India of a drug intended for market in the country were allowed to proceed only if all prior had been completed in Western countries[xix].

Now, in contrast, only Indian companies are permitted to conduct first-in-human studies and involve substantial effort to gain approval. There is lack of state-level regulation in certain areas. The state regulatory authorities suffer from staffing problems as they have very limited technical experience on regulatory issues, which is a serious handicap[xix]. India has recognized the importance of these issues for some time, but it has taken the country several years to go from the talking stage to the implementation of solutions. Although, the Government of India is making efforts to promote biotech sector by various incentives and by eliminating loopholes by making sound regulations but still there are persistent chronic problem which is difficult to eliminate.

Now, we will discuss specifically the adequacy of Governmental initiative to stimulate foreign investment as there are certain steps which the Government of India is taking to promote health innovation in long run by adopting pull and push mechanism and also initiative to remove the access and availability of drugs. Then we will talk about the issues which are left with suggestions.

The National biotechnology policy (National Biotechnology Development Strategy Policy, 2007) 2007 provides many fiscal and non fiscal benefits to the industry. These are, 100 percent FDI approved in biotech units, which implies that there will be no restriction on the quantum foreign direct investment (FDI) in biotech companies and there may not be the need for FIPB (Foreign investment promotion board) approval for equity investment in biotech companies. Under this scheme the international patent costs will get R&D weight age. More importantly, two major regulatory initiatives have been taken. The first is the creation of the National Biotechnology Regulatory Authority (NBRA), under the Department of Biotechnology (DBT), as part of India's long-term biotech sector development strategy will be an "independent, autonomous, and professionally led body to provide a single window mechanism for the biosafety clearance of genetically modified products and processes." In other words, the NBRA will replace many of the other bureaucratic processes with which biologics makers in India now will interact for biosafety issues. The NBRA reportedly will also include a training center for its biotech regulators, to enhance their professional competence[xix]. Under NBRA legislation there is provision to establish Central Drug regulatory authority (CDA) with a single, central, FDA-style agency, the Central Drug Authority (CDA). Over a five-year transition period, this new agency will take on nearly all facility-inspection, manufacturing-license, and data-evaluation functions concerning drugs in India. The CDA is expected to have separate, semi-autonomous departments for regulation, enforcement, legal, and consumer affairs; biotechnology products; pharmacovigilance and drugs safety; medical devices and diagnostics; imports; quality control; and traditional Indian medicines. It will set up offices throughout India and will be paid for by inspection, registration, and license fees[xix]. Although it is a very good idea to have a single contacting authority for all matters concerning modern biotechnology but there are still need of clarity with regard to the definition of genetic engineering and on the functions of the DCGI and NBRA which is blurred as biotechnology products are related to many areas and there can be conflicts between regulatory authorities for the same.

Apart from this, the Government has made policies for the promotion of the biotech Parks: The DBT will promote and support at least 10 biotech parks by 2010. In the biotech parks, concession will be given to the biotech companies located in biotech parks like tax holiday U/S 10B of Income Tax Act 1961, duty free or import of equipments, instruments etc. The Government has devised industry friendly policies without compromising safety aspects of biotech products, incentives like customs duty exemption equipments and consumable required for modern biotechnologies tools will further boost infrastructure capability of Indian biotech industry.

Government has come up with a bill on the line of Bayh Dole Act, in which the academia, national labs and industry collaborate with each other and the IP generated pursuant to that is shared by those who contribute to the generation of IP. Under this scheme university or non profit research organization receives money from Government agency. As per the bill the researcher or scientists who will come up with the invention will get at least 30% as the royalty stemming from the licensing of the patent. If the grantee University or research institution does not choose to retain title over the invention, the Government can step in and patent the said invention. Thus, it can be said that this is a very good move on the part of Government of India by making it easier to patent the invention of a scientist or grantee institution as this particular support may motivate scientist to do research on the drugs for neglected diseases. Also there will be increase in general drug innovation. At the same time this scheme of the Government of India seems to be a good move to promote innovation but incentive for IP protection is not the only factor to stimulate innovation there are other factors which are also important in order to make this scheme successful such as universities need to be modernized with all modern scientific equipment. Also these universities need to be more active and aware. Also modernization of patent office's is barriers in the path of the success of this legislation as already patent offices are overloaded with backlogs. Efficiently trained manpower is required to dispense cases on time.

On the same line, Small Business innovation and Research Initiative (SBIRI) scheme launched by Department of Biotechnology to provide early stage funding to scientists in private industries for high risks, innovative product proposal is an important initiative to boost for early stage funding to scientists. At the same time there is need for schemes to support financially the development process such as clinical trials which is the longest and most expensive phase of drug innovation. Government should provide such scheme.

Ethical trials are very important. There are instances of litigation for unethical trials without requisite permissions, or on ill-informed patients. On the functioning of Indian drug regulatory authorities the World Health Organisation has taken a strong objections to the new drugs are approved in the country without proper documentation on pre clinical, clinical and toxicological studies.

On the issue of clinical trials the regulatory officials have announced stricter enforcement to international good clinical practice (GCP) and World Health Organization protocols. The DCGI also will conduct regular inspection of ongoing trial. The CDA will be responsible for increasing the criminal penalties for illegal clinical trials by a new law (National Biotechnology Department Stratgey Policy, 2007).

The introduction of ethical guidelines for biomedical research on human subjects 2000, acceptance of good clinical practice (GCP) guidelines, revision of schedule Y (multicentric phase II and phase III trials) are increasing India's attractiveness as a place to conduct clinical trials. In the manufacturing area, an amendment to the regulations, "Schedule M" of the Drug and Cosmetics Act, now specifies the good manufacturing practice (GMP) requirements for factory premises and materials, modeled after US FDA regulations.

Apart from these, the National Science and Technology Policy of the Government and the Vision Statement on Biotechnology issued by the Department of Biotechnology have directed notable interventions in the public and private sectors to foster life sciences and biotechnology.

Recently a landmark amendment is being sought by the Union health ministry in the drugs and Cosmetics Act to prevent the practices of drug counterfeiting by Drugs and Cosmetics (Amendment) Bill, 2005 which is cleared by Union Cabinet on 17th July 2008. As per the amended bill the Central or the state Government in consultation with the chief justice of the High court, shall for trials of offences relating to adulterated or spurious drugs designate one or more courts of session as a special court for such cases. The Act has made the provision of minimum ten years of imprisonment which can be extended to life imprisonment with a fine of not less than Rs. Ten lakhs or three times the value of the drugs confiscated whichever is higher in case of selling adulterated, misbranded or spurious drugs. Such kind of stick is a good move on the part of the Government of India to control the counterfeited drugs supply which will boost the confidence of foreign investors to invest in India. This is one of the main causes of reluctance for the foreign investors to invest in India where 10 to 20% of medicines are counterfeited (World Health Organisation Report, 2006).

Although looking at the recent reforms in drug regulations in India it is apparent that it has gained importance and been streamlined in efficient manner to attract more investment but still there is demand for 'Patent-Registration' linkage system. Pharma multinationals are now pushing the Government to introduce guidelines for linking drug registration for marketing approval to the patent system in the country. At present there is no provision under the drugs and cosmetics Act to establish patent linkage, where drug regulators checks the patent status of a drug before granting marketing approval to the company. The regulator gives marketing approval to the company. The regulator gives marketing approval after establishing its safety and efficacy. The patents are handled by ministry of commerce and industry. On the demand by the foreign companies 'Patent-Registration' linkage system has been proposed by the DGCI. Under which DCGI is preparing a document outlining a system to implement the linkage system. The Central Drugs Standard Control Organization (CDSCO) has planned for e-governance project to digitize all their records in a bid to ensure better co-ordination amongst their various offices in order to ensure better accountability and transparency. Such policy will prevent the infringement of patents at the initial stage which is generally sorted out after a long and time consuming judicial process. This is a good move of Government of India to gain confidence of the investors in Indian regulatory system.

It seems that very soon the weaknesses will be removed as the Government is working on the loopholes of the system. These initiatives of Government of India can be looked as incentives as push factors to pull foreign investment for biopharma innovation in India. So, it can be said that in times to come the regulatory system will be quite efficient in order to attract foreign investment. Although it is hard to say that bureaucratic inefficiency which is the major problem of Indian system will be removed so easily or not.

Fiscal Policies

Another important push factor which is important to pull foreign investment and technology is sound tax system, with moderate rates and a broad base.

According to Morisset and Pirnia (Morriset, 1999), Tax policies are obviously capable of affecting the volume and location of FDI. Fiscal policy of a country comprise of mainly two kinds of taxes, direct tax and indirect tax.

Direct Taxes

Corporate tax is a form of direct tax, which refers to a tax imposed by various jurisdictions on the profits made by companies or associations. The tax rate may, however, vary between jurisdictions. In India, corporate tax rate for a company depends on the origin of that particular company. A company based in India will have to pay a flat tax rate of 30%. However, for a foreign company, the tax rate is dependent on a variety of factors. The companies that are domicile to India pays taxes based on the criteria of global income, whereas the foreign companies in India are taxed on the basis of their income generated out of Indian operations. (see Table 4)

There is controversy with regard to the higher corporate tax levied on the foreign corporation than Indian companies. It has been alleged that despite of the various tax treaties which India has made with a large number of countries the corporate tax on foreign companies are higher. As per the foreign multinational companies the non-discrimination clause provides that a foreign enterprise in India will not be subjected to any taxation which is more burdensome than the taxation to which Indian enterprises are subjected to under similar conditions.

Table 4. Tax rates in India

Domestic Corporate Income Taxes Rates		
	Tax Rate	**Effective Tax Rate with surcharge**
Domestic Corporations	30%	30%
Foreign Companies Tax Rates		
	Withholding Tax Rate for non-treaty foreign companies	**Tax Rate for US companies under the treaty**
Dividends	20%	15%1
Interest Income	20%	15%2
Royalties	30%	20%2
Technical Services	30%	20%2

Note: A surcharge of 10% of the income tax is levied, if the taxable income exceeds Rs. 1 million. All companies incorporated in India are deemed as domestic Indian companies for tax purposes, even if owned by foreign companies. Source: Finance Act 2008-09

Despite the above internationally accepted position, explanation to Section 90 of the Income-tax Act provides that the charge of tax in respect of a foreign company at a rate higher than the rate at which a domestic company is chargeable, shall not be regarded as less favourable charge of levy of tax in respect of such foreign company.

The aforesaid non-discrimination clause is on the lines of an article contained in the UN Model of tax treaties, which reads as under: Nationals of a Contracting State shall not be subjected in the other contracting state to any taxation or any requirement connected therewith which is other or more burdensome than the taxation and connected requirements to which nations of that other state in the same circumstances are or may be subjected. This provision shall, notwithstanding the provisions of Article 1, also apply to persons who are not residents of one or both of the contracting states.

On the issue of discrimination by imposing higher rate of tax and India commitment to the International tax treaties made with other sovereign states, it is interesting to discuss recent case judgment in this regard. In Chohung Bank vs. Deputy Director of Income-tax (2006) Korean Chohung bank which had a branch in India filed a case claiming that the tax rate as applicable to Indian companies carrying on similar business should be applied for the Chohung bank too relying upon Article 26 of the Agreement for Avoidance of Double Taxation between India and Korea. Accordingly, the foreign bank income was taxed at higher rate of 48% instead of at the rate applicable to a domestic company i.e. 35%. The branch (called assessee-company) was involved in normal banking activities including financing of foreign trade and foreign exchange transactions.

The ITAT Mumbai after considering Section 90[20] of the Income-tax Act 1961 and the non-discrimination clause in the Tax Treaty with Korea ultimately held that the provision of non-discrimination has nothing to do with the rate. Looking at the decision of the case and other incentive provisions of Income tax Act 1961 it can be said that the judgment of the Mumbai tribunal is justified as foreign companies invest and do businesses which are not available to the domestic companies. For example, foreign exporters are generally not liable to tax if the sale is concluded abroad and the proceeds are received when they have business connection or permanent establishment or any activity in India. Also, a foreign corporation is not liable to tax on operations confined to the purchase of goods in India for the purpose of business. This could be an incentive if a foreign biopharma company were planning on purchasing raw ingredients in

India for making of the Hepatitis A Vaccine for distribution in India.

There is another issue with regard to higher corporate tax in India which is a cause of reluctance for the foreign companies to do business in India. India charges the second highest corporate tax from foreign companies, next only to the US. "India and Japan retain the second and the third highest corporate tax rates in the world. India taxes foreign companies at a maximum rate of 42 per cent while Japan levies a corporate tax of 41 per cent. New York City imposes the highest corporate tax of 46.2 per cent. This is a barrier to establish business in India which needs to be addressed efficiently.

Now coming to indirect taxation, in indirect tax customs tariffs, excise duties are the prominent ones. In order to know the exact amount of the applicable customs duty on a product, it is therefore necessary to add: the basic customs duty surcharge plus countervailing duties. As mentioned before that Government imposed tax and tariff raise the price of the life saving drugs specially the combined tax and tariffs imposed on drugs is 55% which is quite high which needs to be addressed. We will see the recent effort of Government of India to reduce the same.

The Union Budget 2008-09 has reduced of customs duty on raw materials for ELISA kits to 18.72 percent and select vaccines and selects bio-therapeutics to 9.36 percent. An 8 percent reduction in export duties from 16% to 8% along with export duty exemption for indigenous life saving drugs and 5 percent from 10% customs duty reduction for imported Life Saving Drugs will certainly reduce drug prices. Many vaccine companies will also benefit due to reduction in custom duty. This will benefit many MNC who are importing the various combination vaccines into the country. As in case of all drugs (formulations), excise duty rate has been reduced to 8 percent. This is good for biopharmaceutical manufacturers in the country. In general, the budget is a very good sign.

This is a better way to address healthcare costs along with the drug price control. Perhaps the most important tax benefit announced for the sector is the 125 percent weighted average tax deduction on outsourced R&D, which sends a strong signal of the potential that discovery research holds for the Indian pharma and biotech industry.

Despite of all incentives to pull foreign investment still medical technology is left as an industry, which is reeling under unprecedented tax burden (36 percent). There was no special mention made for the industry and no reduction in custom duties for the Import of medical equipments/ devices. The government has not given any incentives like capital grants and subsidies for local manufacturing of medical devices as well.

The Finance Minster has proposed to increase its healthcare allocation by 21.9 percent and has extended a 5 year tax holiday for setting up hospitals in non- urban cities, but this however will not reduce the cost of healthcare in the country, until the cost of medical equipments are reduced. Health of the common man has been neglected and so has been the government's attitude to the Indian medical equipment manufacturer

More incentives are needed for R&D as 10 year holiday of 200 percent weighted tax deduction for both in-house and outsourced R&D is desired to boost investment for R&D and move up the value chain.

Although there is general decline of customs tariffs within the framework of their WTO participation but still India remains one of the countries where the customs duty is the highest. There are 4 basic rates: 5%; 15%; 25% and 35. To these basic duties, a 10% surcharge can be added. India also applies some "countervailing duties" to compensate for the loss of excise duty, which it would have earned on raw materials, components and ingredients if the product had been made in India.

Also despite of India proposal of a six-year phase-out to the WTO for removing quantitative restrictions on imports of some 2,700 items, including consumer goods, which have been maintained since the 1950's on balance of payments grounds. Import licenses are still required for most pharmaceuticals and chemicals, and products reserved in India which serves as an effective ban on importation. India's strict trade policy and complicated import polices and licensing could be limiting factors in an effective policy making process. India's import policy is administered by the means of a negative list. The negative list would make export of pharmaceuticals difficult, as certain pharmaceuticals and medicines could fall into the gray area or restrictive items.

Thus, we have seen that Indian trade policies and tariffs including import duties, and tax structure etc. have been rationalized and liberalized in tune with the WTO requirements in order to make them attractive to attract more investment but still the tariff levels are very high by international standards which need to be lower.

The above discussion is to pull foreign investments in India for the biopharma innovation to make drug available to the needy population of India. We have seen that few incentives such as higher patent protection will raise the prices of the drugs steeply high due to higher monopolies, which will affect the access to drugs. So, a balanced policy needs to look at the side affect too and cure the same. As health is the fundamental right of the citizens thus state must guarantee access to healthcare including drugs for neglected diseases. For access to drugs it is important to have availability in the form of adequate supply of the drugs at first and for the same adequate investment in R&D for the drugs of that particular disease. Secondly, cost also can be a major obstacle for access as if the drugs are available at higher price which is beyond the purchasing power then that creates access problem. There are two ways of making drugs more financially accessible: (1) lowering prices; and (2) adopting system of financing which distribute the cost. Price control mechanism and effective insurance system are important measures to check high prices. As in countries where a substantial part of the population is covered by health insurance systems and where patients generally only pay part of the cost of their drugs on the other hand, in countries where health insurance is less widespread and consumers pay for most of their pharmaceuticals from their own pockets (79). Policy promoting generic drugs will also help to improve financial access to drugs. The mechanism to regulate the prospective abuse of excessively high prices is "price control" this is quite prevelant in India to check the prices of the drugs and make the drug available at affordable price in India. Other than compulsory licensing (which is also granted if the price of the patented drugs is excessively high) and competition law (which see excessively high price of the drugs as the abuse of the dominant position in the market), price control legislation is quite important to curb excessive price and provide access.

Now about the price mechanism in India, with the introduction of Indian Patent Act 1970, the Government of India also instituted a Drug Price Control Order (DPCO) under the essential commodities Act 1955 to control the drug prices to ensure access to drugs to the public. Under DPCO, the Government fixed the prices of the bulk drugs and their formulations. In 2002 pharmaceutical policy was introduced to relax the control and some life saving drugs were excluded which was later on challenged in the Supreme Court is still pending. In the wake of the Patent Amendment Act 2005, the Government of India is considering a new Drug Pricing (regulation and management) Act to strengthen the price control regime. The DPCO will function as the Act will not take away the functioning of DPCO but perform additional function of price monitoring and negotiation of prices of new patented drugs. The Drug Policy Control Order 2006 declared that bulk drugs developed by Indian companies would be exempt from price control for five years or, in the case of new drugs, 10 years. Goods developed in India and patented in the U.S., Japan or E.U. enjoy a three-years waiver from excise duty, and companies get a 10-years tax holiday for income stemming from qualifying R&D. Recently the ministerial review panel on pharmaceutical policy and pricing which has recommended for 354 drugs to bring to the ambit of price control which account for about 7000 formulations packs in the market. Looking at the price control mechanism in India it can be said that it is an important tool to ensure access in India but at the same time it is apprehended that such strong price control may become barrier to foreign investment in Indian biopharma sector as strong price control combined with weak patent system makes Indian investment environment unfavourable. This apprehension is ruled out as India is the destination for low cost research and development of drugs where drugs are developed at the fraction of the investment of US and Europe which is attracting the foreign companies to invest in India.

If IP regime is strengthened for the general drugs along with effective price control system then there will be increased flow of investment for innovation and access to drugs as well. Access to drugs also needs availability which needs proper mechanism developed to ensure essential medicines reaches consumers in remote areas. Medical insurance for all the people who are below poverty line is a good way to cover the health and drugs cost. The same can be a good push incentive by the Government in this regard. Recent national health insurance scheme – Rastriya Swasthya Bima Yojna (RSBY) of the Government of India seems to be a good scheme in this regard. The objective of the scheme is to recognize the diversity with regard to public health infrastructure, socio economic conditions and administrative network and aims to facilitate launching of health insurance projects in all the districts of the states of India all the people living below poverty line (BPL). The beneficiaries will be issued smart cards for the purpose of identification and the scheme covers four people of the family of the insured who live below the poverty line. The unorganized worker and his family (unit of five) will be covered for a total sum of Rs 30,000. There is cashless attendance to all covered ailment. As the smart card provided under the scheme contains Rs 30,000 which tracks expenses day today in the hospital and the money is deducted automatically after each procedure. No need of pre approval or reimbursement.

This scheme is specially beneficial for those who keep on moving from one place to another for their work and the smart card empowers. Hospitalization expenses, taking care of most common illness with a few exclusions is possible in which all pre existing diseases are covered. Also include transportation costs (actual with maximum limit of Rs 100 / visit) within overall limits of Rs 1000. Under the scheme there will be 75% of the estimated annual premium of Rs. 750 contributed by the Government of India subject to a maximum of Rs 565 per family per annum. The cost of the smart card will also be borne by the Government of India where as 25% of the annual premium as well as any additional premium will be borne by the respective state Governments. The beneficiary would pay Rs 30 per annum as registration/ renewal fee. The administrative and other related cost of administering the scheme would be borne by the respective state Governments. Smart card is designed to prevent fraud because of unique types of embedded software also can dramatically cut corruption in all Indian social program. It can also carry data on payments for rations (PDS) or earnings from employment schemes (National Rural Employment Guarantee Scheme) and it can expose corruption very easily. This seems to be a very good support from the Government of India in the direction of supporting the access to drugs problem. If it goes according to the plan 30 core people or one third of India will be covered in five years at annual costs of Rs 4500 core- a tiny sum compared to the money wasted in dozens of other schemes. The main cause of failure of previous state health insurance is insistence to the people to use public hospitals and public insurers whereas in this scheme, insurance companies, hospitals and patients all have skin in the game.

Towards the access of drugs, the Government of India has also planned to open a national chain of generic drug stores, one in each district to sell inexpensive unbranded medicines with the help of Red Cross society, non Government organization, charitable bodies and consumer groups[21]. The Government might finance operating costs of these stores for four in five years. This is a nice practical on the part of the Government but all these policies will not be successful unless they are implemented properly and for the proper implementation system free from corruption is required which is major problem in India due to which the benefit never reach the needy for which the policies are made. So the mindset of the people should be changed who are in the system.

CONCLUSION

As mentioned earlier, technology and technical know-how are essential for improving productivity, promoting export growth and attaining the development aspirations of countries. Therefore, technology transfer is seen as an important tool to narrow the growing technological gap between developed and developing countries, for integrating developing countries into global economy and enabling developing countries to meet their international obligations. TRIPs Agreement contains statements about the importance of technological innovation and the role of transfer of technology along with the stronger patent protection. So, according to TRIPs text, intellectual property protection has a direct link with technology transfer (Article 7). Foreign direct investment can be an important means of transferring technology to developing countries (Blomstrom & Wolff, 1989) it is widely recognized that both policy makers and analysts require a better understanding of the effect, if any, that a developing country's system of intellectual property rights protection has on the transfer of technology.

Mounting costs of drug discovery and development, extended timelines for bringing new drugs to the market, fierce competition, pricing pressure and funding challenges have combined to prompt pharma companies in the developed countries to look to countries beyond their own borders on multiple fronts. This places India at a particular advantage. At the same time, there are certain factors which need to be strengthened in order to attract foreign investment especially on regulatory front. As per John Dunning's (Dunning, 1979) ownership location and internalization (OLI) approach ownership specific location, location specific and internalization specific ownership advantage create monopolistic advantages which can be used to prevail in market abroad. The paradigm shows that market size and structure, prospect for market growth and degree of development, the cultural, legal, political and institutional environment and Government legislation and policies of the host country governs the decision of foreign investors to invest. The paper has evaluated the Government IP policies and other drug legislations which govern the foreign investment as well as strength of other stimulant factors.

This analysis evaluates current scope of patent policy along with drug regulations and fiscal policies to attract foreign investment in biopharmaceutical sector for the new drugs innovation as well as providing accessibility of the same to the needy. Based on this analysis, the paper concludes that optimal design of patent for biopharma innovation in India should confine to trade off between patent length and breadth taking into account the flexibility provisions of TRIPs. The fine tuned patent policy can be achieved by allowing broader biotech patents to set incentive to attract investment for the innovation, as patents with longer lives allow the patent holder to set a higher market prices for the patented product, patents with longer lives allow patent holders to obtain revenues for longer period of time. Larger breadth makes it more difficult to imitate or improve upon the protected invention, whereas increasing the duration of patent protection enhances the incentives to improve the invention (Bostyn, 2004/2005) by increasing R&D investment for the innovation of biopharmaceutical drugs in general and neglected diseases drugs through foreign investment in particular carrot in terms of data exclusivity and higher form of patent protection should be given along with the stick of properly framed compulsory licensing provision and parallel importation to balance the need of the access that are applicable to the. In biotechnology, the surge in innovation, notably by start-ups, benefited greatly from the possibility of obtaining patent protection, which attracted the capital needed in this area. At the same time, Government subsides can directly promote the development of biopharma technologies. Government's funds significant research and development through basic grants for medical research. Governments also seek to spur innovation indirectly, especially in the environmental and occupational health and safety fields, through regulatory reforms.

While the crisis of access to medicines in poor countries has multiple determinants, intellectual property protection leading to high prices is well-established as one critical element of the access gap. The study believes that one of the major causes of high price of life saving drugs is the high prices of the Government posed taxes and tariffs as the combined taxes and tariffs on imported medicines. Higher intellectual property protection cannot be blamed solely for the high prices of the drugs. At the same time an 8 percent reduction in export duties along with export duties exemption for indigenous life saving drugs and 5 percent customs duty reduction for imported life saving drugs by the budget of 2008-09 will definitely make the drugs price lower. In fact this is a better way to address drugs costs rather than drugs price control. Given the current international political climate, systemic, Government driven reform of intellectual property protection seems unlikely in the near term.

This paper suggests policymakers to consider targeted policies taking support of pull and push mechanism for neglected diseases drugs for innovation and access. By effective implementation of open source mechanism combined with public private partnership and Advance Market Commitment these drugs innovation can be accelerated along with push factor by the Government in the form of early stage research grants and clinical development tax credits. They would address the funding gap and above average development risks associated with neglected diseases. The initiative of inter Governmental working group of WHO to develop a global plan of action on intellectual Property, innovation, and public health is a good initiative. While this is undoubtedly a useful initial step, true reform of intellectual property protection can only be achieved through domestic, government driven reform or binding international agreements along the lines of the TRIPS regime. Therefore, the study proposes India should adopt a modest balanced policy that works within existing patent law and drug development paradigms in order to proactively circumvent both innovation and access of general biopharma drug. Whereas the case of innovation of the biopharma drugs for the neglected diseases can be treated differently by adopting alternate method such as open source and other suggested mechanism.

Thus it can be said that if the patent, regulatory and fiscal policies are framed on a model balancing both the interest groups and implemented properly, then it can be said for sure that TRIPs agreement leads to greater technological innovation and transfer of technology for the mutual advantage of producer and user of technological knowledge in a manner conducive to the social and economic welfare too.

So, at the moment the Indian biotech sector is needs a strong international partnering model to step forward towards the path of innovation for general biopharma drugs and even for the neglected diseases drugs innovation. Partnership model is all about using cost arbitrage to deliver value arbitrage. Through international partnering the Indian biopharma sector will attain DUI mode of innovation which will create access to explicit codified knowledge at the moment. This international collaborations and investment will bring interaction among both the partner's employees and such interaction will lead to learning by doing which will generate actual drug innovation gradually on its own strength. These will typically involve organizational frameworks and relationships between employees that utilize implicit knowledge and promote interactive learning (Lundvall, 2004).

Partnership model and foreign technology is a strategy for the present which will enable India to learn biotech innovation but to become a knowledge superpower in biotechnology should be the important mission for India finally. As per Dr. Abdul Kalam knowledge society has two dimensional objectives of societal transformation and wealth generation. The societal transformation in respect of education, healthcare and governance leads to employment generation, higher productivity and prosperity whereas which eventually generates wealth. For the same suitable policy and administrative procedures, changes in regulatory methods, identification of partners are and most important, creations of young and dynamic leaders are essential. For the same Government policy should promote entrepreneurship along with citizen centric approach to shape business policy user driven technology generation and intensified industry lab and academia have to be established along with the focused approach to Intellectual Property Rights and related issues. Such policy should be made and properly implemented by the corruption free system. With such dynamic approach only India will be able to reap the benefit of globalisation and implementing WTO norms.

REFERENCES

Aoki, R., Kubo, K., & Yamane, H. (2006). Patent and public health under countries lessons from Japan. *Bulletin of the World Health Organization, 84*(5).

Bashir, S. (2005). India's tryst with TRIPs: The Patents (Amendment) Act, 2005. The Indian Journal of Law and Technology, volume 1.

Basu, I. (2004; Jul 23). *India's clinical trials and tribulations*. Asia Times Online; Accessed 2008; January 16 at www.atimes.com/atimes/South_Asia/FG23Df03.html

Bengt-Åke Lundvall. 2004. *Why the New Economy is Learning Economy*. DRUID Working Papers 04-01, DRUID, Copenhagen Business School, Department of Industrial Economics and Strategy/Aalborg University, Department of Business Studies.

Bioventure for Global Health (BVGH). Retrieved from http://www.bvgh.org/innovation/crisis/default.asp accessed on 29th Oct 006 NACO Report(2003). HIV Sentinel Survey, Government of India Bioventure on Global Health (BVGH). Retrieved from http://www. Bvgh.org accessed on 29th Oct 007

Blomstrom, M., & Wolff, E. (1989). [Starr Center for Applied Economics, New York University.]. *Multinational Corporations and Productivity Convergence in Mexico*, *C*, V.

Bostyn, S. J. R. (2004/2005). Narrow trousers and narrow patents, a health risk? Product protection or purpose bound protection for biotechnological inventions. *Bio-Science Law Review*; 2 BSLR 95.

Chambers, J. (2002). Patent Eligibility of Biotechnological Inventions in the United States, Europe and Japan: How Much Patent Policy is Public Policy? 34 GEO. WASH INT'L L. Rev. 223.

Chohung Bank vs. Deputy Director of Income-tax (2006) 102 ITD.

Commission on Intellectual Property Rights. *Innovation and Public Health Report, World Health Organisation* (2003). London.Correa, C. (1999). *Intellectual property rights and the use of compulsory licenses: options for developing countries*. Trade-Related Agenda, Development and Equity, Working Papers, South Centre, Geneva.

Correa.C (1998). *Implementing the TRIPs Agreement; General context and implications for developing countries*. Penang, Malaysia: Third World Network: De Boeck & Larcier (2001). *Pharmaceutical patents and accessibility to drugs, German velasquez, essential drugs and globalisation*. Revue Internationale de Droit Economique. A new way to get medicines to the poor; www.spicyip.com; May 13, 2008

Deep, M. (2008). *Data protection: Emerging trends*Accessed on 3rd March 2008.

Di Masi, J. A., Hansen, R. W., & Grabowski, H. G. (2003). The price of innovation: New estimates of drug development costs. *Journal of Health Economics*, *22*, 151–185. doi:10.1016/S0167-6296(02)00126-1

Diamond v. Diehr, 450 U.S. 175, 185 (1980). *Excluded from such patent protection are laws of nature, natural phenomena.*

Dimminaco A.G. v. Controller of Patents, Designs & Trade Marks, (2002) I.P.L.R. July 255,269 (Calcutta H.C.)

DNDi submission to WHO IGWG (15th Nov 2006). Building a global strategy for policy change in neglected disease research. Geneva.

Dunning, J. (1979). Explaining changing pattern of international production. In defence of Eclectic theory," 41 Oxford Bulletin of econometric statistics, 269-275.

Encaoua, Guellec & Martinez. (2006, November). Patent systems for encouraging innovation: Lessons from economic analysis. *Research Policy*, *35*(9), 1423–1440. doi:10.1016/j.respol.2006.07.004

Erramouspe, M. (1996). Staking Patent Claims on the Human Blueprint: Rewards and Rent-Dissipating Races. *43 UCLA L. Rev.*, *961*, 988–990.

Evenson, R. E., & Westphal, L. E. (1995). Technological change and technology strategy. In Behrman, J., & Srinivasan, T. N. (Eds.), *Handbook of Development Economics*. Amsterdam: Elsevier Science.

Gerhardsen, T.I.S. (12 March 2007), "Thailand Presents Report On Compulsory Licensing Experience," Intellectual Property Watch.

Hardin, G. (1968). The Tragedy of the Commons. *Science*, *162*, 1243–1248. doi:10.1126/science.162.3859.1243

Hati, M. (2006). *Does the new intellectual property regime impede innovation in developing countries: A case study of the Indian pharmaceutical industry*. http//www.globelicsindia2006.org/Moutushi%20 hati.pdf, accessed on Feb17th 2008-05-29

Heller, M., & Eisenberg, R. (1998, May 1). Can Patents Deter Innovation? The Anticommons in Biomedical Research. *Science*, *280*(5364), 698–701. doi:10.1126/science.280.5364.698

Hoffman La Roche Ltd Vs. Cipla ltd(2008);

I.A 642/2008 IN CS (OS) 89/2008; High Court Of Delhi Hoffman La Roche Ltd Vs. Cipla ltd(2008);

I.A 642/2008 IN CS (OS) 89/2008; High Court Of Delhi http://www.vidhiglobe.com/Data_Protection_1_.pdf

Investment Commission Report. (2006). *Investment strategy for India*. Ministry of Finance Government of India.

Keyala, B. K. (2000). TRIPs patent system; implication for health care and pharma industry. *Centre for the study of global trade system of development*. New Delhi, India.

KPMG Report on Indian Biotechnology industry (2006). 18, India.

Lall, S. (2003). Indicators of the relative importance of IPRs in developing countries. [New York: Elsevier.]. *Research Policy, 32*(9), 1657–1680. doi:10.1016/S0048-7333(03)00046-5

Love, J. (2003). *Evidence regarding research and development investments in innovative and non-innovative medicines*. Consumer Project on Technology.

Love, J. (2003). *Evidence regarding research and development investment in innovative and non innovative medicines*. USA: Consumer Project Technology.

Malerba, F., & Orsenigo, L. (2001). *Towards a history friendly model of innovation, market structure and regulation in the dynamics other pharmaceutical industry: The age of random screening*. Working Papers with number 124 CESPRI, Centre for Research on Innovation and Internationalisation, Universita' Bocconi, Milano, Italy.

Matter, P. S. accessed from http://www.southcentre.org/publications/publichealth/publichealth-04.htm#P327_38330#P327_38330 on 12th Jan 2008

Merges, R., & Nelson, R. R. (1990). On the complex economics of patent scope. *Columbia Law Review* 90, Morriset, J.P. (1999), *How tax policy and incentives affect foreign direct investment: A review*. World Bank Policy Research Working Paper No. 2509; World Bank Group - Foreign Investment Advisory Service (FIAS)NEDE PIRNIA World Bank.

Munos, B., (2006). *Can open source R&D reinvigorate drug research*? Nature Review drug discovery.

Nair, M.S. (21/09/2004). Data exclusivity: *The Indian perspective*. Lex Orbis; Mondaq Publication.

Nandan, K. (2001-2002). *Operation successful but patient dead; MSc thesis in Economics and social history*. Oxford, UK: University of Oxford.

National Biotechnology Development Strategy Policy. (2007). Department of Biotechnology, Ministry of Science and Technology; Govt of India, available at http://www.dbtindia.gov.in/biotechstrategy.htm.

Organization for Economic Cooperation and Development Report. (2004). *Patents and innovation: Trend and policy challenges*. OECD Publication.

Ott, R. (2004).Applying the patent to living material in India. 2 Okla J.L. & Tech. 17.

Parker, T. (2001). *Biotech companies call for patent reform*. Managing Intellectual Property. TRIPS and Public Health, Paper submitted to TRIPS Council, 29th June 2001, IP/C/W/296

Pharma (2001). India: A cautionary tale on critical importance of IP protection. Retrieved from at http://www.Phrma.org/publications/documents/backgrounders/2001-03-01.202/html

Rank hovis Mc Dougall case (1978). FSR, 588

Rashmi, R. (Sept 2007). Indian biopharmaceutical industry in the era of globalization: Path for novel drug industry. In *Genetically Modified Organisms: Emerging Law and Policy in India.* New Dehli, India:TERI Press.

Reddy, P. March 31, 2008; Foreign pharma may adopt differential pricing; Spicy IP; Accessed at http://spicyipindia.blogspot.com/2008/03/foreign-pharma-may-adopt-differential.html

Report, U. N. C. T. A. D. (2004). The Biotechnology Promise Capacity-building for Participation of Developing Countries in the Bioeconomy, UNCTAD/ITE/IPC/2004/2,Geneva

Roche vs NATCO: India's First "Doha Style" Compulsory License? January 16, 2008 http://spicyipindia.blogspot.com/2008/01/roche-vs-natco-indias-first-doha-style.html accessed on February 18 2008

Rose, C. M. (1998). *The several futures of property: Of cyberspace and folktales, emission trades and ecosystems.* 83 MINN. *L. REV.*, *129*, 179.

Singham, S.A. (2000). Competition Policy and the Stimulation of Innovation: TRIPS and the Interface Between Competition and Patent Protection in the Pharmaceutical Industry. 26 BROOK. J. INT'L L.363, 388-89.

Srinivas, K. R. (2006), *TRIPs, access to medicines and developing nations: Towards an open source solution*. Working paper no.248; Indian Institute of Management, Bangalore.

United States Supreme Court (June 16, 1980). 447 U.S. 303, 206 USPQ 193

Van Dijk, T. (1996). Patent Height and Competition in Product Improvements. *The Journal of Industrial Economics*, *44*, 151–167. doi:10.2307/2950643

Vanessa Bradford Kerry. Kelley Lee (24 May 2007), "TRIPS, the Doha declaration and paragraph 6 decision: what are the remaining steps for protecting access to medicines?" Globalization and Health, 3:3.Wilson Tim, "In defense of patents," Times of India, New Delhi, April 26, 2008

World Health Organisation Report (November 2006);

Counterfeit medicines, Fact sheet N°275; Revised 14, WHO/. available on http://www.who.int/mediacentre/factsheets/fs275/en/

ADDITIONAL READING

A list of District Court websites is available at http://indiancourts.nic.in/sitesmain.htm (last visited Oct 12, 2008).

A list of various High Court websites is available at http://indiancourts.nic.in/sitesmain.htm (last visited Oct 12, 2008).

A new way to get medicines to the poor; www.spicyip.com; May 13, 2008.

Agreement on Trade-Related Aspects of Intellectual Property Rights, Apr. 15, 1994,Marrakesh Agreement Establishing the World Trade Organization, Annex 1C, Legal Instruments—Results of the Uruguay Round, 33 I.L.M. 1125, 1208-11 (1994)

Basu, I. (2004; Jul 23); India's clinical trials and tribulations. Asia Times Online. Accessed 2008; January 16 at www.atimes.com/atimes/South_Asia/FG23Df03.html. Business Memo from Belgium; publication of the Embassy of Belgium, Washington DC, 2/2005 http://www.diplobel.org/usa/NewsPublications/BusinessMemo/BM022005.pdf accessed on 10/06/07.

Biogen vs. Medeva(1997); RPC; 1.

BiospectrumApril, 2006.

BiospectrumJanuary; 2008.

Biospectrum, February 2005.

Biospectrum; June 2008.

Biospectrum, March 2008.

Biospectrum; May, 2008.

Biospectrum, September 2006.

Bioventure for global health(n.d.). ;

Retrieved from http://www.bvgh.org/innovation/crisis/default.asp.

Diamond Vs Chakrabarty; United States Supreme Court, June 16, 1980, 447 U.S. 303, 206 USPQ 193.

Chohung Bank vs Deputy Director of Income-tax (2006) 102 ITD

Data Protection: Emerging Trends, Deepu M. Accessed on 3rd March 2008 http://www.vidhiglobe.com/Data_Protection_1_.pdf.

Department of Biotechnology. (n.d.). *Government of India*, Retrieved from http://dbtindia.nic.in.

Diamond v. Diehr (1980), 450 U.S. 175, 185.

Dimminaco A.G. v. Controller of Patents, Designs & Trade Marks, (2002) I.P.L.R. July 255,269 (Calcutta H.C.)

Finance Act 2006, Government of India.

Finance Act 2007, Government of India.

Florey/relaxin case(1995); O.J.E.P.O, 388.

Foreign pharma may adopt differential pricing, Prashant Reddy; Spicy IP; March 31, 2008; http://spicyipindia.blogspot.com/2008/03/foreign-pharma-may-adopt-differential.html.

Hoffman La Roche Ltd Vs. Cipla ltd (2008);

I.A 642/2008 IN CS (OS) 89/2008; High Court Of Delhi

Indian government to set up biotech regulator; Wednesday (Nov 14,2007);

The Hindu. Business Daily http://www.thehindubusinessline.com/2007/11/14/stories/2007111452301000.htm Accessed on January 16, 2008.

Jain Jayant and Panchal Dharmesh,(November 1-15, 2006);

Taxing Indian pharma, Express Pharma.

Mashelkar, R. A., et al. Report of the Technical Expert Group on Patent Law Issues 1-56 (Dec. 2006), available at http://patentoffice.nic.in/ipr/patent/mashelkar_committee_report.doc. 35 U.S.Code.

Moutshi hati (2006); Does the new Intellectual Property regime impede innovation in developing countries: A case study of the Indian pharmaceutical industry; http//www.globelicsindia2006.org /Moutushi%20 hati.pdf, accessed on Feb17th 2008-05-29.

National Biotechnology Development Strategy policy (2007);

Department of Biotechnology, Ministry of Science and Technology. Govt of India, available at http://www.dbtindia.gov.in/ biotechstrategy.htm.

(n.d.). Retrieved from http://www.ipwworldonline.com.

(n.d.). Retrieved from http://www.southcentre.org.

(n.d.). Retrieved from www.who.edm.org.

NovBiospectrum, 2005.

Pedro Roffe and Taffere Tesfachew; Revisiting the Technology Transfer Debate: Lessons for the New WTO Working Group, http://www.ictsd.org, accessed on 20th April 2006.

Pharma 2001, India; A cautionary tale on critical importance of IP protection at http://www.Phrma.org/publications/documents/backgrounders/2001-03-01.202/html.

Rank hovis Mc Dougall case(1978); FSR, 588.

re Thomas F. DEUEL(March 28, 1995);

No. 94-1202; United States Court of Appeals, Federal Circuit.

Roche vs NATCO: India's First "Doha Style" Compulsory License? January 16, 2008 http://spicyipindia.blogspot.com/2008/01/roche-vs-natco-indias-first-doha-style.html accessed on February 18 2008

Shri Justice N. Rajagopala Ayyangar, Report on the Revision of the Patents Law (September 1959).

The Patents Act, No. 39 of 1970, § 5(1) (Universal 2005) (amended 1999 & 2002) available at http://www.indiacode.nic.in.

The Patents (Amendment) Act. No. 17 of 1999 (1999), *available at*http://indiacode.nic.in.

The Patents (Amendment) Act. No. 38 of 2002, *available at*http://indiacode.nic.in.

The Patents (Amendment) Act, No. 15 of 2005 (Universal 2005); The Patent Act 2005 was officially published in the Gazette of India on April 5, 2005

The Patents (Amendment) Ordinance. No. 13 of 1994, *available at*http://www.wipo.int/clea/docs_new/en/in/in001en.html. The 1994 Ordinance was officially published in 81 THE GAZETTE OF INDIA (Dec. 31, 1994

Tim, W. In defense of patents, Times of India, New Delhi, April 26, 2008.

World Trade Organization. *Declaration on the TRIPS Agreement and Public Health* (WT/MIN(01)/DEC/2) (Nov. 14, 2001), *available at*http://www.wto.org/english/thewto_e/minist_e/min01_e/mindecl_trips_e.htm [hereinafter *Declaration on the TRIPS Agreement and Public Health*].

World Trade Organization. Dispute DS199, Brazil—Measures Affecting Patent Protection, available at http://www.wto.org/english/tratop_e/dispu_e/cases_e/ds199_e.htm.

World Trade Organization. (n.d.).Retrieved from www.wto.org.

WTO General Council. Implementation of Paragraph 6 of the Doha Declaration on the *TRIPS Agreement and Public Health*, WT/L/540 (Aug. 30, 2003, *available at*http://www.wto.org/english/tratop_e/trips_e/implem_para6_e.htm.

ENDNOTES

1 Rakhi Rashmi is working as Research Fellow with CNRS GREDEG, University of Nice, France, A PhD candidate of Centre for the study of Law Governance, Jawaharlal Nehru University, New Delhi, who has the degrees of LLB (Delhi University), MBL (National Law School of India University, B'lore), MFM (Pondicherry University. Contact: GREDEG-CNRS, University of Nice Sophia Antipolis, 250, rue Albert Einstein, 06560 Valbonne, France.Email: rakhi.rashmi@gmail.com

2 35 U.S.C 102 Conditions for patentability; novelty and loss of right to patent

3 35 U.S.C. 101 Inventions patentable; Whoever invents or discovers any new and useful process, machine, manufacture, or composition of matter, or any new and useful improvement thereof, may obtain a patent therefore, subject to the conditions and requirements of this title.

4 1. The following shall not be patentable: (a) plant and animal varieties; (b) essentially biological processes for the production of plants or animals

5 *Article 4.2* The following shall not be patentable: (a) plant and animal varieties; (b) essentially biological processes for the production of plants or animals. 2. Inventions which concern plants or animals shall be patentable if the technical feasibility of the invention is not confined to a particular plant or animal variety. 3. Paragraph 1(b) shall be without prejudice to the patentability of inventions which concern a microbiological or other technical process or a product obtained by means of such a process.

6 1. The human body, at the various stages of its formation and development, and the simple discovery of one of its elements, including the sequence or partial sequence of a gene, cannot constitute patentable inventions. 2.

An element isolated from the human body or otherwise produced by means of a technical process, including the sequence or partial sequence of a gene, may constitute a patentable invention, even if the structure of that element is identical to that of a natural element.

7 1997; RPC; 1.

8 1995; O.J.E.P.O, 388

9. European patents shall not be granted in respect of: (a) inventions the publication or exploitation of which would be contrary to "ordre public" or morality, provided that the exploitation shall not be deemed to be so contrary merely because it is prohibited by law or regulation in some or all of the Contracting States;

10 Members may also exclude from patentability: (a) diagnostic, therapeutic and surgical methods for the treatment of humans or animals; (b) plants and animals other than micro-organisms, and essentially biological processes for the production of plants or animals other than non-biological and microbiological processes. However, Members shall provide for the protection of plant varieties either by patents or by an effective *sui generis* system or by any combination thereof. The provisions of this subparagraph shall be reviewed four years after the date of entry into force of the WTO Agreement

11 Section 3 (d) reads: 'The mere discovery of a new form of a known substance which does not result in the enhancement of the known efficacy of that substance or the mere discovery of any new property or new use for a known substance or the mere use of a known process, machine or apparatus unless such known process results in a new product or employs at least one new reactant.

12 *'Salts, esters, ethers, polymorphs, metabolites, pure form, particle size, isomers, mixtures of isomers, complexes, combinations and other derivatives of known substance shall be considered to be the same substance, unless they differ significantly in properties with regard to efficacy.'*

13 and held that "*...this court has no jurisdiction to decide the validity of the amended section, being in violation of Article 27 of TRIPS, we are not going into the question (of) whether any individual is conferred with an enforceable right under TRIPS or not. For the same reason, we also hold that we are deciding the issue namely, whether the amended section is compatible (with) Article 27 of TRIPS or not*".

14 Members when requiring as a condition of approving marketing of pharmaceutical or of agricultural chemical product which utilize new chemical entities, the submission of undisclosed test or other data, the originator of which involves considerable efforts shall protect such data against disclosure, expect where necessary to protect the public or unless steps are taken to ensure that the data are protected against unfair commercial use".

15 Satwant Reddy Committee Report 2007, on Steps to be taken by Government of India in the context of Data Protection Provisions of Article 39.3 of TRIPS Agreement

16 any act of making, constructing, using, selling or importing a patented invention solely for uses reasonably related to development and submission of information required under any law for the time being in force, in India, or in a country other than India, that regulates the manufacture, construction, use, sale or import of any product.

17 Other Use Without Authorization of the Right Holder where the law of a Member allows for other use of the subject matter of a patent without the authorization of the right holder, including use by the government or third parties authorized by the government

18 Here the condition of obtaining compulsory licence is expanded, (in case of LDCs having no Patent Law or provision for compulsory

licence) to include an 'authorisation' or notification from such a country. This is done by modifying sub-section (1) of section 92A as follows: Adding the following words after the words "provided compulsory licence has been granted by such country" "or such country has by notification or otherwise allowed importation of the patented pharmaceutical products from India."

19 Indian government to set up biotech regulator; Wednesday (Nov 14, 2007); The Hindu, Business Daily; At http://www.thehindubusinessline.com/2007/11/14/stories/2007111452301000.htm; Accessed on January 16, 2008

20 AGREEMENT WITH FOREIGN COUNTRIES. (1) The Central Government may enter into an agreement with the Government of any country outside India - (a) For the granting of relief in respect of income on which have been paid both income-tax under this Act and income-tax in that country, or (b) For the avoidance of double taxation of income under this Act and under the corresponding law in force in that country, or (c) For exchange of information for the prevention of evasion or avoidance of income-tax chargeable under this Act or under the corresponding law in force in that country, or investigation of cases of such evasion or avoidance, or (d) For recovery of income-tax under this Act and under the corresponding law in force in that country, and may, by notification in the Official Gazette, make such provisions as may be necessary for implementing the agreement. (2) Where the Central Government has entered into an agreement with the Government of any country outside India under sub-section (1) for granting relief of tax, or as the case may be, avoidance of double taxation, then, in relation to the assessee to whom su

21 A new way to get medicines to the poor; www.spicyip.com; May 13, 2008

APPENDIX A: INTERNATIONAL EXAMPLES OF COMPULSORY LICENSING (SOURCE: BASED FORMATION COLLECTED FROM VARIOUS SOURCES)

Country	Year	Detail
Argentina	18 Oct 2005	Argentina announced intention to grant CL for Tamil Flow (Oseltamivir) It later found that the patent was never granted in Argentina
Brazil	6th July 2007	Brazil grants CL for Abott's Kaletra
Brazil	4 May 2007	Brazil grants CL for Mercks HIV/AIDS drug Efavirenz
Equador	2003	Petitions by local manufacturer (Acoomax) for CL for Combivir Lamivudine and AZT) refused, appealed and refused again. GSK agrees to supply drug at a discount
Eritrea	5th June 2005	Eritria issues CL for importation on Eritria of generic HIV / medicines.
Zimbabwe	May 2002	Zimbabwe declares Period of emergency which enables it to ignore antiretroviral drug patents for 6 months via a CL. In 2003 the period of emergency was extended by 5 years (Until 31st Dec 2008)
Thailand	April 2007	After weeks of negotiations and public debate (and statements about with drawing drugs from Thailand, abbott agrees to provide discounted AIDs drugs (Kaletra / Aluwin) to Thailand
Thailand	Feb 2007	Thailand announced two compulsory licenses on Plavix (heart diseases drug) made by BMS and Sanofi-Aventis and Kaletra for HIV by Abott
Thailand	Dec 6 2006	Thai Government issues CL on Merck's AIDS drugs Efavirenz.
Zambia	21st 2004	CL for Lamitudine, Claudine and Nevirapine
Ghana	Oct 2005	CL for the importation into Ghana of Indian generic HIV/AIDS medicines
Guinea	18th April 2005	CL for importation on patents on drugs to treat HIV –AIDS
Indonesia	5th Oct 2004	Indonesia issues act to manufacture generic versions of Lumi vidine and nevirapine (both Hive /AIDS) drugs)
Israel	Jan 1942	Cl to manufacture Bop Hep B under Biogen patent. Biogen's appeal was unsuccessful. The patent expired in 1999 before the SC ruled on the dispute.
Italy	21 march 2007	Italy grants CL in relation to Finas teride and related generic drugs for two years prior to the 2009 expiration of patent monopoly
Republic of Korea	June 2002	Korea rejects application for CL of Novartis Glivec
Malaysia	29th Sept 2004	Cl to import from India dida nosine (ddl) Zidovudine (AZT) and Lamivudine (Combivir)
Rwanda	19th July 2007	Rwanda becomes the I country to notify the WTO that intends to use the paragraph 6 system (of DOHA Declaration). Paragraph 6 is designed for those countries that do not have the facilitates to manufacture pharmaceuticals and also allows importations
Mozambique	5th April 2004	CL for Lamivudine, Slavudine and Nevirapine
Canada	4 Oct 2007	The WTO received from Canada on 4th Oct 2007, the first notification from any Government that it has authorized a company to make a generic version of patented medicine for export under special WTO provisions agreed in 2003. The triple combination AIDS therapy drug, Tri Avir, Can now be made and exported to Rwanda, which is unable to manufacture the medicine itself. Earlier on 17th July, Rwanda informed the WTO that it intentends to import 260,000 packs of Tri Avir – a fixed dose combination product of Ziol Ovudine, lamivudine two years. The drug so to be made in Canada by Apotex the and is called Apo Triavir by the manufacture
Canada	14th May 2004	Canada amends its patent law to allow Canadian manufacturer to export to countries which lacks the ability to manufacture pharmaceuticals but only in respect of drugs listed in schedule, of the Act which include, Lamivudine Nevirapine Zidovudin tablets etc.
Canada	Oct 18, 2001	CL in relation to the Bayers ciprofloxacin and authorized generic manufacturer to build manufacturer to build stockpiles protection against and attack and anthrax

Chapter 8
Embracing Emerging Technologies:
The Case of the Singapore Intelligent Nation 2015 Vision

Pak Tee Ng
Nanyang Technological University, Singapore

ABSTRACT

Many countries are attempting to develop themselves as intelligent nations, equipped and empowered by advanced info-communications technology (ICT). This chapter describes and analyses how Singapore attempts to develop an intelligent nation under its umbrella vision and strategy of Intelligent Nation 2015 (iN2015). It examines the substance of the vision and highlights some of the initiatives implemented to realise the vision. It also discusses the different roles of the government in implementing the iN2015 vision and some of the challenges ahead.

1. INTRODUCTION

As info-communication technology (ICT) rapidly advances, many governments around the world launch initiatives to get their country wired and connected, both nationally and internationally, in order to navigate and compete more effectively in the knowledge age and economy. Many countries are reviewing and restructuring their national network infrastructures, information and media regulations, and fiscal policies to match the speed of these ICT drivers. If they do not match up or transform themselves into intelligent societies, they will be left behind in a connected global economy. Many areas of the society are affected by such technological advancement: basic communications, education, health care, business, and even lifestyle (read, for example, Drucker, 1993, 1998; Greider, 1997; Ohmae, 1990, 1995; Poster, 1990; Postman, 1993; Webster, 1995). These areas will benefit from the development of an integrated network of national information super-highways.

This chapter describes and analyses how Singapore attempts to develop an intelligent nation under its umbrella vision and strategy of Intelligent Nation 2015 (iN2015). It examines the substance of the vision and highlights some of the initiatives implemented to realise the vision. It also

DOI: 10.4018/978-1-61692-846-9.ch008

Copyright © 2011, IGI Global. Copying or distributing in print or electronic forms without written permission of IGI Global is prohibited.

discusses the different roles of the government in implementing the iN2015 vision and some of the challenges ahead.

2. INTELLIGENT NATION 2015 (IN2015)

Launched in 2006, the Intelligent Nation 2015 (iN2015) vision is a 10-year master-plan to grow the info-communications sector and build a well-connected society, thus fulfilling the aim of becoming an intelligent nation, empowered by the latest technologies. According to the Info-communications Development Authority of Singapore (iDA), iN2015 will propel Singapore into the ICT forefront to ensure continued international competitiveness (iDA, 2006). This 10-year master-plan culminates in 2015, which coincides with Singapore's 50th anniversary of its independence and nation-building.

According to the Finance Minister, Tharman Shanmugaratnam, the info-communications industry today contributes 6% to Singapore's Gross Domestic Product (GDP). Info-communication technology usage in Singapore households ranks among the highest in the world, with at least one computer in 74% of Singapore homes, more than 40% dial-up Internet penetration, household broadband penetration at 42%, 99% island-wide broadband coverage and mobile penetration rates at 92% (Tharman, 2008). But beyond physical ICT provisions, the iN2015 vision aims for a state of complete ease among citizens with technology, where technology enables new businesses and lifestyles.

A high-level iN2015 Steering Committee was formed at the inception of iN2015 to oversee the project. In particular, its aim was to identify areas of opportunity where ICT could create new value propositions and innovative services, thereby enhancing the competitiveness of key economic sectors in Singapore. These sectors include Digital Media and Entertainment, Education and Learning, Financial Services, Healthcare and Biomedical Sciences, Logistics and Hi-Tech Manufacturing, and Tourism, Hospitality and Retail. The committee also examined how ICT and other emerging technologies (e.g. sensor technology, bio-computing and nanotechnology) could transform the way people live, work, learn and play, the ICT infrastructure to support the future ICT needs, the ICT manpower development strategies to support new high-growth industry clusters, and government policies and regulatory frameworks to support the ICT edifice. Efforts to develop iN2015 include (iDA, 2006):

- seeking views and perspectives of local and international info-communications players;
- seeking views and perspectives of business players in key economic sectors such as manufacturing, logistics, healthcare and education;
- environment scanning and analysis of future global and local trends;
- examination of Singapore's (and the info-communications industry's) strengths and weaknesses, as well as possible opportunities and threats;
- review of the current state of info-communications development; and
- identification of the long-term info-communications industry structure and manpower profile.

In a way, the iN2015 vision is not totally new. Rather, it is the confluence of various efforts made in the past, for example:

- **1981–1985:** Singapore embarked on the National Computerisation Plan in the Civil Service to develop automated work functions and improve internal operational efficiencies. This was a massive exercise and laid the foundation for subsequent ICT development in other parts of the economy.

- **1986–1991:** The National IT plan, which aimed to take the extension of the computerisation efforts beyond the civil service, was launched in 1986. In particular, networks such as TradeNet, LawNet and MediNet were developed to facilitate business in these three domains. During this period, three data hubs, Land, People and Establishment, were created to facilitate information-sharing across public agencies for better service delivery. The aim was to provide one stop services.
- **1992–1999:** During this period, the most significant national initiative was the IT2000 master-plan. Building upon the foundation laid by the National IT Plan, the IT2000 masterplan aimed to position Singapore as a global IT hub by expanding the focus of IT infusion to improving the quality of life in Singapore, boosting the economic engine and linking communities locally and globally. In particular, Singapore ONE (One Network for Everyone) was developed to provide an integrated network for everyone in the country.
- **2000–2002:** As telecoms and IT began to converge in many applications, the Infocomm 21 blueprint was launched in 2000 to develop Singapore into a global Infocomm Capital. The vision was a thriving and prosperous e-economy and e-society. As part of Infocomm 21, the first e-Government Action Plan was developed to place all public services online.
- **2003–2005:** The push for an e-society was refocused with a new vision of 'Connected Singapore', first launch in 2003. Connected Singapore focused on further developing ICT in the private and public sector, and promoting ICT among the public. In particular, it aimed to create new lifestyle ideas, produce new consumer value and open up new business possibilities. As part of Connected Singapore, the second e-Government Action Plan was launched with an aim to delight customers and connect citizens through infocomm. A significant event during this period was the establishment of the Media Development Authority (MDA) on 1 January 2003, whose remit was to spearhead the transformation of Singapore into a global digital media hub.

3. IN2015 INITIATIVES

A suite of initiatives followed the launch of the iN2015 vision in 2006. A few recent projects from various sectors in Singapore serve to illustrate the efforts put into realising the iN2015 vision.

In the area of ICT infrastructure, one of the first island-wide projects is Wireless@SG, a wireless broadband programme developed by iDA and launched in December 2006 as part of its Next Generation National Infocomm Infrastructure (NGNII) initiative. Run by three local wireless operators, users enjoy free indoor and outdoor seamless wireless broadband access with speeds of up to 1 Mbps at public areas (free access until 2013). It facilitates people on the move, especially those who want or need to be connected while away from their homes, schools and offices, including students, tourists, business travellers and corporate executives out for meetings. In 2009, there are 7,500 hotspots covering Singapore and 1.3 million subscribers (iDA, 2008a).

In the area of education, ICT has always been a strategic thrust, with the implementation of three master-plans hitherto. The first Masterplan for ICT in Education (1997-2002) laid the foundation for schools to use ICT, by providing the basic ICT infrastructure and basic level teacher training. The second Masterplan (2003-2008) strove for a more effective and pervasive use of ICT in education by establishing baseline ICT standards for students and integrating a certain level of ICT into the curriculum. The third and current Masterplan, launched in 2009, continues and enhances the vision of the first and second Masterplans by emphasising the transformation of learning environments through ICT and equipping students with the competencies to succeed in a knowledge economy. The third Masterplan aims to integrate ICT extensively into the planning, design and implementation of curriculum, assessment and pedagogy. Students will be required to use ICT to find information, generate reports, give feedback on one another's work and collaborate with people within and outside school (Ng, 2008).

In the longer term, the government aims for all schools to be ICT-intelligent by 2015. 15 to 20 percent of schools are designated as Experimental Schools to trial innovative applications of ICT in teaching and learning. Another five per cent are identified as Schools of the Future, sites which are models of ICT-empowered education and leaders in the innovative use of ICT in Singapore and beyond. Local ICT enterprises are invited to work with like-minded schools in developing, piloting and deploying innovative products and services in education, and subsequently market successful innovations to overseas markets (iDA, 2008b).

In the area of commercial business, one of the latest projects is 'TradeXchange', in which the government invites companies to form consortia and integrate their key IT systems, business processes and data flow so as to achieve greater business efficiency. The government also plans to develop the next generation e-payment infrastructure through the large scale deployment of contactless point-of-sale terminals to promote e-payment adoption using Near Field Communication technology. Singapore also aims to be the wealth management hub of Asia though its 'i-wealth Management' project by providing high technology front-end customer service, middle office management and back-end office operations. Through advanced technology, customers can securely obtain a complete view of their financial positions and perform personalised transactions anytime and anywhere (iDA, 2008b).

In the area of health care, advanced ICT has already allowed partial integration of medical records so that different hospitals and some clinics can provide holistic care, especially for those suffering from chronic diseases. However, in the middle term (2009-2011), the aim is to allow patients to manage their health at home through remote monitoring solutions, especially for patients with chronic diseases. In the long term (2011 and beyond), the public in general will be able to access on their electronically-accessible personal health records and related health information (iDA, 2008b).

In the area of defence, the Singapore Armed Forces (SAF) has begun its transformation into a high technology third generation fighting forces, with advanced ICT capabilities. For example, a notable development in the Army was the deployment of the Advanced Combat Man System (ACMS). Boasting an array of portable sensors, computers and communications equipment, this system allows Command Headquarters to link seamlessly with SAF assets and soldiers on the ground. Commanders can view real-time satellite images of battle situations, communicate with all friendly forces within its vicinity as well as call for support fire from air and sea counterparts. Minister for Defence Teo Chee Hean explained that this was an example of how "the SAF is ready to confront current and future operational challenges because we have embarked on the transformation to the 3rd Generation SAF. Such investments have enabled the SAF to restructure and streamline its training and operational requirements, leading to more effective use of our National Servicemen". (Teo, 2009)

4. THE ROLES OF THE GOVERNMENT FOR IN2015

The government is usually the initiator and key player in most national level projects. But, it is not so common to find a case like the Singapore's iN2015, where government presence is so pervasive, obvious and instrumental. The iN2015 is a story of a wired intelligent nation engineered by the government. The Singapore government is the single most influential body to realise the iN2015 vision, playing the roles of:

- Visionary
- Investor
- Developer
- Regulator

The Government as Visionary

The government is currently the main visionary for the nation, taking on the leadership role in defining Singapore's future direction of growth and allocating appropriate resources to realise the iN2015 vision for Singapore. Indeed, the iN2015 forms part of a greater vision to transform Singapore into a Learning Nation and develop intellectual capital for the country (Goh, 1997a; Ng, in press). In the Singapore Learning Nation, the continuous development of intellectual capital – human capital, infrastructural capital and relational capital (Roos, Edvinsson and Roos, 1998) – initiated by deliberate policies and facilitated by fiscal investment, becomes the strategy to gain a competitive edge in the global economy.

Indeed the iN2015 vision reflects the government's pragmatic style of governance. The political leaders continually exhorted the nation that the priority was economic growth linked to national survival and that social order was essential for the nation to meet international economic competition (Chan, 1971; Chua, 1995). At every level of the socio-political and economic arena, pragmatism, meritocracy, anti-corruption and government leadership permeated. In the rhetoric of economic survival and expediency (Hill and Fee, 1995), the intelligent nation discourse fits in comfortably. The iN2015 vision does not merely allude to a state of technology within Singapore but more importantly incites an idea of high technology urbanity and world class living environment (Goh, 1997b). It is symbolic of what the country should believe in and strive for. It is part of manipulating the symbolic content of national identity, but in Singapore's case never too far from its utilitarian objectives (Hill and Fee, 1995).

The Government as Investor

The government is the main ICT investor in the country. The various national level projects mentioned in the previous section illustrate the extent of the investment on the part of the government. Indeed, the biggest investor in the ICT industry and customer of the ICT companies can be said to be the government itself. The industry has benefited from the government's continual ICT investment. Local companies were awarded a total of 67% of government contracts, while multi-national companies captured 33%, a ratio that has been consistent for the last four years (Lui, 2009). Such investments range from the investment in the Next Generation National Info-communications Infrastructure, comprising the ultra-high speed Next Generation National Broadband Network (NGNBN) and the pervasive Wireless Broadband Network, to the tax incentives and special grants for technology investment offered to local MNCs, SMEs and individual entrepreneurs.

The clearest illustration of the ICT investment commitment of the government was found during the collapse of the US sub-prime mortgage market in 2008, which cumulated in a full blown global financial crisis in 2009. During the severe recession, which hit Singapore very hard, the government kept faith with its iN2015 vision. Despite the tight fiscal situation, the iN2015 masterplan marched on to transform Singapore into an intelligent nation powered by advanced info-communications. The government invested S$1.73 billion worth of new ICT tenders in financial year 2009, up from S$1.60 billion in 2008 (Lui, 2009). Major projects continued. Wireless@SG was enhanced. The deployment of NGNBN began as scheduled and was expected to cover 60% of all homes and offices by 2010 and 100% nationwide by 2012 (Lui, 2009).

The Government as Developer

Other than the creating the vision and providing the funding, the government is the primary iN2015 developer in the country. It plays the role of a strategic integrator by ensuring that the various programmes and projects, past and present, such as IT2000 and iN2015, are cohesive and well integrated. It spearheads the development of the physical infrastructure that provides the means for citizens and companies to enter the information age. Concurrently, it is also the main player in developing policies that will entice multinational ICT companies to invest, develop and promote Singapore as an ICT hub. It also sees to the development of manpower needs. Through its Talent Management Programme, the iDA grooms a sustainable pool of industry-ready, globally competitive ICT professionals and offer them a head-start in an ICT career (Lui, 2008).

But the government's role as a developer goes beyond the physical infrastructure to the softer issues of an ICT society. Here, it plays the role of an educator and life-style engineer. The government goes all the way out to educate its citizenry of the benefits of ICT and the lifestyle conveniences that would bring. The iDA conducts a broad-based outreach programme to help people in Singapore leverage ICT for a vibrant digital lifestyle. Going beyond imparting basic knowledge such as Internet connection and email usage, the programme covers more advanced uses such as Web 2.0 and Internet collaboration tools. The aim is to provide opportunities for growth, fun and fulfilment for young and old alike (Tay, 2009). The government also attempts to enable citizens a greater voice in public policy-making through the virtual medium. The REACH (Reaching Everyone for Active Citizenry @ Home) portal incorporates Web 2.0 and new media technologies such as blogs, social networking, online chats and short text messaging for government-citizen exchange (Lee, 2007).

The Government as Regulator

The government is also the primary regulator of the ICT industry and ICT usage in Singapore. As ICT opens up a new world of possibilities, there are now more players to be managed and issues of regulation to be considered. Through the iDA, the government tries to enact policies and regulations that are pro-consumer and pro-business (iDA, 2006). The government is the central body that formulates and develops standards, codes of practices and advisory guidelines pertaining to issues such as licensing, inter-connection, and resource and competition management. As an example, one of iDA's roles as a regulator is to enforce the regulatory requirements on licensees of the tele-communications sectors and ensure that they do not engage in anti-competitive behaviour. The iDA also ensures that licensees comply with an established price control framework and service quality standards. Because the ICT industry is a dynamic one, therefore the iDA also has to regularly and progressively review these policies and regulations. ICT policies and regulations are built on the principles of:

- Promoting effective and sustainable competition;
- Promoting facilities-based competition to the greatest extent possible;
- Relying on market forces;
- Adopting proportionate regulation;
- Remaining technology-neutral; and
- Providing a transparent and reasoned decision-making process.

5. CONCLUSION: FUTURE CHALLENGES

While Singapore marches on purposefully in its iN2015 journey, there are still challenges along the way that it has to address. Firstly, it has to develop a comprehensive system of managing intellectual property and computer security. As Singapore becomes a highly-networked society that is connected globally, issues such as policy and legal frameworks, which are developed in parallel to the industry, will need to take into account evolving national values and international expectations. It has to maintain the trust of international companies and trading partners that want their intellectual property to be protected. However, among its big international clientele such as the US, Europe and China, each has a different standard and expectation of intellectual property rights and virtual medium trading regulations. So, the challenge is to find the right common ground to work with international partners while defining its own code of practice and ICT laws which reflect local conditions.

Secondly, ICT brings with it the challenge of the digital divide, the segregation of a society into the ICT literate and illiterate. Those with ICT knowledge and skills will be able to access more information and services because most information and services are going to be put on the Internet. The challenge of the digital divide is real. The older generation, which has live through an era without these advanced ICT, will have a problem navigating in a paperless society. The poorer sections of the society may not be able to afford constant upgrades of technology. Singapore has been trying hard to unify one citizenry in and through one place as a response to Singapore's roots in linguistic diversity and diasporas (Yeoh and Chang, 2001). But such homogenising efforts could be parodied by the digital divide.

The government responded early and with some urgency to this issue. Various interventions were launched to address the problem. For example, a three-year initiative "eCelebrations Singapore" was launched in 2000 to raise awareness on how technologies could enhance the quality of life and standard of living in Singapore, especially among those who were inexperienced in Internet usage so that they could stay connected and be relevant in Singapore's new economy. The government committed S$25 million to bridging the digital divide, paying special attention to senior citizens, home-makers, blue-collared workers and special interests groups. The government, through iDA, also worked with self-help groups and grass-roots groups to offer used computers, together with free Internet access and basic training to low-income households (Yeo, 2000). In addition, the government launched the CitizenConnect Centres in 2006 to help citizens or residents who do not have access to, or need assistance in using the computer. Such centres are located in neighbourhood community clubs. From 5 centres in 2006, this grew to a network of 28 centres across Singapore in 2008 (iDA, 2008).

Thirdly, the Singapore government has played an overwhelming role in the iN2015 journey so far. While the government has demonstrated a resolve to continue the iN2015 vision, shown clearly in its response to the 2008/9 financial crisis, the transformation needs to be more organic. In going forward, a critical challenge is to bring up the level of involvement from the citizenry and the private sector. The success of bringing Singapore into the vision of iN2015 depends as much on people's attitudes and motivations as on the government. The average Singaporean has to willingly allow the permeation of such ideologies to their private domains and the crux of their daily lives. The more the vision is to permeate the whole society, the more the various parties have to collectively own the vision. This involves the government, the citizens, the public sector and the private sector. The implementation of the projects also requires the participation of, and has direct impact on, many governmental departments, statutory boards, commercial companies and civil organisations. Singapore's iN2015 vision can be better achieved through synergistic efforts, adopting a highly organic co-evolutionary dynamic (Ng, 2009) at all levels of the country. There needs to be more ground-up innovations in ICT. Digital space, whether private or public, is embedded in actual social structures and power dynamics (Sassen, 2000). The challenge is to find the soft edge to hard technology.

ICT has changed the way people live, work and play. It has reshaped businesses worldwide, revising business structures and eliminating companies and economies that could not keep up. In a way, the iN2015 initiative is not just an internal affair of being an intelligent nation. It is good global publicity for the investment grown and driven city. So, it is not just a matter of being an intelligent hub. It is also a matter of being perceived globally as an intelligent hub. In an environment of global competition, this is an extremely important dimension and contribution of iN2015. Singapore's iN2015 vision aligns well with its intended future. With ICT advancement, many developing countries will be trying to develop themselves into similar 'intelligent nations' to connect more effectively to the global economic pie and compete for a bigger slice of the pie (Yergin and Stanislaw, 1998). The experience of Singapore will serve as a mirror for those that wish to pursue this path.

REFERENCES

Chan, H. C. (1971). *Nation-building in Southeast Asia: The Singapore case*. Singapore: Institute of Southeast Asian Studies.

Chua, B. H. (1995). *Communitarian ideology and democracy in Singapore*. London: Routledge.

Drucker, P. (1993). *The Post-Capitalist Society*. New York: Harper Business.

Drucker, P. (1998). The future that has already happened. *The Futurist*, *32*(8), 16–18.

Goh, C. T. (1997a). Shaping Our Future: Thinking Schools, Learning Nation. Speech by Prime Minister Goh Chok Tong at the Opening of the 7th International Conference on Thinking, 2 June.

Goh, C. T. (1997b). Singapore 21 - A New Vision for a New Era. Speech by Prime Minister Goh Chok Tong at the Debate on President's address in Parliament, 5 June.

Greider, W. (1997). *One World, Ready or Not: The Manic Logic of Global Capitalism*. New York: Simon & Schuster.

Hill, M., & Fee, L. K. (1995). *The Politics of Nation Building and Citizenship in Singapore*. London: Routledge. doi:10.4324/9780203424438

Info-communications Development Authority of Singapore. (2006). *Innovation. Integration. Internationalisation. Report of the iN2015 Steering Committee*. Singapore: Info-communications Development Authority of Singapore.

Info-communications Development Authority of Singapore. (2008a). *Fact Sheet: Wireless@SG*. Singapore: Info-communications Development Authority of Singapore.

Info-communications Development Authority of Singapore. (2008b). *Fact Sheet: Updates on Singapore's 10 Year Infocomm Masterplan – Intelligent Nation 2015 (iN2015)*. Singapore: Info-communications Development Authority of Singapore.

Lee, B. Y. (2007). Singapore's e-Government – Sharing Our Experience. Speech by Dr Lee Boon Yang, Minister, Information, Communications and the Arts at the 4th Ministerial e-Government Conference on "Reaping the Benefits of e-Government" at Feira Internacional de Lisboa, Lisbon, Portugal, 20 September.

Lui, T. Y. (2008). Infocomm talent as a key driver of Singapore's economic competitiveness. Speech by RAdm (NS) Lui Tuck Yew, Senior Minister of State for Information, Communications and The Arts, at the National Infocomm Scholarship 2008 Award Ceremony, 5 August.

Lui, T. Y. (2009). Speech by RAdm (NS) Lui Tuck Yew, Acting Minister for Information, Communications and The Arts, at the Official Opening Of imbX Event 2009, 16 June.

Ng, E. H. (2008). Opening Address by Dr Ng Eng Hen, Minister for Education and Second Minister for Defence, at the International Conference on Teaching and Learning with Technology at the Suntec Convention Hall, 5 August.

Ng, P. T. (2009). Examining the Use of New Science Metaphors in Learning Organisation. *The Learning Organization*, *16*(2), 168–180. doi:10.1108/09696470910939224

Ng, P. T. (in press). The Singapore learning society: Intellectual capital development strategies and its response to the 2008/9 financial crisis. *International Journal of Learning and Intellectual Capital.*

Ohmae, K. (1990). *The Borderless World.* New York: Harper Collins.

Ohmae, K. (1995). *The End of the Nation State: The Rise of Regional Economies.* New York: The Free Press.

Poster, M. (1990). *The Mode of Information: Poststructuralism and Social Context.* Cambridge, MA: Polity.

Postman, N. (1993). *Technopoly: The Surrender of Culture to Technology.* New York: Vintage.

Roos, J., Edvinsson, L., & Roos, G. (1998). *Intellectual Capital: Navigating in the New Business Landscape.* New York, NY: New York University Press.

Sassen, S. (2000). Digital Networks and the State: Some Governance Questions. *Theory, Culture & Society*, *17*(4), 19–33. doi:10.1177/02632760022051293

Tay, R. (2009). *Fostering a Vibrant Infocomm Ecosystem.* Speech by RAdm(NS) Ronnie Tay, Chief Executive Officer, Infocomm Development Authority of Singapore, at the Singapore Infocomm Technology Federation 2009 ICT Business Outlook Forum at the Raffles City Convention Centre, 18 February.

Teo, C. H. (2009). Speech by Minister for Defence Teo Chee Hean, at Committee of Supply Debate in Parliament, 12 February.

Tharman, S. (2008). Budget Statement 2008 by Finance Minister Tharman Shanmugaratnam at the Singapore Parliament, 15 February.

Webster, F. (1995). *Theories of the Information Society.* London: Routledge.

Yeo, C. T. (2000). Speech by Mr Yeo Cheow Tong, Minister for Communications and Information Technology, at the launch of eCelebrations Singapore at Suntec City, 1 March.

Yeoh, B., & Chang, T. C. (2001). Globalising Singapore: Debating Transnational Flows in the City. *Urban Studies (Edinburgh, Scotland)*, *38*(7), 1025–1044. doi:10.1080/00420980123947

Yergin, D. A., & Stanislaw, J. (1998). *The Commanding Heights: The Battle between Government and the Marketplace that is Remaking the Modern World.* New York: Simon & Schuster.

Chapter 9
Dynamics in a Non-Scale R&D Endogenous Economic Growth for Chinese Development

Qiong He
Xiamen University, China

ABSTRACT

By introducing nonlinear technology gap into Jones (1995b), this chapter constructs an R&D non-scale growth model that includes endogenous human capital and technological progress. The goal is to take the model's implications to the data to explaining the Chinese economic development experiences at period 1979-2004. Our model suggests that the technology gap has the block neck effect on the economic development. The market competitive equilibrium solution shows that R&D and technology spillover can enhance the steady state growth rate. The mode's transitional dynamics is also analyzed on the effects of human capital, capital, technological progress and intersectional labor movements on economic growth, technological progress has the most effect on the economic development, and more human capital shift into R&D sector from final goods sector.

1. INTRODUCTION

'Development miracles' constitute one of the most intriguing phenomena associated with modern economic growth. The fast-growing economy in China is truly staggering. China averaged output growth rate over 9 percent per year during 1978-1999. Figure 1 illustrates the growth experience of the miracle country. Figure 1 reveals an interesting feature of miraculous experience: the sharp increase of output per work was characterized by growth rates that did not peak at the beginning of the convergence process but later on, thus giving way to a hump-shape growth path.

Since the influential paper by Lucas (1993), there has been surging interest in theoretical research attempting to explain development miracles. Since Lucas (1993),there has been surging interest in theoretical research attempting to explain development miracles. The underlying characteristics of China economic growth are focused widely. Liu, et al.(2004) consider that

DOI: 10.4018/978-1-61692-846-9.ch009

Copyright © 2011, IGI Global. Copying or distributing in print or electronic forms without written permission of IGI Global is prohibited.

Figure 1. The growth experience of China during 1979-2004

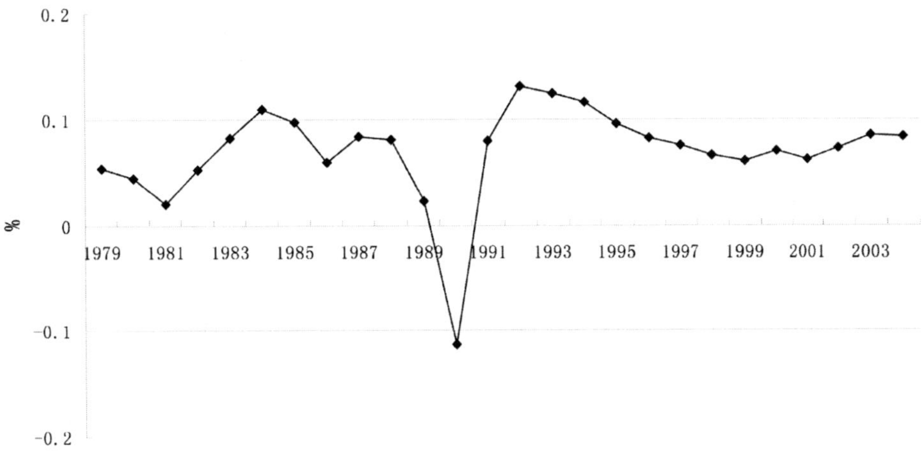

rapid economic growth in China is not isolated at the end of the 20th century, which is a pattern of global economic growth in the whole of Asia, especially East Asian economies. Stiglitz (1996) mainly discussed phenomenon of rapid economic growth of East Asian countries from the perspective of the Government's policy. Maddison (1998), Bhattasali (2001), World Bank (1997) consider that the Chinese rapid growth mainly rely on the Total Factor Productivity. However, some scholars believe that China's economy is a high growth, low-efficiency model. Young (1995b) consider that East Asian new-industrialized countries are not high productivity growth, although output and manufacturing export growth are very fast. Sachs, Woo (1997) point out that Chinese economic growth in reform period of is consistent with East Asian economic model, which there is no technological progress.

However, if China's economic growth is purely for non-efficiency material input as the pillar, then, as Krugman (1994) predicted, China's economic growth will also as 'rabbit's tail' as other East Asian countries. A number of papers are able to reproduce. However, growth models have not in general been able to predict the variable convergence speed needed to generate the adjustment path of output growth rate.

In this chapter, we propose a model in which the complementarities between human capital and endogenous technology, and technology are able to replicate and explain development miracles. Despite human capital and technical progress are indeed complementary, there have been few attempts in the theoretical literature to explore growth models with these engines. Our model is a hybrid R&D-based model a Jones (1995) and Papageorgiou, Perez-Sebastian (2006) in which technical progress is enhanced through innovation and imitation, and human capital is enhanced through formal education. We choose a schooling-based human capital technology following the approach suggested by Klenow, et al. (1997). Our choice of schooling technology is based on the Mincerian approach (Mincer, 1974) that has recently been revives by Papageorgiou, Perez-Sebastian (2006).

The hybrid R&D-based non-scale growth model has been explored by Eicher and Turnovsky (1999a, 2001) and Perez-Sebastian (2000). Unlike us, those authors do not consider human capital. Further, there are a few literatures that investigate the relationship between human capital accumulation and technical progress. Lloyd-Ellis and Roberts (2002) develop models with both endogenous human capital and technical innovation, but they are only concerned with steady-state predictions. Further, Keller (1996) and Funke and Strulik (2000) study dynamics in a model of human capital. Nevertheless, they do not take the predictions of their models to the data. Perhaps closer to the scope of our paper is Papageorgiou, Perez-Sebastian (2006) who focuses on the dynamics in a model of human capital and technical progress, which technical progress depends on imitation and innovation. However, Papageorgiou, Perez-Sebastian's work reproduces postwar miraculous development through endogenous changes in the barriers size by considering human capital accumulation and intersectional labor movements. Similarly, we adopt the same way to simulate the adjustment path, but the technical advance in different way.

The remainder of the paper is organized as follows. Section 2 presents the basic model and examines its steady-state properties. Section 3 and section 4 we choose China as the country in our investigation and explain the parameterization and calibration of our model to fit the China's growth process, and obtain the adjustment paths implied by our improved model. Section 5 concludes; discuss the main findings and limitations of our work.

2. MODEL

This section presents an economic growth model with endogenous human capital and technical progress. We start by describing the model economy's environment, and improve the technical progress according to the China's fact, then set up and solve the central planner's problem. Finally, we derive and discuss the steady-state solutions of the model. Our analysis is based on the central planner's solution.

2.1 Economic Environment

We consider an economy with stocks of human capital and physical capital. Both types of capital are used to produce a final output that can be consumed or added to the stock of physical capital. The population in this economy consists of identical infinitely lived agents, and grows exogenously at rate n. Agents are engaged in three types of activities: consumption-good production, R&D, and human capital attainment. Each period, consumers are endowed with one unit of time that is allocated between working and studying, consumers obtain their human capital through formal education; human capital can be used for good-production in output sector and R&D work in R&D sector. We assume that agents have preference only over consumption, and obtain utility through consumption.

Assume that at period t, output (Y_t) is produced using human capital (H_{Yt}) and physical capital (K_t) according to the following aggregate Cobb–Douglass technology:

$$Y_t = A_t^{\xi} H_{Yt}^{1-\alpha} K_t^{\alpha} \tag{1}$$

Where A_t is the economy's technology level, ξ is the technology-output elasticity, and α is the share of capital. And $\xi > 0$, $0 < \alpha < 1$.

In this model, innovative activities are introduced, similar to Romer (1990), assume that independent R&D output depends on human capital inputs and the existing stock of technical knowledge of R&D sector, which enhance its own absorptive capacity of advanced technology from outside world. Ideas created anywhere in the world can be copied by local researchers at a cost that diminishes with the country's technological gap. Owing to denying the existence of scale in empirical studies, according to Jones (1995b), we assume that there is non-linear relationship between independent R&D output and R&D inputs. In this paper, we take into account only human capital and knowledge accumulation, but also the technology imitation and spillovers. Caniëls, Verspagen (2001) stress that spillover is a product of knowledge economy, their research is considered to be great significance for region economic development. Technological progress is the engine of economic growth through knowledge spillovers and the spread of technology transfer, which can increase productivity, and then promote economic growth. Wang, Zhang (2001) find that the spillover number has index with technology gap. Wang, Ma (2003) further improve the Verspagen (1992) model and confirm that there is index relationship between knowledge spillovers and knowledge gap. Papageorgious (2003) regard that the relationship between technology spillovers and technology gap is non-linear, and depend on its human capital input of R&D sector, which introduce actually the spillover idea. The economy's technology level evolves according to the following equation:

$$A_{t+1} - A_t = \mu A_t^{\phi} H_{At}^{\lambda} e^{\frac{A_t^*}{A_t}} - \delta_A A_t \qquad (2)$$

Where $A_t^* \geq A_t$, $\phi < 1$, $\lambda > 0$。 δ_A represents the technology depreciation rate, H_{At} is the number of human capital employed in the R&D sector at time t, A_t^* is the worldwide technology frontier that grows exogenously at rate g_A^*, A_t is the own technology level. μ is a technology parameter, ϕ measures the effect of the existing technology stock on the R&D product. λ measures decreasing returns to R&D effect. Equation (2) is the modification of Jones (2002a), the own country technology output is composed by own existing technology stock and technology spillover, which depends on the following factors: the portion of human capital inputs in R&D sector and technology gap A_t^* / A_t. The greater the technological gap between the technology frontier and the follower, more significant the possible of the technology spillovers to catch up through imitation of existing technology. The assumptions are consistent with Findlay (1978).

The pioneering study of Borensztein, et al.(1998) confirm that human capital investment in the host country is important for technology absorption. Formal education, training are some ways to increase human capital. In order to facilitate the collection of data for quantitative analysis, we use the average formal education years to estimate human capital stock; the economy's output function evolves according to the following equation:

$$H_{jt} = e^{\eta S^{\beta}} L_{jt}, j \in \{Y, A\} \qquad (3)$$

Where S is the average formal education years of labor at period t, similar to Papageorgiou, Perez-Sebastian (2006), we can derive the law of motion of the average educational attainment as follows:

$$S_{t+1} - S_t = (\frac{1}{1+n})(\frac{L_{Ht}}{L_t} - nS_t) \qquad (4)$$

Where L_{jt} is the total amount of labor allocated to sector j at period t. Y, A represent output sector and R&D sector, respectively. η and β are fixed parameters, and $\eta > 0$, $\beta > 0$. The derivative $f'(S)$ represents the return to schooling estimated in a Mincerian wage regression: an additional year of schooling raises a worker's efficiency by $f'(S)$. The evolution of S across time depends on the share of people in education L_{Ht}/L_t and the growth rate of population, with the latter inducing a dilution effect.

2.2 Central Planner's Problem

The central planner's problem is of particular interest in our model and deserves careful consideration. We focus on a centrally planned economy for simplicity and for comparability to existing models although there are external effects in the model. The ultimate aim of production is to satisfy the increasing consumption demand, it means that all individual consumption utility maximized. Assume now that a benevolent planner maximizes the following utility function:

$$\max_{\{C_t,S_t,A_t,K_t,L_{Yt},L_{At},L_{Ht}\}} \sum_{t=0}^{\infty} \rho^t \left[\frac{(C_t/L_t)^{(1-\sigma)}-1}{1-\sigma}\right] \tag{5}$$

Where C_t is the consumption at time t, ρ is discount factor, σ is marginal utility elasticity, which is the inverse of the inter-temporal elasticity of substitution, and $\rho > 0$, $\sigma > 0$.

A central planner chooses the sequence $\{C_t, S_t, A_t, K_t L_{Yt}, L_{At}, L_{Ht}\}_{t=0}^{\infty}$ to maximize the lifetime utility subject to the following feasibility constraints, the initial values of S_0, A_0, K_0 and L_0 are given.

Subject to (1), (2), (3), (4) and

$$I_t = K_{t+1} - (1-\delta_K)K_t = Y_t - C_t \tag{6}$$

$$L_t = L_{Yt} + L_{At} + L_{Ht} \tag{7}$$

$$L_{t+1}/L_t = 1+n \tag{8}$$

$$A^*_{t+1}/A^*_t = 1 + g^*_A \tag{9}$$

L_0, S_0, K_0, A_0 are given,

Where I_t is the investment at period t, δ_K is capital discount rate. L_{At} is the number of labor employed in the R&D sector, L_{Ht} is the number of labor studying in the school, L_{Yt} is the number of people employed in the output sector. Eq. (6) is the economy's feasibility constraint combined with the motion law of physical capital stock; it states that, at the aggregate level, domestic output must equal consumption, C_t, plus physical capital investment, I_t. Eq. (7) is the population constraint; the number of labor employed in the output and the R&D sectors – plus the number of individuals in school must equal total population L_t.

Optimal control problem requires that constraints (1)-(4) and (6)-(9) hold, and can be stated as the following equations:

$$\begin{aligned} V(A_t,K_t,S_t) = \max_{\{L_{Ht},L_{At},I_t\}} & \frac{\{A_t^{\xi} H_{Yt}{}^{1-\alpha} K_t^{\alpha} - I_t)/L_t\}^{1-\sigma} - 1}{1-\sigma} \\ & +\rho V\{A_t(1-\delta_A) + A_t^{\phi}[H_{At}]^{\lambda}(\varphi \frac{A_t^*}{A_t})^{\theta}; \\ & K_t(1-\delta_K) + I_t; S_t + \frac{1}{1+n}(\frac{L_{Ht}}{L_t} - nS_t)\} \end{aligned} \tag{10}$$

where $V(\cdot)$ is a value function, L_{Yt}, L_{At} and I_t are control variables, S_t, A_t and K_t are state variables. According to steady-state solving rules, we solve the optimal control problem and get the Euler equations that express the optimal allocation of population in human capital investment, in R&D investment, and in consumption/physical capital investment as follows:

$$(\frac{C_t}{L_t})^{-\sigma} \frac{(1-\alpha)Y_t}{L_{Yt}} = (\frac{\rho}{1+n})(\frac{C_{t+1}}{L_{t+1}})^{-\sigma}[f'(S_{t+1})(\frac{L_{Y,t+1}+L_{A,t+1}}{L_{Y,t+1}})+1] \tag{11}$$

$$\begin{aligned} (\frac{C_t}{L_t})^{-\sigma} \frac{(1-\alpha)Y_t}{L_{Yt}} = & (\frac{\rho}{1+n})(\frac{C_{t+1}}{L_{t+1}})^{-\sigma} \frac{\lambda[A_{t+1}-(1-\delta_A)A_t]}{L_{At}} \\ & \times \left\{ \frac{\xi Y_{t+1}}{A_{t+1}} + \left[1-\delta_A + (\phi-\theta)(\frac{A_{t+2}-(1-\delta_A)A_{t+1}}{A_{t+1}})\right] \right. \\ & \left. \times \frac{(1-\alpha)Y_{t+1}/L_{Y,t+1}}{\lambda(A_{t+2}-(1-\delta_A)A_{t+1})/L_{A,t+1}} \right\} \end{aligned} \tag{12}$$

$$(\frac{C_t}{L_t})^{-\sigma} = \frac{\rho}{1+n}(\frac{C_{t+1}}{L_{t+1}})^{-\sigma}[\frac{\alpha Y_{t+1}}{K_{t+1}} + (1-\delta_K)] \tag{13}$$

In equilibrium, the central planner must be same between investing labor in schooling, R&D and final output production. The LHS of equations (11) and (12) represent the return from allocation an additional unit of labor to output production.

The RHS of equations (11) is the discounted marginal return to schooling, taking into account population growth. The RHS of equations (12) is the return to R&D investment. Equations (1)-(4), (6)-(9) and (11)-(13) constitute the equilibrium dynamics system of the model.

2.3 Steady-State Growth

We now derive the model's balanced-growth path according to above dynamic system. We first solve the interior solution. According to the relationship of consumption, investment and output, the variables Y, C and K have the same growth rate, it is the steady-state condition $g_{Y,ss} = g_{K,ss}$. Owing to equation (1) we can deliver the gross growth rate of output as a function of the gross growth rate of technology and the population growth rate as

$$G_{Y,ss} = (1+n)(G_{A,ss})^{\xi/(1-\alpha)} \tag{14}$$

Where $G_{xt} = 1 + g_{xt}$. Equation (7) implies that L_{Yt}, L_{At} and L_{Ht} must all increase at the same rate as L_t in order for labor allocations to grow at constant rates. This means that the ratio L_{Ht}/L_t is invariant along the balanced-growth path. Hence, equation (4) implies that S_{SS} is constant and given by

$$S_{ss} = v_{H,ss}/n \tag{15}$$

Where $v_{H,ss} = L_H/L\big|_{SS}$, it means that the proportion of labors employed in R&D sector of all labors in the steady-state. Equation (15) implies that along the balanced growth path, the economy invests in human capital just to provide new generations with the steady-state level schooling. This is consistent with Papageorgiou, Perez-Sebastian (2006), $G_{A,ss}$ is constant, according to equation (2), we can derive the growth rate of technology in the host country as a function of the growth rate of world frontier technology as

$$G_{A,ss} = [(1+n)^{\lambda} G^{\theta}_{A^*,ss}]^{1/(1+\theta-\phi)} \tag{16}$$

Equation (16) expresses the relationship between technology growth rate and world frontier technology growth rate. Because the technology growth rate A_t in one country is impossible to bigger than the world frontier technology growth rate A^*_t at period t, $G_{A,ss}$ cannot be bigger than $G^*_{A,ss}$ according to the front assumptions, otherwise A_t will eventually be bigger than A^*_t. Owing to the ratio $\theta/(1+\theta-\phi) < 1$, equation (16) is a concave function, then the technology growth rate A_t in one country has a unique point with the world frontier technology growth rate, it means that the unique point is

$$G_{A,ss} = G^*_{A,ss} = (1+n)^{\lambda/(1-\phi)} \tag{17}$$

For simplicity, we focus on the special case in which all countries grow at the same rate at steady state; that is, we assume that $G^*_{A,ss}$ is given by expression (17) and so is $G_{A,ss}$, which be input into equation (16), and we can derive the following expression:

$$G_{A,ss} = G^*_{A,ss} = (1+n)^{\lambda/(1-\phi)} \tag{17}$$

Then we can derive that output growth rate in steady-state can be expressed by the population growth rate.

$$G_{Y,ss} = G_{C,ss} = G_{K,ss} = (1+n)^{1+\lambda\xi/[(1-\phi)(1-\alpha)]} \tag{18}$$

Equation(18) explain the economic growth rate at steady-state depend on population growth rate, parameter ϕ, R&D return decrease parameter λ, technology's output elasticity ξ and human capital in output sector's output elasticity $1 - \alpha$.

3 DATA AND PARAMETER

For calculating, this paper adopt *China Statistical Yearbook* (1984-2005) and *New China 50 years statistical information* as data sources during the period 1978-2004 of China, gross domestic product (GDP) and China fixed capital stock express capital K and output Y, respectively, which are based on 1978 year constant prices. In accordance with Ye (2002) we assume that capital annual depreciation rate is 6%, and estimate capital stock. Technology stock have not direct time-series data in the Statistical Yearbook, according to Xia (2002)'s studies, the average annual rate of China's technological progress is 2.78 percent during 1979-2000. Another view is that China's pure technological progress rate is 2%. This article adopts Xia (2002)'s estimation, assume that average technological progress rate is 2.78 percent during 1979-2004 in China. Because U.S. economic and technological power lie in the world's leading level, this paper uses the technology stock of the United States as the world's leading technology. Denison (1962) studies that the U.S. economy average growth rate is 3.3 percent and technological progress rate is 1.4 percent during the period 1950-1962, economic growth has a positive correlation with technological progress. This article assumes that the technology growth rate of U.S. remains unchanged, it is 1.4%. We adopt directly Wang and Fan (2000)'s human capital stock data as data during the period 1978-1999 years, the human capital stock of 2000-2004 years used in this article are measured in accordance with the method of Wang and Fan (2000).

Parameters directly affect the results of model, therefore, so the determination of the parameter values need to have a strong background research to support, they can come from the output of other models, or may also the results of the other analysis other than models. According to population data on the China Statistical Yearbook, we can calculate the average population growth rate during 1979-2004 years is 1.17%. Technology will be depreciation like as physical capital. Caballero, Jaffe (1993) estimate that the technology depreciation rate is at least 0.1, it indicates that the life span of a new technology is 10 years. In this article $\delta_A = 0.1$. Bils, Klenow (2000) calculate that economic output steady-state growth rate is 1.6% according to 91 national panel data including China. At this paper the steady-state output growth rate, $g_{y,SS}$, is 1.6%. The values of parameter ξ, α in the production function by regression are 0.20, 0.574, respectively. According to equation (18) and the value of $g_{y,SS}$, we can derive 0.828 for external parameter of R&D ϕ. Other parameters values keep the same with Papageorgious, Perez (2006), then the parameter values obtained as shown in Table 1.

4. ANALYSIS ON ECONOMIC GROWTH ENGINE IN CHINA

We simulate the China economic experience on steady state in accordance with the model and the given parameter values, whose simulate output result and the actual output shown in Figure 2. In Figure 2, we can see that the model simulated output values can reflect in reality, the fitness between analog output and actual output can be broadly divided into three stages. From the actual development of China's economic situation, it mainly because the stock of human capital and R&D activities in China during this period lag behind the necessary need for economic development, the stock of knowledge is not enough for economic growth, and China still remain in the resource-exhausting growth, and there is no through technological progress, improve product quality to achieve intensive development. Hu, Xiong (2008) show that the portion of scientific and technological strength in China account for

Table 1. Parameter values

Values			
$\delta_K = 0.06$	$\rho = 0.96$	$n = 0.0117$	$\alpha = 0.574$
$\lambda = 0.5$	$\xi = 0.20$	$g_{y,SS} = 0.016$	$\delta_A = 0.10$
$\eta = 0.69$	$\beta = 0.43$	$\phi = 0.828$	

Note: subscript SS express steady state.

0.81% of the world's total in 1980, the relative gap with the United States 32 times, the relative gap with Japan 20 times, China fall behind the world's scientific and technological revolution who is also a 'new entrant.' The early stage of China's reform, there were inadequate education investment and inefficient funds use, which not only was lower than the level of developed countries and the world average, but also was lower than India, Egypt and other developing countries.

Productive forces have developed rapidly; the key role of economic growth is not a simple labor, but a certain amount of knowledge and technology workforce. The contribution of labor inputs to economic growth will be primarily explained by the quality of labor, rather than the quantity of labor. The high-level education population proportion is too low in China, and the lower secondary education proportion of the population is too large. By 2000, industry staff in China who accepts more than high school and high school education accounted for only 17.31 percent, of which the industry staff who accept more than junior college accounted only 4.66%. According to the Organization for Economic Cooperation and Development (OECD) statistics, nearly 80% of the labor forces in nearly 30 countries have high school and more than high school education, in which junior college and university account for more than 25%. Hong (1999) consider that the worst input factor in our country is the quality of human capital. In the meantime, the proportion of the state funding for science and technology in GDP account for a significant decline, which result in China's share of world scientific and technological strength in the world total stock reduce to 0.82

Figure 2. Simulated GDP and real GDP

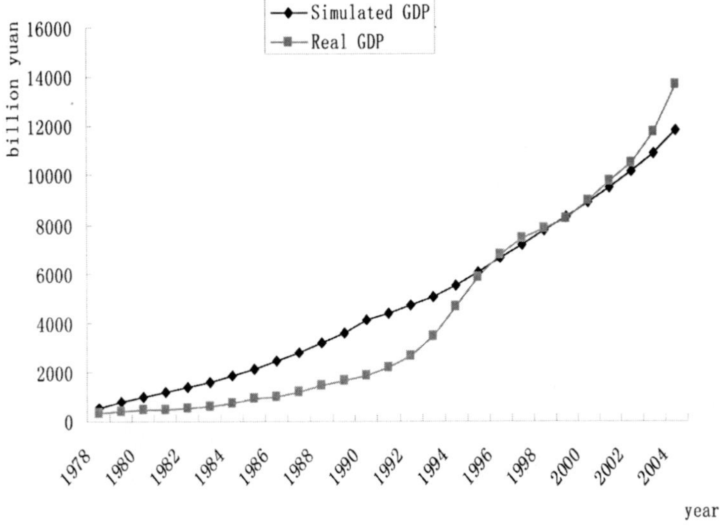

percent during 1985-1990 period, the relative strength of science and technology gap between China and the United States expanse from 22 times in 1985 to 48 times in 1990, China may be behind once again in the information technology revolution, because of lower human capital stock, inadequate R&D investment, weak independent research and development capability, the technology gap between the world technology frontier and China is so serious that it affect the ability of transformation and absorbability the new frontier technology, and then cannot to achieve its 'catch-up' effect. The technology gap has a large extent constraints for economic growth in China, technology gap will increasingly reveal its role in economic bottlenecks in the beginning stages of rapid economic growth.

The simulation results accord basically with the actual data during 1994-2001 period. Because of the gradual opening to the outside world, then afford some opportunities for introducing, learning and re-development new technology through foreign trade. China put first forward the country science and education strategy in 1995, which put technology and education in an important position for economic and social development, promote vigorously the development of education, focus on the consolidation of compulsory education, improve the quality of higher education and speed up the cultivation of high-quality workers.He, Wang (2006) find that abundant human capital in China is one of the decisive factors to attract multinational corporation R&D institutions. In 1996, staff employed in R&D sector of China is 804,000 person, R&D expenditures is 40.45 billion yuan, which account for 0.6 percent of GDP. In 2004, staff employed in R&D sector of China reach 1.153 million person year, R&D expenditures is 245 billion yuan, which account for 1.23% of GDP. China's rapid introduction of the Internet, mobile phones and other information technology in early 1990s, the rise of new economy in China would also contribute greatly to technological progress (Wang, He, 2007).

At the same time, China import directly the 'hard technology' through foreign trade, that is, import technology products and technical capital goods, especially zero import tariff for information technology and low import tariffs for technology products. According to the Statistics data of National Bureau, imports of high-tech products accounted for 2.0% share of GDP in 1990, which increased to 4.47% in 2000. According to World Bank data, China's direct imports of 'soft technology', that is, international copyright and licensing of imports, China imported 540 million U.S. dollars in 1997, accounting for only 0.9% of the world's total, which increased to 3.7% in 2004. The purpose of the use of foreign capital in China has also changed from making up 'funding gap' to making up 'technology gap'. More than 310 multinational corporations of world top 500 multinational corporations have invested in China until the first half of 1999, whose technology used in China is generally higher than the level of similar enterprises, but also a substantial proportion of multinational corporations in China provide advanced technology to fill gaps. By introduction of new technologies, continuous innovation and R&D, China's scientific and technological strength in the world total rose from 0.82% in 1990 to 1.33% in 1995, which further rose to 3.97 percent in 2000 and ranked fifth in the world, and the relative gap behind the United States and Japan significantly reduced. The enhancement of creation and absorption capacity of new technologies will be increasingly help to economic development. After 2002 the simulation results began to appear less than actual results. In fact, China's economy was unexpectedly rapid recovery the second half of 2002, the GDP annual growth rate exceeded 9 percent. The reason are mainly following areas: China accession to the WTO at the end of 2001 gave a clear signal to foreign investors, which increased security of the foreign direct investors in China's investment, coupled with the labor costs inherent advantages, which resulted in directly the foreign direct investment rapid incensement. This is a direct result of the foreign direct investment increased rapidly. As Liu, Zhou (2005) proved, a large influx of non-direct foreign investment in China promote greater economic growth after China's accession to the WTO. Owing to the efficient flow of foreign direct investment and international capital, which drive the domestic large-scale investments demand, which promote the potential role of 'environmental factors' on economic growth into a reality. The simulation model in this paper has not considered the impact of foreign investment, then result in the simulation results small than actual results after 2003.

In fact, foreign direct investment is one of the main channels for China to access to international advanced technology. Huang, Yang (2007)'s empirical study of China find that technology spillover of foreign direct investment is one of the main impacts for the host country's economy. According to the United Nations Committee on Trade and Development statistics, the stock of FDI flows in China accounted for 0.5 percent of GDP in 1980, and which rose to 5.8 percent in 1990, further to 17.9 percent in 2000, respectively. FDI stock inflow in China has reached 6,919 million in 2006, and which accounted for 26.3% of GDP. At the same time, Chinese rapid development of science and technology accounted for 7.09 percent of the world's total in 2004 and became the world's third scientific and technological power, and the scientific and technological strength more closer with the United States and Japan, the relative science and technology gap between China, the United States and Japan was 3.4 times and 2.1 times, respectively in 2004. Technological gap theory suggests that, due to the skill levels gap between economies, those with a relatively low level of economic development and technological level economy can imitate production technology of leaders and emerge catch-up phenomenon.

Some empirical researches also show that the technology ability increase of laggard economies indeed can contribute to narrow the technology gap. Once the new technology spread, apply, popularize rapidly in the world, the proportion of scientific and technological strength of imitative countries in the total will increase, at the same time those countries begin to transform to the creative country. The unprecedented increase of learning ability and innovative ability in the country will effectively promote economic stability growth in China.

In order to find the motivation of China's economic growth and achieve sustainable economic growth, according to the idea of Papageorgious, Perez (2006), we put equation (3) into equation (1), then derive per capita output of labor force, adopt logarithm on both sides to derive output growth rate pre labor force:

$$g_{Yt}^{w} = \xi g_{At} + (1-\alpha)\frac{df(S_t)}{dt} + \alpha g_{(K/L),t} + (1-\alpha)[g_{v_{Y,t}} - g_{(1-v_H),t}] \tag{19}$$

Where, g_{Yt}^{w} is output growth rate pre labor force, which can be decomposed into four factors: technological progress, the economic growth changes owing to change of average education years of each labor force, the economic growth changes owing to the net contribution of capital and labor.

Based on simulated output data and equation (19) we can calculate the effect of these four different factors on per capita output growth rate, which is showed by Figure 3. From Figure 3, the largest contribution to per capita output growth rate of China is capital in period 1982-1997 years, that is to say China's economic growth during this period mainly depends on the accumulation of capital, high saving rate and high investment rate are important contribution factors to economic growth, and capital has a decisive role for economic growth. Qiu, Zheng (2006) find that capital investment incensement is the most important source of China economic growth by establishing economic growth model with combination factors. After 1995, the contribution role of capital to economic growth is declining, which is

Figure 3. The effect of different factors on pre labor output growth rate

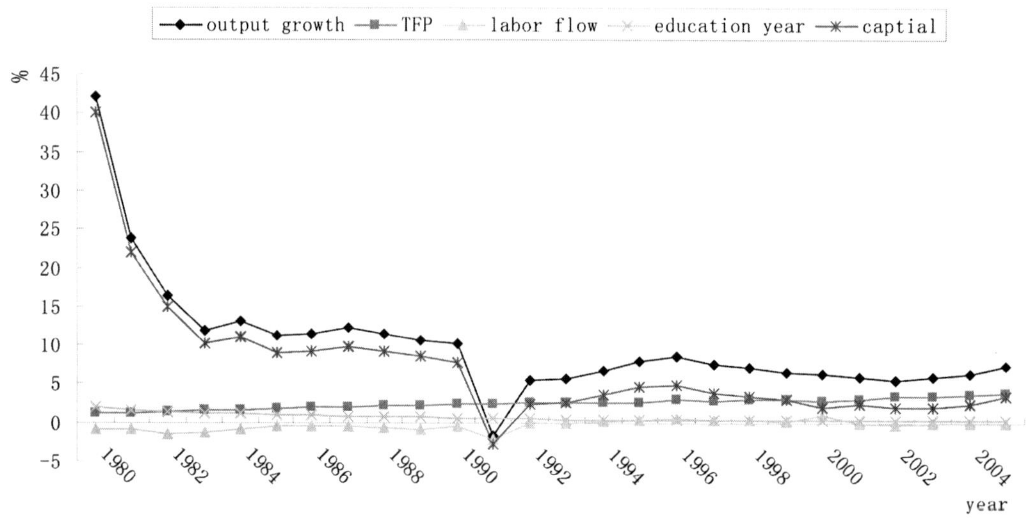

the same with the conclusion of foregone scholars (Yang, 2006).

The contribution of technological progress to China economic growth is only behind the capital, technological progress has always been a positive contribution to economic growth rate in the years 1979-2004, which show a steady growth, but contribution of technological progress is not significantly distinct in the years 1979 -1990. Hu, Khan (1997) point out that the accumulation of capital play a major role for China economic growth in the years 1952-1994, but also technological progress play a major role, The contribution of technological progress to China's economic growth is over 40% in the 1979-1994 years. After 1997, The contribution of technological progress to China's economic growth is more than the contribution of capital, and become the greatest impact factor on China economic growth, which is directly related with science and education strategy, and vigorous introduction of advanced technology to catch up 'technology gap'.

The contribution of labor education average years to China economic growth is positive and small, but its direct impact on economic growth is relatively small, which is consistent with Cao (2004), Hu, Liu (2004). Qiu, Zheng, et al (2006) find that the contribution of labor input increase is relatively weak, which is relative with China labor supply relative surplus and low labor marginal efficiency. The contribution of average education years to the per capita output growth rate was reducing in the years 1979-2004, mainly because schooling years accepted by workforce have diminishing marginal utility. Psacharopoulos (1994) point out that the marginal human capital decline in with the increase of education year, education. It is to say that the number of average educated years of labor force is increasing, but the direct impact on economic growth is not increasing year after year. The declining contribution of human capital in fact is caused by our declining human capital investment. Wu, Wang (2004) measure the human capital inputs of China provinces and find that the number of human capital in China is increasing, but investment in human capital and effective use of human capital are low, in fact, the human capital stock need to have sufficient payment in order to become effective human capital. Expansion recruit of college in recent years can increase rapidly the number of human capital, but it reduces the effective investment in human capital.

It is worth noting that labor has a small net positive contribution to economic growth after 1992. Net contribution of the workforce reflects the cross-sector labor mobility. As a result of labor into the final product sector will eventually increase the production output, but the additional labor will reduce per capita labor output. Thus the net contribution of workforce will reduce as students enter the labor market after graduation and result in part of R&D labor re-allocation to the final product decrease effectiveness of R&D. Owing to wage rates of investment in human capital employed in R&D sector is higher than in the final output sector, which led to the human capital in final output sector will transfer to the R&D sector.

5. CONCLUSION

In this paper, we build a non-scale R&D model including human capital base on endogenous growth theory to analyze China economic growth during the period 1979-2004, and analyze technological progress, human capital, and physical capital and inter-sector labor mobility factors on the impact of China economic growth, and provide a basis for analyzing motivation of China economic growth. The equilibrium solution results of the model competitive market show that R&D activities and technological spillovers can promote balanced growth path when the country lie in steady-state growth path. Simulation results show that the technology gap has large extent constraints to China economic growth. When Chinese economy is entering the phase of rapid development, technology gap reveal increasingly the bottleneck role for economy. Only increasing scientific and technological inputs to improve the quantity and quality of human capital, and vigorously the introduction of innovative technologies to enhance the ability of creating and absorbing new technologies, and reduce the technology gap between the world technology frontier and host-country, then which can promote economic intensive development.

We analyze the causes of by decomposing the four factors about China economic growth, the contribution of capital to China economic growth has a decisive role, but the contribution of capital shows a downward trend to economic growth. The contribution of technological progress to China economic growth is smaller than that of capital in the early, and has a positive contribution to economic steady growth. The implementation of China science and education strategy has greatly decreased the technology gap between the world frontier countries and China, and which led to technological progress in China by imitating and learning, the contribution of technological progress to the economic growth become gradually more than that of capital, and become the greatest impact factor of China economic growth. Due to the existence of diminishing marginal utility and low effective utilization of human capital, the contribution of average number of education years accepted by labors to economic growth is positive, but the positive contribution has a decline trend in year-on-year, and has relatively a direct small impact on economic growth. As a result of the wage rates of the final output sector and R&D sector are different, which will result in human capital inflow into R&D sector to gain higher payment.

REFERENCES

Angang, H., & Wenzhi, X. (2008). Quantitative evaluation on science in China(1980-2004). *Journal of Tsinghua University Philosophy of the Social Sciences*, *23*(2), 104–119.

Bhattasali D., (2001). *Sustaining China's development:Some issues*. Presentation to Tsinghua University 90th Anniversary Celebrations Seminar Series.Beijing, People's Republic of China, April 24.

Bils, M., & Klenow, P. (2000). Does schooling cause growth? *The American Economic Review, 90*(5), 1160–1183. doi:10.1257/aer.90.5.1160

Bing, W., & Zheng, W. (2004). A study on measurement of human capital across provinces of China. *Science Research Management, 25*(4), 60–65.

Borensztein, E., Gregorio, J. D., & Lee, J.-W. (1998). How does foreign direct investment affect economic growth? *Journal of International Economics, 45*, 115–1351. doi:10.1016/S0022-1996(97)00033-0

Caballeros, R. J., & Jaffe, A. B. (1993). How high are the giants' shoulders? In Blanchard, O., & Fischer, S. (Eds.), *NBER Macroeconomics Annual*. Cambridge, MA: MIT Press.

Caniëls, M. C., & Verspagen, B. (2001). Barriers to knowledge spillovers, and regional convergence in an evolutionary model. *Evolutionary Economics, 11*, 307–329. doi:10.1007/s001910100085

Denison, E. F. (1962). United States economic growth. *The Journal of Business, 35*(2), 109–121. doi:10.1086/294483

Eicher, T., & Turnovsky, S. (1999a). A generalized model of economic growth. *The Economic Journal, 109*, 394–415. doi:10.1111/1468-0297.00454

Eicher, T., & Turnovsky, S. (2001). Transition dynamics in non-scale models. *Journal of Economic Dynamics & Control, 25*, 85–113. doi:10.1016/S0165-1889(99)00070-6

Findlay, R. (1978). Relative backwardness, direct foreign investment, and the transfer of technology: S simple dynamic model. *The Quarterly Journal of Economics, 92*(1), 1–16. doi:10.2307/1885996

Funke, M., & Strulik, H. (2000). On endogenous growth with physical capital, human capital and product variety. *European Economic Review, 44*, 491–515. doi:10.1016/S0014-2921(98)00072-5

Hu, Z., & Khan, M. S. (1997). Why China growing so fast. *IMF Staff Papers, 44*, 103–131.

Huangsong, L., & Zan, Z. (2005). Cause research on new round economic growth. *Management World, 11*, 152–153.

Jiechang, X. (2002). Positive analysis on technical progress and economic growth and the related fiscal and tax policy. *Research on Financial and Economic Issues, 11*, 53–59.

Jinwen, C. (2004). Demonstration research on human capital and economic growth in China. *Research On Financial and Economic Issues, 9*, 9–13.

Jones, C. I. (1995b). R&D-based models of economic growth. *The Journal of Political Economy, 103*, 759–784. doi:10.1086/262002

Jones, C. I. (2002a). Sources of US economic growth in a world of ideas. *The American Economic Review, 92*, 220–239. doi:10.1257/000282802760015685

Keller, W. (1996). Absorptive capacity: on the creation and acquisition of technology in development. *Journal of Development Economics, 49*, 199–227. doi:10.1016/0304-3878(95)00060-7

Klenow, P., & Rodrı′ guez-Clare, A. (1997). Economic growth: A review essay. *Journal of Monetary Economics, 40*, 597–617. doi:10.1016/S0304-3932(97)00050-0

Krugman, P. (1994). The myth of Asia's miracle. *Foreign Affairs (Council on Foreign Relations), 73*(6), 62–78.

Lengyun, H., & Wen, Y. (2007). Economic growth: FDI spillover effect. *Economic Review (Federal Reserve Bank of Atlanta), 5*, 72–76.

Lloyd-Ellis, H., & Roberts, J. (2002). Twin engines of growth: skills and technology as equal partners in balanced growth. *Journal of Economic Growth*, *7*, 87–115. doi:10.1023/A:1015633528122

Lucas, R. (1993). Making a miracle. *Econometrics*, *61*, 251–272. doi:10.2307/2951551

Maddison, A. (1998). *Chinese economic performance in the long run*. France: OECD Publication.

Papageorgiou, C. (2003). Imitation in a non-scale R&D growth model. *Economics Letters*, *80*(3), 287–294. doi:10.1016/S0165-1765(03)00140-X

Papageorgious, C., & Perez, F. S. (2006). Dynamics in non-scale R&D growth model with human capital: Explaining the Japanese and South Korean development experiences. *Journal of Economic Dynamics & Control*, *30*(6), 901–930. doi:10.1016/j.jedc.2005.03.005

Perez-Sebastian, F. (2000). Transitional dynamics in an R&D-based growth model with imitation: Comparing its predictions to the data. *Journal of Monetary Economics*, *42*, 437–461. doi:10.1016/S0304-3932(99)00055-0

Psacharopoulos, G. (1994). Returns to investment in education:A global update. *World Development*, *22*(9), 1325–1343. doi:10.1016/0305-750X(94)90007-8

Qiong, H., & Zheng, W. (2006). Study on the location factors of overseas R&D investment in China. *China Soft Science*, *7*, 113–120.

Romer, P. M. (1990). Endogenous technological change. *The Journal of Political Economy*, *98*(5), 71–102. doi:10.1086/261725

Sachs, J. D., & Woo, W. T. (1997). Understanding China's economic performance.*NBER Working Paper Series, (Working Paper 5935)*.

Stiglitz, J. (1996). Some lessons from the East Asian miracle. *The World Bank Research Observer*, *11*(2), 151–177.

Verspagen B. (1992). *Uneven Growth between Interdependent Economies*.Data Wyse:Maastricht Press.

Wei, L., Xianchun, X., & Zhizhou, C. (2004). China economic growth from long term tactic. *Management World*, *7*, 6–15.

World Bank. (1997). *China 2020: development challenges in the new century*. Washington, D.C.: The World Bank.

Xiaohua, Q., Jingping, Z., & Donghua, W. (2006). Factor analysis and prospects of China's economic growth. *Economic Research Journal*, *5*, 4–12.

Xiaolu, W., & Gang, F. (2000). *Sustainability of China's economic growth*. Beijing: Economic Science Press.

Xiaoping, Y. (2006). Staggered analysis on China's economic growth. *Market Modernization*, *62*, 200–201.

Yingxing, H. (1999). Basic meaning on transformation economic growth mode. *Management World*, *4*, 15–24.

Yongyuan, H., & Zhiyong, L. (2004). Effects of Different Types of Human Capitals on Economic Growth. *Population & Economics.*, *2*, 55–58.

Young, A. (1995b). The tyranny of numbers: confronting the statistical realities of the East Asian growth experience. *The Quarterly Journal of Economics*, *110*(3), 641–680. doi:10.2307/2946695

Yuling, W., & Shiying, Z. (2001). Studying on decomposition of technological innovation effects spillover. *China Soft Science*, *8*, 53–56.

Yuming, Y. (2002). Analysis and calculation on TEP of China and provinces. *Economist*, *3*, 115–121.

Zheng, W., Ma Cuifang. (2003). A geographical investigation into knowledge spillovers between regions. *Acta Geographica Sinica*, *58*(5), 773–780.

Zheng, W., & Qiong, H. (2007). An analysis to the possibility of regional new economy industry development in China. *Science Research Management*, *28*(2), 105–111.

Chapter 10
Analysis of Online Game Distribution in China's Internet Cafés

Qun Ren
Bournemouth University, UK

Philip Hardwick
Bournemouth University, UK

ABSTRACT

Internet Cafés are the most popular locations for internet users to enjoy online game playing in China. At the beginning of this chapter, the authors analyzed reasons for Internet cafes' popularity as well as listing difficulties the Internet Cafés are confronting. Applying the distribution strategy put forward by Kasper, the authors analyzed how five of the top ten most popular games are distributed in the Internet cafés by their game operators. In the conclusion, the authors suggest game operators of different market dominance should utilize different distribution strategies (intensive or selective one) to match the game product with the players' demand. At the end of this chapter, some recommendations are put forward for the Internet cafés to take into account.

1 INTRODUCTION

Online game operators in China are the game suppliers. Due to the limited use of online payment system in China, cash-based point card sales (both virtual and physical cards) have been the most popular payment method. Online game players can buy the point cards through game operators' official game websites or through distributors and retail outlets, including Internet cafes, software shops, supermarkets, bookstores, newspaper stands, and convenience stores.

1.1 How Prepaid Point Card Works in an Internet Cafe?

In order to play online games or to pay for in-game value-added services, game players have to buy pre-paid cards, which are sold in both virtual and physical forms. Each pre-paid card

DOI: 10.4018/978-1-61692-846-9.ch010

Copyright © 2011, IGI Global. Copying or distributing in print or electronic forms without written permission of IGI Global is prohibited.

contains a unique access code and a password so that users can add value to their accounts for one or more specific online games. Online game operators, such as Shanda, NetEase etc, generally sell offline physical pre-paid cards to a group of regional distributors who resell the cards to sub-distributors. In turn, these sub-distributors sell the cards directly to Internet cafes and other retail points of sale.

Online game operators always offer sales discount to the distributors and the discount percentage also depends on the tier the each distributor belongs to. In 2006, Shanda offered average sales discounts of approximately 19.3% and 16.6% to e-sales distributors and offline distributors. Here is another example. Shanghai-based online game operator, The9, chose Junnet as its exclusive national distributor for the game titled MU. Junnet got a discount offer of 30% because it was responsible for the all the logistics of dealing with lower-tier distributors. In addition, about 500 city-wide distributors contracted with NetEase directly and gained a discount of 14%.

Physical cards are "scratch" cards with pass-codes printed on them. Virtual cards are essentially a list of pass-codes. Presently, majority of urban Internet cafés adopted the E-sales distribution systems which are always developed by online game operators for buying the pre-paid virtual cards directly and then sell to the customers.

Here is an example of virtual pre-paid card purchase in an Internet Café. The customer should pay in advance at the counter usually RMB 10 and the cashier will give him or her one slip or a card with the login and password. Sometimes it is only a login number with the password blank. For instance, 00452917 is the login and 789 is the password. Once registered, the customer may access the Internet, select and activate the desired games he or she wants to play, and then his or her account will be charged from the prepaid card or prepaid online points he or she just bought. After the customer has finished his or her session, the cashier will return the change to the customer. Large operators usually explore their own "E-Sales" systems which are used for monitoring virtual card transactions from operators to different levels of distributors, and including finally sales in Internet cafés.

1.2 Internet Cafe: A Big Revenue Generator

Internet cafés are not limited to work as point card sellers. Following homes, they are the second most popular places to enable game products to finally reach the end users – game players. "Internet cafés are critically important to China's online game market," said Lisa Cosmas Hanson, the managing partner of Niko Partners (2008). It is the key factor for online game operators to distribute their online games and is expected to generate 40 percent of $2.5 billion service revenues in 2008.

In order to sustain the competitive advantage, game operators usually have to manage channels to meet the customer demands with supply through the efficient exploration of each element in the value chain (See Figure 1). At the same time, each element in the value chain has its own specific needs, which should be carefully taken into consideration by each sensible online game operator. Current situations and problems which are encountered by the Internet Cafés in China are displayed in the second part of this paper.

2 RESEARCH BACKGROUND

2.1 Popularity of Internet Café in China

It is reported by CNNIC (2008) that 67.3 percent of the 220 million Chinese internet users rank home as the most convenient place of Internet access. Internet café is chosen by over 33.9 percent of the Chinese internet users for online access. In

Figure 1. Online game supply chain (Source: APRP, 2007)

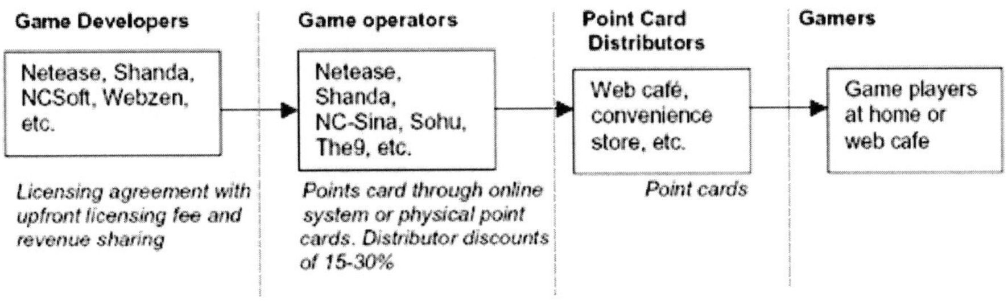

2007, the number of people surfing the Internet at Internet café increased by 60.9% over 2006. The popularity of Internet access in Internet café is due to the following reasons.

- CNNIC (2008) statistics shows that it will take RMB 900 for a family per year for the average cost of internet access in China. Not all families can afford internet access expenses and the cost of possession of internet accessible computers and other equipments. Internet café is the popular location for the low income people or internet users without computers or internet access at home. By comparison, the average expense of an Internet user at the Internet café is RMB 51.6 per month (p34, CNNIC 2008) and 1/3 of them spend less than RMB 15 per month.
- Internet cafés may be the only source of internet access in China's rural areas and smaller cities. 48% of the rural internet users go to Internet cafes for surfing online (CNNIC January, 2008). It is amazing to know that Internet café is widely located in China. They are not limited to Eastern developed areas, but scattered in most small and medium-sized cities rural areas and even can be found in Kashgar, close to China's border with Pakistan.
- Internet cafés are attractive places to the young game players whose parents show their disapproval in online game playing. Thomas and Lang (2007) noted that internet cafés "have emerged as the place for urban Chinese youth to be youth and as one of the few places young urban Chinese can escape from the pressure of schooling, work and their parents." Besides, most game players in the Internet cafe enjoy the presence of other game players' presence and are bored at playing online game at home.
- Despite owning personal PCs at home, lots of young internet users still enjoy going to Internet cafés for playing online games, meeting friends, watching network films, videos and other socialization activities. According to Lindtner and Nardi etc (2008), some game players in China are always friends in the physical world and some of them maybe live very close with each other. They enjoy playing in the Internet cafés together for a couple of hours and then going out for dinner or tea. Lindtner and Nardi et al conclude that playing in the Internet cafe with friends is regarded by the game players as "a very stimulating social experience comprised of physical and digital elements."
- Lots of game players prefer to sit beside experienced game players and learn how to enhance their skills in the Internet cafe through observation and advice from them, especially from those who lead the raids (one of the hardest activities in the game).

2.2 Internet Café's Development and Government Regulations

The development of the Internet café is not smooth in China at all. Here is a brief overview of its progress in China.

- Between 1995 and 1998, it is the period of fast development for China's Internet Cafés. During that period of time, few families have own PCs. The main function of Internet cafés in China is to offer unconnected games to café users with the price of RMB 15-20 per hour.
- Between 1998 and 2002: It is the flourishing era of Internet cafes. Competition became more and moresevere. Nine people were killed in an Internet café fire in Beijing in June, 2002. Since then, Internet cafés got the close concerns from the Chinese government. Following the concerns are the regulations released constantly by the Chinese government.
- At the end of 2004, more than 70,000 Internet cafés were closed due to the safety reasons and the game contents which are thought to be harmful to the young adolescents' growth.
- In March 2007, 14 Chinese ministries jointly announced not to approve any new Internet cafe open in China till 2008. "The regulations govern the estimated 185,000 Internet cafés nationwide in China, 71,000 of which are unlicensed by the appropriate regulatory authorities (Niko Partners[2], 2008)." According to different media, this activity taken by the Chinese government aimed to drive the illegal cafes out of business and made it easier for the Internet cafes' reconsolidation.
- In early 2008, Chinese governmental agencies announced the lift of the one-year long ban and new Internet café applications started again. By comparison, the new license approval procedures are more convenient. More preferential policies are released for supporting the development of the Internet cafe industry. The policies are especially favourable for the development of franchised or the chain of Internet cafes.

Also, since 2008, all visitors to internet cafés in Beijing have been required to have their photographs taken when using Internet cafés. It is said that a city-wide database run by the Cultural Law Enforcement Taskforce is responsible for storing the personal information. By mid-December all internet cafés in the main 14 city districts installed cameras to record the identities of their internet users, who must by law be 18 or over; although a survey by the internet version of the *People's Daily* indicated that "72 per cent of respondents were opposed to the measure, calling it an infringement of their rights. Just over 26 per cent supported the photographing because it would benefit children." (Macartney, 2008) This reminded people think of the Anti-fatigue system which was issued two years ago and regarded as a failure by many internet users.

2.3 Current Situations Facing the Internet Cafés

2.3.1 What People Do in Internet Cafés?

If you go to an Internet café in China, you can see lots of people sitting in front of terminals to satisfy their needs. According to the iResearch investigation (See Table 1), over half of internet users play online games. 18% of them like to chat online with friends and others enjoy watching films or listening to music downloaded off the Internet (iResearch, 2007).

In a licensed Internet café, it will take about RMB 2 to 3 ((between 26¢ and 40¢) to download a film legally. However, it is usually free to watch due to the illegal downloads. This can explain

Table 1. Aims for surfing in the Internet Café (Source: iResearch Inc. Nov 2007. Sample numbered =2000)

Aims for surfing in the Internet Café	Percentage of the Internet users
Online game playing	59%
Online chatting	18%
Watching films and listening to music	18%
Information searching	4%
Others	1%

why 71,000 illegal internet café still exist and are hard to be eradicated. These unlicensed Internet cafés offer their customers lower-priced internet services by evading taxes which were enforced to take effect in April 2001 by the Chinese Government.

2.3.2 Current Difficulties in the Internet Café Industry

The stringent government regulations did not bring the shrink of the Internet café market. Instead, the entry of more and more capitals in 2007 and 2008 made the industrial competition more severe.

- Mr Zhao Zhiqiang, the secretary of Beijing Internet Café Association, recalled that the Internet access fee was RMB 8-10 per hour. However, with the wide broadband penetration in households, in order to retain the customers, Internet cafés have to lower the price of the internet access fee to RMB 2-3 or even below RMB 2 per hour. Majority of Internet Cafés are involved into a price war which hurts them and the rivals (see Table 2). Nearly all distributors complain about the low profit margin they have got. According to the survey by Tianxia Netcafe Association (2008), half of the investigated targets confessed that the attendance rate of the Internet users in their Internet Cafés cannot reach 60%.

Table 2. How Internet cafés charge the Internet access? (Source: iResearch Inc. Nov 2007. Sample numbered =2000)

Rate of Internet access (RMB / per hour)	Percentage of Companies who adopted the rate
Over RMB 3 per hour	15%
RMB 2-3 per hour	59%
Less than RMB 2 per hour	26%

- Facing the endless price war, nearly all Internet cafés target the value-added services that could be resorted to, such as selling snacks, beverage and print-paid point cards, etc. However, these services are easily duplicated by almost all Internet cafés and cannot distinguish one Internet café from others.
- In order to guarantee the quality and consistency of broadband connectivity, Internet Cafés try their best to update their computers and related equipments, decorate the internal environment, which is also a huge sum of expense.

3 LITERATURE REVIEW

Groucutt (2005) states that the way a company distributes its products can help itself gain a competitive advantage over its competitors. Channel management is the form through which a product or service is distributed to the end user. Blythe (2006) supports the above idea by regarding choosing the right distribution channel as a strategic issue. Segmentation of the market, product characteristic and the statement of competition are factors to be taken into account when choosing the channel. In his opinion, an unconventional distribution channel can bring the company a competitive advantage which cannot be provided

Figure 2. Steps in developing a communication plan: from target audience to determination of the communication mix (Source: Service Marketing Management, by Kasper, H. 2006)

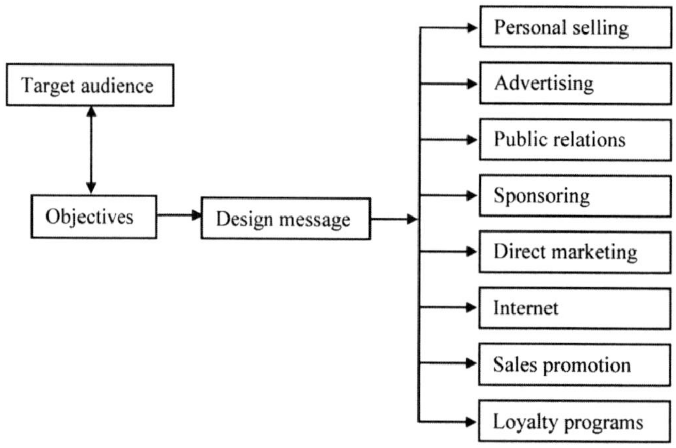

by the product alone. Kasper etc (2006) reminds the success of a company will not only rely on the company strategy but also the performance and competitive capabilities of its marketing partners in the distribution channel. That is to consider whether the distribution partners can deliver excellent quality and therefore value to customers. Distribution channels can thus have a number of levels. Kotler defined the simplest level is the direct contact with no intermediaries involved, as the 'zero-level' channel. IT product manufacturers (or software publishers) work directly with the dealers. A one-tier or two-tier channel means that product manufacturers work directly with distributors who sell the products.

Before choosing the right distribution partners and distribution channel, the company should have a clear understanding of the marketing segmentation. Doole etc (2005) define market segmentation as the "process of dividing the market up into distinct subgroups of customers who display common characteristics and then selecting one or more subgroups which company action targets with a unique strategy." He further added that some segmentation variables, such as geographic (Physical location) and demographic (sex, age, ethnic groups, etc) are important to be identified. Without the scanning of the marketing environment where the customers exist, how a company understand the customers and deal with the customers and compete against the competitors?

Only when taking consideration of the environmental forces, market opportunities and distribution strategy together, the provider can position the service offer clearly. Distribution strategies are usually summarized as intensive distribution, selective distribution and exclusive distribution by scholars (Blythe, 2006). Intensive distribution means that large distribution coverage is desirable and the producer can be available as many outlets as possible; selective distribution indicate that a limited number of specialist outlets are used for stock the products. Exclusive distribution means that the retailers are given the unique rights to sell the product and cut out the competition from other distributors within their area.

Internal and external communications of services are important in creating excellent service quality and also are related to creating a firm's desired reputation among its customers. The details of the general communication plan are listed in Figure 2.

Presently, limited deep academic study mentions the distribution management in the online game industry. Some research available focus on the business plan for Internet Café[3].

Lindtner and Nardi et al from UCI conducted comparative studies of WoW players in China and the USA. Their research is not only based on but also to extend the hybrid ecology approaches put forward by Crabtree and Rodden (2007). Lindtner and Nardi et al indicate that "hybrid ecologies are spaces that merge the physical and digital to facilitate collaboration." They underscore the "cultural, social and contextual assemblies and infrastructures that shape and are being shaped by the digital environment." In the detailed scene of the Chinese online gaming, they found that Internet cafe is a physical environment where friends enjoyed meeting and playing a virtual game with others online through peer-based learning. The fact indicates that "Internet cafe is a place where people transitioned between the virtual and the physical in many ways, where technological, cultural, political and social-economic factors are entangled together."

4 RESEARCH QUESTIONS

Combining the concepts of marketing environment, distribution strategy and communication of services within the context of online game distribution management in the Chinese internet café, this study aims to analyze how the dominant game operators utilize their different distribution strategies and specific communication plans in the Internet cafés to create their desired images among their customers – game players.

5 RESEARCH METHODS

Primary data are consisted of 20 semi-structured online interviews with sampled Chinese online game operators and Internet café owners. Each interview lasted 20 and 30 minutes. Through these interviews, we discussed contributory factors that promote the development of Internet café industry. The main interviews proceeded with the discussion of the following questions:

- What are the main factors that are critical for development of Internet cafés in China?
- How do you think of the price war engaged by lots of Internet cafés?
- What kind of value-added service you think can bring the Internet café satisfying profit?
- How do you think of the cooperation between Internet café and online game operators?

In addition, secondary data are widely used, such as CNNIC report in 2008, the survey of iResarch in 2007 and the Internet Café survey by Tianxia Netcafe Association in 2008. These three investigations are based on the study of large amount of samples, which offer quite useful and reliable data for this study.

6 DATA ANALYSIS

Top ten most popular online games in China's Internet cafés are listed in Table 3. The names of each game's developer, operator are listed in the table as well. The information of game style and game's market share (i.e. revenue rank) are also shown in this table.

Homes and Internet Cafés are the two main locations for game players to play online games. When playing item-based online games, game players can either pay for nothing if he enjoys playing the basic functions, or they can spend money in buying virtual items for enhancing their competitiveness. Hence, some popular games maybe have a huge percent of game players who would like to pay nothing but still quite loyal to the game. In addition, contributions from Internet cafés account for 40 percent of the whole online game industry revenues; and 67.3 percent of in-

Table 3. Top ten most popular online games in China's Internet cafés (Sources: authors)

Popularity rank in Internet cafés in 2007	Revenue rank in Online game market in 2007	Online Game Name	Name of Game Developer	Date of Commercial Launch in China	Name of Game Operator
1	3	World of Warcraft	(USA)Blizzard Entertainment	June 2005	The9
2	5	Crazyracing Kart Rider	(Korea) Nexon	June 2004	Tiancity
3	4	Audition	(Korea) T3	May 2005	9you
4	10	Street Basketball	(Korea)JC Entertainment	March 2006	T2CN
5	2	ZT Online	Giant	Jan 2006	Giant
6	Out of Top 10	Counter Strike(or: CS)	(USA) Valve Software	May 2004	Tiancity
7	8	Zhu Xian	Perfect World	May 2007	Perfect World
8	Out of top 10	SUN	(Korea)Webzen	May 2007	The9
9	1	Fantasy Westward Journey	NetEase	January 2004	NetEase
10	Out of top 10	Woool	Shanda	September 2003	Shanda

ternet users prefer to access Internet at home. This explains why some quite popular games in the Internet cafés fail to bring the huge revenues as it is expected.

7 FINDINGS: GAME OPERATORS' DISTRIBUTION STRATEGIES AND GAMES' POPULARITY

In order to have a good understanding of how Chinese online game operators adopt their distribution and marketing strategies in the Internet Cafés, the authors interviewed 20 Internet café owners and game operators. How game operators adopt different marketing strategies and distribution management to enable their games' popularity in Internet Cafés are elaborated as follows. The authors list five of the ten most popular games cafés which are entitled with their special distribution characteristics by their game operators in the Internet Cafés.

7.1 World of Warcraft (3D Realized Style MMORPG)

- **Distribution strategy:** The9, operator of *World of Warcraft* (WoW), adopts selective distribution but multi-segment strategy because they target at the game players in the eastern costal, economically-developed cities. With more than 10 million players worldwide, half of them are in China. WoW game players represent a wide range of ages and different demographics.
- **Competitive advantage in game content:** Many Internet café owners and game players agree that WoW is specifically designed for the game players in the Internet cafés. Instead of emphasizing on individual heroism, it underscores the team work and collaboration. Mr Tan, an Internet café owner in Dalian city stated in the interview that "Majority of the guild activities are at night, so lots game players who always know each other show up late in the Internet café. They enjoy exchanging ideas

and playing together." In addition, WoW does not demand full concentration at all times, that is why some friends can play for a period of time and then go out to the nearest restaurant for dinner and or a chat. Its activities involved are so varied and some are violent and some are peaceful, which can meet the demands of different customers. Its unique content excellence is widely imitated by Chinese online game developers. However, till now, no games surpass its significance in the global game industry.

- **Direct marketing:** Internet Café owners and The9's marketing team staff are always available to help players learn WoW in the Internet Cafés.
- **Promotion strategy:** The top one game content not only brings the Internet cafés customers, but puts forward the Internet cafés challenges and promotes the upgrading of the hardware industry. The9 collaborates with Intel, the global leading innovative chip manufacturer and Blizzard, the developer for promoting Blizzard Entertainment's upcoming expansion pack, *World of Warcraft: The Burning Crusade.* A number of Internet cafés who have high-quality computers were chosen as numerous "The Burning Crusade" themed Internet cafes. After that, The9 organized a series of nationwide gaming competitions for Chinese players. Mrs Zhang, an Internet café owner in Changchun city indicated during the interview that game players would regard an Internet Café fall short of the standard if this café's computer can not meet the demand of running WoW smoothly. In 2007, many Internet Cafés upgraded their computers by installing driving power for high density games and applications, which can work as an excellent platform for playing games such as *World of Warcraft: the Burning Crusade.*

Presently, playing WoW in the Internet café is a fashion and necessity, especially to the youth in the eastern urban cities.

7.2 ZT Online (2D Realized styleMMORPG)

- **Distribution strategy:** Intensive distribution strategy is adopted by Giant, the game's developer and operator. Mr Shi Yuzhu, CEO of the Giant, tried to target all game players (excluding adolescents who are under 18) as his customers. It is him who firstly focused on the customer segment in the Internet cafés of the second and third tier of cities. Mr Shi grouped game players into two groups, rich game players, who accounted for 15 percent of the entire game players of ZT online. They are deep-pocketed and ambitious to be leaders, bosses and kings no matter in the virtual world or the physical world. The other group consists of non-RMB players who accounted for 85 percent of the total game player amount and are willing to spend lots of time building their skills but are reluctant to pay the money. It is the 15 percent of the rich game players that created the net income generation of RMB 290.2 million for Giant in the third quarter of 2007 (i.e. 13.8% of the entire Chinese online game market in the third quarter of 2007). At the same time, he realizes the same importance to retain the rich players and non-RMB players. Without the non-RMB players' participation in the game, the rich players have no chance to enjoy the excitement of winners. Therefore, Mr Shi invented some stimulating activities to keep the non-RMB players, such as issuing in-game player wage payout (with virtual money), red envelopes, and special events.
- **Competitive advantage in game content:** As a super game player, Mr Shi Yuzhu in-

teracted with hundred of game players to collect ideas during the development of ZT Online. He managed to make ZT Online appeal to game players with different tastes. Game players can attack or kill monsters; and they can have fun when doing casual activities such as boating, fishing and biking etc. The 2D content of ZT Online doesn't require high demand to the computers, which enlarges its customer segments from the urban cities to the rural areas.

- **Marketing network (Including advertising and promotion):** Giant had a 3,000 market network personnel in 2007 and Giant is planned to expand it to 5,000 in 2008. Nearly every Internet café in the second-tier and third-tier city or rural areas can be seen posters of ZT online on the wall, the handle and the toilets. Another unique activity initiated by Giant is the ZT-Online-exclusive incentive to the Internet Cafés. That means, Giant would like to pay the Internet Café the Internet access fee for a fixed period of time on the premise that ZT-Online is the only game can be played during this specific period of time. This kind of activities is especially welcome in the Internet cafés in the less-developed rural areas.
- **Sales promotion:** It is told by Mr Wang, an Internet café owner that currently all Internet café owners are trying to get the profit from the value-added services. Giant is the company who always give a higher discount offer for the pre-paid point card sales of ZT Online, which is another reason why ZT Online is welcome in Internet cafés.

7.3 Fantasy Westward Journey (2D Cartoon-style MMORPG)

- **Distribution strategy:** Intensive distribution strategy is adopted by NetEase, the game's developer and operator. This game aims to cover all customers in the country.
- **Competitive advantage in contents:** The top one market share in 2007 indicates its popularity. Fantasy Westward Journey has over 4 years' long lifecycle. The success is mainly attributed to its constant release of an expansion pack which can always bring game players fantasy. Another reason is that the game is based on one of the four famous four Chinese classic novels *Journey to the West*.
- **Direct marketing:** NetEase has a strong distribution team whose members are easily seen in Internet Cafés. Instead of putting advertisements here and there, the marketing staff are famous for their distribution skills in the Internet Cafés. One Internet owner Mrs Hu expressed in the interview the marketing staff usually issue the game players their business cards firstly and introduce the core services of their game and teach the new game players patiently. Excellent word-of-mouth communication is applied successfully in NetEase's game distribution in the Internet Cafes and enables NetEase to migrate the existing players to the upgraded version.

7.4 Zhu Xian (3D realistic style MMORPG)

- **Distribution strategy:** Selective distribution strategies are applied. Target customers are the youth in urban areas.
- **Competitive advantages in the content:** Developed and operated by Perfect World, the game content is based on the extremely popular novel that is dominating the

top ten most popular novels of Baidu [1]. Considering the popularity of this novel, lots of Internet cafés would like to be the "Zhu Xian" themed Internet cafés.

- **Advertising:** Perfect World and Nestle jointly launched an advertising campaign in 23 provinces and regions across China, which brought the Internet cafés more new customers. The advertising campaign is available in thousands of supermarkets and shops across the country, within a one-month period since its launch which has helped increase sales of Nescafe nationwide. A wide range of Nescafe products now contain game top-up cards for "Zhu Xian Online," integrating the two influential products of Nescafe and Beijing Perfect World.
- **Marketing strategy:** Perfect World usually offer dedicate painting of Zhu Xian to the Internet cafés. In addition, lots of big parties are held in the Internet cafés which not only issue the prizes to some game players but also create a good communication between the game operator, game players and the Internet cafés.

7.5 Crazyracing Kart Rider (2D Cartoon-style online car racing game)

- **Distribution strategy:** Developed by the Korean Nexon, Crazy Racing Kart Rider is operated by Shanghai-based Tiancity with the intensive distribution strategy. This game content has a very low demand to the computer configuration, which means that it can be played in all Internet cafés. It is especially very popular by girls because the characters are very cute.
- **Competitive advantage in its simple play rules:** Crazy Racing Kart Rider, the most popular online game in China, with the market shares following WoW. Its competitive advantage is that it can be learned and operated easily. Many Internet café owners agree one reason for the popularity of Crazy Racing Kart Rider is that leisure online games have gradually favoured by people in the recent years.
- **Sales promotion:** Players can race alone or in teams with different modes. There are endless national tutorials and driving tests held every year. Many options with the maps, special tools make the game players never get bored.

8 CONCLUSION

It is become widely recognized that Internet cafés become popular for the online game players because they are the social locations where "game players assembled a variety of digital and physical artefacts and interrelated them in terms of individual preferences, requirements of the social group players belonged to (for example, guild or friends from college), societal norms and in-game goals. (Lindtner and Nardi, et al, 2008)." Whether game operators can adopt their distribution strategies to operate their games in the Internet café depend on their communications with the Internet café owners, the understanding of needs from both owners and the customers in the Internet café. It can be seen clearly from the above analysis; the success of the games distribution should appeal to the preference of the Internet café owners and the take mutual efforts between game operator's marketing teams and Internet café owners.

Prior to the game operation, the game operators should segment the customers geographically and demographically and then position the Internet cafés they need for working together. Following this, the game operators need to distinguish what is the customers' basic requirement and how to achieve added value through the supplementary service ((Bennett et al, 2003). Bennet et al pointed out that "It is important that customers, not mangers, define customer-perceived value." The quick huge profit generation of Giant is not by accident but due to the deep and thorough understanding of their CEO, Mr Shi Yuzhu. Before establishing Zhengtu Network, Shi Yuzhu was an experienced game player, who always spent more than 10 hours or even 15 hours each day in playing online games that different operators offered. While playing, he usually attended two MSN chat groups. One is the community for players and the other for Giant's developers. In building ZT Online, he managed to work as a bridge and transferred the ideas and demands of players to the game's development team timely for.

In addition, intensive distribution strategy is the second factor that guarantees Giant's success. That is to say, the large scale employment of frontline staff was firstly adopted by Shi Yuzhu in the service delivery, which is imitated by other dominant domestic game operators too. However, not main game operators have the capability to gather a 2000-people well-trained distribution team. Mr Shi's unparalleled marketing team is based on the nationwide marketing network he had built for his previously successful business, the sale of a health supplement product named Naobaijin (melatonin). Therefore, selective distribution strategy combined with unique value-added services (which is based on the customers' perceived value) is recommended to consider by most less dominant Chinese online game operators when operating the game products in Internet Cafés.

9 RECOMMENDATIONS TO INTERNET CAFÉ OWNERS

In the end of this study, the authors realized that lots of middle-level and low-level Internet café owners show their confusions about whether Internet café is still a good business. Confronting the low attendance rate of the game players and the endless price war, lots of them worried the industry future and the survival of his or her business. The following recommendations are based on information from some successful Internet café owners, representatives from the game operators and important lessons from some Internet café owners who suffered from the business failure. Hopefully, they can offer some clues for the people who have ambitions in Internet café industry.

- **Segment your target customers and then choose your Café location:** Make sure of the targeted customers who you would like to deliver your services because the tastes of the university students differ a lot from the game players from the congested living areas. Pick a desired location for your café and you can earn money expectedly with your business strategy. If you choose the wrong place, it would be hard to satisfy yourself.
- **Keep the broadband connectivity with high speed and consistency**: No matter whether you would like to offer the 3D games or 2D games to your customers, you must enable the game players to enjoy the smooth running of the games. No matter how excellent your face-to-face service is and how splendid the Internet café environment you have created, your customers will feel frustrated if you can not handle the internet connectivity.
- **Not encourage to take the price war:** Most Internet café owners expect to win customers by charging the lowest price. In fact, exceeding the lower boundaries can drive you and your competitors to bankruptcy. A prosperous industry is a one in which a trade union is established with a regulated pricing policy to use. Players in the same industry can take the strategic alliances, franchising and M&A approaches to create a healthy environment which can benefit the growth of the industry.
- **Offer variety of value added services:** Office related services are available in

many Internet cafés, such as scanning, desktop publishing, encoding, and printing. If you can find one or two reliable and qualified technicians, then you can deliver the handy services like PC repairs, electric parts sales.

REFERENCES

Asia Pacific Research and Positioning. (2007). *The Chinese Video Game Market: Market Entry for Canadian Firms*. Retrieved from www.asiapacificresearch.com

Bennett, R. (2003). *Services marketing: a managerial approach* (1st ed.). Milton: John Willey & Sons Australia.

Blythe, J. (2006). *Principles & Practice of Marketing* (1st ed.). London: Thomson Learning.

CNNIC. (2008). *21st Statistics Survey Report on the Internet Development in China*. Beijing: CNNIC.

Crabtree, A., & Rodden, T. (2007). *Hybrid Ecologies: understanding cooperative interaction in emerging physical-digital environments*. Personal and Ubiquitous Computing.

Doole, I. (2005). *Understanding and Managing Customers* (1st ed.). Harlow: Pearson Education Ltd.

Groucutt, J. (2005). *Foundations of Marketing* (1st ed.). New York: Palgrave Macmillan.

IResearch Co. Ltd. (2007). *The report of current situation of Internet Cafés in China*. Shanghai: Shanghai iResearch Co., Ltd

Kasper, H. (2006). *Services Marketing Management: A strategic perspective* (2nd ed.). West Sussex: John Wiley & Sons Ltd.

Lindtner, S., Nardi, B., et al. (2008). A *Hybrid Cultural Ecology: World of Warcraft in China*. CSCW 08, November, San Diego, USA.

Macartney, J. (2008). China watches over internet café customers in web crackdown. *Times*. Retrieved from http://technology.timesonline.co.uk/tol/news/tech_and_web/the_web/article4958778.ece

Niko Partners. (2008). *PC Games and Console Hardware Sales Surge in CHINA'S Booming Economy*. Retrieved from http://www.nikopartners.com/npr_050208.pdf

Thomas, S., & Lang, T. (2007). From Field to Office: The Politics of Corporate Ethnography. In *Proc. Of Ethnographic Praxis in Industry Conference* (pp. 78-90). 2007.

Tianxia Net café Association (2008). *The report of current situation of Internet Cafés in China*. Retrieved from http://www.txwm.com/special/research/

ENDNOTES

1 Baidu (NASDAQ: BIDU) is the leading Chinesesearch engine for websites, audio files, and images. Baidu offers 57 search and community services including an online collaboratively-built encyclopaedia (Baidu Baike), and a searchable keyword-based discussion forum.

2 Niko Partners is the leading market intelligence firm for China's video game industry,

3 The data is available from: http://www.morebusiness.com/internet-cafe-business-plan

Chapter 11

From Lab to Market:
Issues in Industry-Academy Cooperation and Commercialisation of R&D Outputs in Nigeria

A. A. Egbetokun
Obafemi Awolowo University, Nigeria

W. O. Siyanbola
Obafemi Awolowo University, Nigeria

A. A. Oyewale
Obafemi Awolowo University, Nigeria

ABSTRACT

For nations to compete successfully in the long run, they must innovate; and innovating entails building new competencies, new capabilities and new knowledge. Starting with a brief review of the concepts of Science, Technology and Innovation, this chapter discusses the factors militating against the translation of scientific R&D outputs into innovations in Nigeria. The situation in selected Asian countries are reviewed to highlight the significance of specific factors such as institutional framework, adequate funding and knowledge pool in the process of building indigenous innovation capability for national development. Policies, strategies and structures that could facilitate the effective management of R&D and innovation, and enhance university-industry connections are suggested.

INTRODUCTION

In a remarkably short period of time, economic globalisation has changed the world economic order, bringing new opportunities and new challenges (Commission of the European Communities, 2006). These challenges and opportunities are further amplified by the recent global economic meltdown. In this new economic order, the rules of the game of competition and competitiveness have greatly changed. Nations can no longer compete based on natural resource endowments and locational advantages. Therefore, most countries now devote an increasing proportion of their resources to Science and Technology (S&T), and to the associated Research and Development (R&D), in an attempt to build competitive

DOI: 10.4018/978-1-61692-846-9.ch011

Copyright © 2011, IGI Global. Copying or distributing in print or electronic forms without written permission of IGI Global is prohibited.

advantages, or to catch up with others who have done so. This is basically because of the general belief that Science and Technology (S&T) are the keys to progress and industrial growth in today's increasingly knowledge-driven world (Egbetokun and Siyanbola, 2008a).

Despite the abundance of information in the globalised world economy, a competitive gap exists between nations and organisations. The most significant cause of this competitive gap is knowledge (Prusak and Davenport, 1998). Prusak (1996) made this more explicit when he explained that the only thing that endows a competitive edge on an organisation or a nation, is what it knows, how it uses what it knows and how fast it can know something new. Thus, the advantages of globalisation are only derived through the creation as well as effective management and deployment of knowledge. Within this context, therefore, innovation can be better understood as a process in which the organisation or nation creates and defines problems and then actively develops new knowledge to solve them (Nonaka, 1994).

A significant paradigm in successful innovation is the development of competencies and capabilities for the ultimate practical application of new or re-combined knowledge, through innovation and R&D commercialisation. This chapter sets out to discuss the challenges of this paradigm within Nigeria's National Innovation System (NIS); and to draw lessons from Asia in this regard. This sort of discussion is useful despite the extant argument in the literature that innovation in latecomer settings is not essentially R&D driven (Bell, 1984; Katz, 1987 and Lall, 1987; 1991). For instance, it is very beneficial to for policy-makers to understand how best latecomer economies should engage in R&D in their efforts towards technology catch-up.

The rest of the chapter is structured as follows. We begin by creating a context for the discussion in Sections 2 and 3 which look at Science, Technology and Innovation (STI) and their interconnectedness; and the roles which Research and Development (R&D) play within the NIS to facilitate innovation. Section 4 deals with the specific challenges of commercialising R&D results in Nigeria, drawing extensively from recent research. A review of Asian country cases which offer important lessons in certain areas is presented in Section 5; while specific recommendations are made in Section 6. The paper concludes in Section 7.

SCIENCE, TECHNOLOGY AND INNOVATION (STI) CONCEPTS

Nowadays, it is generally impossible to discuss issues about technology without mentioning science. In fact, science and technology have become so closely related that the one now depends on the other for its development (Ilori *et al,* 2002). The concept of "Science and Technology (S&T)" means, in our time, the totality of activities in a nation that lead to innovation. These include, according to UNESCO, 1982):

1. **Scientific and technological research:** the study, experimentation, conceptualisation and theory testing involved in making discoveries or developing new applications;
2. **Experimental development:** the process of adaptation, testing and refinement which leads to practical applicability;
3. **Scientific and technological services:** a mixed group of activities crucial to the progress of research and to the practical application of science and technology. The services collect, process and disseminate the scientific and technological information needed for such purposes;
4. **Innovation:** the use of a new product, process or invention (resulting from new knowledge, or new combinations of existing knowledge) in the national economy. Innovation can also include the transfer of technology and introduction of products or processes into countries in which they were previously unknown; and

5. **Diffusion:** the propagation of innovations throughout the productive sectors of the economy.

The benefits derivable from S&T are not actually realised in an economy until the innovation and diffusion occur. It is innovation that transforms R&D results into useful products or processes which are placed on the market (commercialised). For this reason, it is now becoming increasingly important to foster a conncetion between the universities and research institutes where knowledge is created on the one hand, and the industry where knowledge is deployed, on the other hand. On its own, innovation is a process which comprises several stages and usually takes a considerable amount of time, conscious efforts and money. The success of innovation depends entirely on the effective management of the individual stages of the innovation process. In the following section, we present a short conceptualisation of the role of research and development in the generation of innovations and the context within which this occurs.

RESEARCH AND DEVELOPMENT (R&D) AND THE NATIONAL INNOVATION SYSTEM (NIS)

Research and innovation are intimate bedfellows. On one hand, there would be little innovation without good science because innovation feeds on knowledge that results from cumulative R&D experience; and it also contributes to this stock of knowledge. On the other hand, the economic and social pay-off of research would be much diminished without innovation. An economy's productivity level, therefore, depends on its cumulative R&D effort and on its effective stock of knowledge, with the two being inter-related (Coe and Helpman, 1993).

Indigenous R&D produces traded and non-traded goods and services that bring about more effective use of existing resources, thereby raising a country's productivity level (Jones and Williams, 1999). Also, domestic R&D enhances a country's benefits from foreign technical advances; and the better a country takes advantage of technological advances from the rest of the world the more productive it becomes (Coe *et al.,* 1997). Therefore, the cumulative R&D effort is intimately related to the national capacity of a country to produce and commercialise a flow of innovative technology over the long term (Stern *et al.,* 2000). However, the success achieved by any nation in exploiting R&D outputs for growth and development depends on the effectiveness of the nation's National Innovation System (NIS). The NIS thus provides a framework for evaluating holistically a nation's attempt at generating and applying knowledge for meeting the needs of her society.

The National Innovation System (NIS) has been described as the network of institutions in the public and private sectors whose activities and interactions initiate, import, modify and diffuse new technologies (Freeman, 1987). Four key elements of the NIS - Education and Research, Industrial Production, Finance, and Public Policy and Regulation, were identified by Tiffin (1997). These elements have some organisations or institutions that perform specific functions within the NIS. The activities within the elements of the NIS include teaching and talent filtering, research and development, industrial production, brokerage, policy, financial and legal support, among others. Thus, innovation is the result of a complex interaction between various individuals and institutions that make up the innovation system (Oyewale, 2005).

The innovative performance of an economy depends on how individual institutions and actors (e.g. education and research organisations; firms; funding institutions) perform in isolation and how they interact with each other as elements of a collective system of knowledge creation and use. Without adequate development of these actors and institutions in the domestic and regional settings the innovation system remains underdeveloped and anaemic (Egbetokun and Siyanbola, 2008b). In what follows, we examine Nigeria's NIS as far as innovation, R&D commercialisation and industry-academic cooperation are concerned.

THE COMMERCIALISATION OF R&D OUTPUTS IN NIGERIA: STATUS AND CHALLENGES

Structure of the Economy

Nigeria, Africa's most populous country with estimated 150 million people, has a land area of 910,768 sq km and is blessed with several mineral resources. The vegetation of the country is diverse, ranging from mangrove forest in the south to arid land in the north. Prior to and immediately after independence, agriculture was the mainstay of the Nigerian economy, with farm products such as cocoa, palm oil, cotton and groundnut the major export commodities. However, since the 1970s, when petroleum was discovered in commercial quantities in Nigeria, crude oil has risen to the forefront, and now provides about 20% of GDP, 95% of foreign exchange earnings, and about 65% of budgetary revenues for the country. The Nigerian economy now largely depends on natural resources, primary processing and manufacturing, and for the most part, on imported technologies (Oyewale, 2005).

The National Innovation System

The key elements of the NIS can all be found in Nigeria but each element has been found to have operated without adequate interaction with the other elements (Oyewale, 2005). The system is also plagued with gross imbalance. Public R&D is carried out in over 70 research organisations controlled by various Ministries such as Science and Technology, Education, Agriculture and Rural Development, Industry, and Health. Other bodies and institutions also play very important parts in the successful prosecution of R&D. As at the first half of 2009, there were 94 Universities where research in various fields is carried out and over 100 polytechnics, monotechnics and colleges for manpower training and research. These are apart from some applied R&D that takes place in the private sector.

With a fairly large number of educational research institutions (collectively known as knowledge centres), a lot of research activities take place in various fields. However, interface or brokerage organisations are mostly absent or inactive. Venture capital provision is nearly absent; and the activities of the patent agency and the incubators remain relatively unknown to the other stakeholders in Nigeria's NIS (Oyewale, 2005). A particularly recent institutional antidote to this challenge was the establishment in 2008 of Intellectual Property and Technology Transfer Offices (IPTTO) in some of the nation's tertiary institutions by the National Office for Technology Acquisition and Promotion (NOTAP). In spite of this, the level of industry-academic cooperation is still rather low. For instance, Egbetokun et al (2009) showed, using data from a premiere industry-wide innovation survey of maufacturing firms in Nigeria, that knowledge centres rank least among actors with whom firms collaborate for innovation.

The Challenges of R&D-driven Innovation: Empirical Evidence

Identifying and understanding the main factors that hinder the commercialisation of research results in Nigeria has been the subject of much research in the last one-and-a-half decade. Among the main issues identified by researchers (Oyebisi *et al*, 1996; Ilori *et al*, 2002; Oyewale, 2005) are:

1. Inadequate research orientations whereby more than 75% of research projects executed

in the educational institutions/research institutes are not demand-driven;

2. Non-availability of information on commercialisable inventions and R&D results to the intended user industries;
3. Poor technological entrepreneurial culture;
4. Inadequate infrastructure;
5. Inadequate motivation for the commercialisation of inventions/research results;
6. Lack of funding and efficient funding structure for innovation;
7. Inadequate patent education;
8. Absence of effective linkage between research organisations and industries, and
9. Preference for foreign technology.

Oyewale's (2005) assessment of the interaction within Nigeria's NIS offers deeper insights into the gravity of the problems that disourage the translation of research results into innovations within the system. He observed, among other things, that a severe information gap exists within the system; and that interactions among the elements are too weak to bring about innovations and knowledge-based industrial development in the country. R&D activities are present but are either not effectively managed or not adequately funded and are generally prosecuted in an unfavourable environment. The situation remains the same in both the public and private sector. From his findings, 50% of educational/ research institutions and researchers had some interactions with their colleagues in other local and foreign institutions/ institutes, but just about 25% of them interacted with industrial firms. These patterns show that interactions of Nigerian universities with other local and foreign institutions/institutes were moderate, but their interactions with industrial firms were low. Furthermore industrial organisations were only sparsely involved in formulating and suggesting research ideas to the institutions/institutes and researchers. This indicates the existence of gaps between industrial needs and the R&D activities of the institutions/institutes.

Over 80% of the researchers and the institutions/institutes assessed by Oyewale (2005) claimed to have some inventions, but less than 30% had patented some of the inventions/research results. Because of information gap between the Nigerian academia and industrial firms, utilisation of the inventions/research results by industrial firms was low, and such were probably for resolution of problems, rather than commercialising new inventions. Research and development outputs of Nigeria's institutions/institutes were not satisfactory enough to the firms, possibly because over 50% of the researchers conducted knowledge-driven researches which are considered as ends in themselves, while about just 38% conducted market-driven researches which may be relevant to industrial needs. Furthermore, about 70% of the industrial firms claimed to have research and development laboratories but less than 60% of these employed qualified research staff.

Regarding private sector R&D, Ilori *et al* (2000), in a study of Nigerian food companies, found that most of the firms were not active enough in R&D as is reflected in their low funding and inadequate staffing for research. About 40% of these firms devoted less than 0.5% of their annual turnover to R&D while only 20% devoted between 1.5% and 2.4% of their annual turnover to R&D. This is in sharp contrast to the situation in Japan where close to 50% of companies spend at least 5% of their turnover; and in Australia where almost 30% of companies spend more than 5% of their turnover on R&D (Liao and Greenfield, 1997a; Liao and Greenfield, 1997b).

In spite of its importance in the NIS, the Nigerian university system has lived below expectations, not only in its research functions but also in its capacity building role. University education in Nigeria has declined in quality over the last decade or so, owing to several factors acting in tandem. Key among these factors, as identified by Oyelaran-Oyeyinka (2005) are:

1. Unstable political situation and declining support from governmental agencies: With declining economic and research facilities, employment opportunities and diminishing value of earned income, the brain and manpower drain from Nigerian universities to universities abroad has been immense.
2. Structural Adjustment Programme: Reduced educational expenditure, one of the pronounced negative effects of the Structural Adjustment Programme (SAP), has contributed to a threat to scientific and technological development in Nigeria. According to the National Economic Intelligence Committee, about 0.6% (250,000 persons) out of the economically active population of about 40 million constitute the scientific and engineering manpower of Nigeria, when compared with about 10% in China, 11% in South Korea, 12% in Malaysia, 15% in Japan and about 25% in Russia, USA, Germany, UK and France.
3. Lack of cooperation with other actors within and outside African countries: regional politics, lack of capacity and lack of coordination amongst African universities has also contributed to declining quality of research in Nigerian universities.
4. Expanding university enrolment and graduate output: Universities within Nigeria have had expanding enrolments with no or little by way of a commensurate increase in academic staff, research funding and other infrastructure. Figure 1 shows the student-teacher ratio in 2 selected Nigerian universities between 1984 and 2000. As against intended capacities, the present potential of university departments and centres of excellence has been stretched beyond limits due to the explosion in enrolment. This has resulted in sub-optimal student-teacher ratios (over 30:1 in 2005) that have adeversely affected university performance.

Figure 1. Student/teacher ratios in selected Nigerian universities, 1984-2000 (Source: Oyelaran-Oyeyinka, 2005)

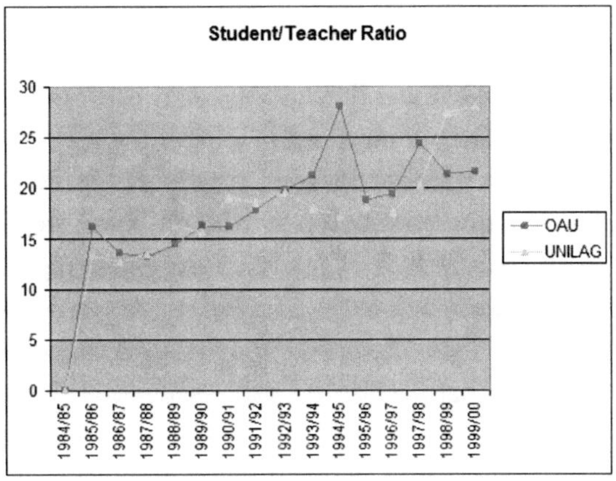

The lack of research endowments also forces university research to place enormous reliance on funding and collaboration from foreign agencies. Most activities – like those of the Malaria Group of the University of Ibadan and the facilities of the Central Science Laboratories of the Obafemi Awolowo University– are totally or significantly dependent on foreign grants. Added to this, the focus only on publication as an end of research programmes is a major disincentive to translate research into product innovation coupled with a great lack of mobility and research links between university researchers and private firms.

Oyelaran-Oyeyinka (2005), speaking of biotechnology innovation, observed that there seem to be two major reasons affecting the capacity among Nigerian universities to perform well. The reasons, as identified by him are, nevertheless, system-wide features; their effects go well beyond the domain of biotechnology. The first main reason is the state of university education in Nigeria and the various pressures that act upon it. Second, there is a lack of incentives to motivate researchers. These have resulted in systemic and institutional weaknesses that clearly affect the capacity of university departments and researchers to perform research and also to collaborate meaningfully with researchers and departments in other Nigerian and foreign universities. Consequently, the few precious results achieved by the researchers remain largely in the pre-commercialisation stage.

From Oyelaran-Oyeyinka's (2005) assessment, three key factors stand out in explaining why the innovation process remains largely at the level of pre-commercialisation in Nigeria. First, there is a significant lack of facilities and financing to move the research to the concluding stages. Second, even in situations where significant research results had been collected, with evidence of possible utility of the process and product, demand by the end-users is almost non-exixtent. As Jolly (1997) correctly observed, technologies and for that products and process inventions fail not so much for the skills of the inventor and the lack of market but because no one promotes or get sufficiently interested in them. Third, failure to commercialise sometimes resulting from institutional rigidity much of which relates to the ways traditional Public Research Institutes (PRIs) and universities are set up. There are two issues that attract attention here: who initiates the process of commercialsiation - the PRI/university or a firm/entrepreneur?; and what form of formal or informal contract guides the process? The non-availability of clearly defined answers to these questions is a deterrent to the commercialisation of research results.

In advanced developing and highly industrialised countries, two broad types of formal contracts are common, which are, academic entrepreneurship, and spin-off companies from public research or universities. Academic entrepreneurship takes several forms namely; (a) Involvement in large-scale externally funded research, (b) consultancy to earn supplementary income; (c) university-industry research and transfer of technology, (d) patents and trade secrets, and (e) commercialisation which might involve holding equity in private enterprises by scientists (Altonen, 1998). The last three options have been proposed consistently by Nigerian researchers (see for example Oyebisi *et al*, 1996; Ilori *et al*, 2002; Oyewale, 2005; Ilori, 2006).

Having so far discussed the Nigerian peculiarities, we seek, in the next section to examine sepcific aspects of the NIS in three Asia countries. These countries have been selected to each highligh a unique set of points.

TECHNOLOGY-LED ECONOMIC TRANSFORMATION: SOME EVIDENCE FROM ASIA

A sound scientific and technological base, from which wealth-creating technological innovations and applications can develop, is essential to economic growth in a competitive international environment. This knowledge base should address the full spectrum of economic accumulation covering resource mobilisation, effective production, knowledge-based marketing, sales, services and distribution of manufactured products. The efforts of the Newly Industrialising Countries (NICs) of East Asia, and their resultant economic growth in record time, are quite illustrative of this. The Asian NICs adopted the explicit technology policy approach. Thus, technological development in these countries took place over a much shorter period when compared to the European countries.

Most East Asian and sub-Saharan African (SSA) countries shared similar economic conditions in the mid-19th century. Today, the East Asian nations have outgrown the SSA countries in no small measure. We therefore consider the cases of two of these nations, India (with a population estimate of 914 million and land area of 3,287,590 sq. km.) and South Korea (with a population estimate of 45 million and land area of 99,020 sq. km.). We note that Nigeria has an estimated population of 150 million people and a land area of 910,768 sq. km. It is a known fact that these rapidly industrialising countries that are poorly endowed with natural resources have become successful economies by encouraging innovation through the development of sufficient human capital; the building of strong institutions and institutional networks; and focused efforts in the appropriate application of S&T.

In this section, we discuss the landscape in three Asian economies: China, South Korea and Malaysia. The rate and manner with which these nations made progress is interesting and has caught the attention of the whole world.

The Role of Technology in China's Emergence

The rate at which China emerged to attain leadership status in science and technology-led development has attracted the attention of the whole world. The achievement was undoubtedly contingent upon aggressive S&T human capital development and strong institutional structures. For instance, there were just about 50,000 science and technical personnel in China in 1949, of which only 1% was involved in scientific research. By 1955, a total of 840 scientific and technological research institutes which engaged over 400,000 researchers had been formed. The number of R&D centres in the high-tech industry sector has been on the increase, being 750 in the year 2004. China's technology-led growth has not only attracted FDI in the real sector, records show that 39% of high-tech R&D centres across the globe prefer to locate in China as against 29% and 28% for US and India respectively. Consequently, China has grown to become a major industrial economy, with the country's manufactured goods as percentage of total exports growing from already high 84% in 1995 to about 92% in 2005. Consequently, between 1981 and 2005, it appeared that much of the poverty reduction in the world came almost exclusively from China. China's poverty rate fell from 85% to 15.9% (over 600 million people) during that period.

Technology and Economic Transformation: the Case of South Korea

The case of South Korea offers us a good understanding of the place of education in national development. The Republic of Korea which lacked indigenous technological capabilities for industrialization in the 1960s sought a more radical solution of establishing educational and R&D institutions. From an illiterate workforce in the 1950s (22.0% in 1953), literacy rate increased to nearly 90% within two decades. Interestingly, the private sector invests more than the government in education. All levels of education are concurrently focused, creating a uniquely balanced system complemented with aggressive training of high calibre scientists and engineers in the West. In addition, technological capabilities are key facilitators of the development process in Korea. In the past two decades, the Korean government has shifted its stance from having a leading hand in R&D activities from the 1960s and 1970s to the encouragement of private firms to perform R&D. In 1980, 62% of R&D was performed by public institutes, along with 9.2% by Universities and only 28.8% by private firms. By 2005, this had dramatically shifted to over 76% of research performed by private firms (an increase of over 48%) whilst the public sector only performed 13.8% (a drop of almost 48%). South Korea now focuses on technology development rather than basic research. Basic research declined from 22.9% in 1970 to 13.2% in 1996. In addition, the Korean state founded companies had reached the technological frontier from reverse engineering and efforts were poured into R&D to gain international competitiveness.

Harnessing Technology Prospects: the Malaysian Experience

Malaysia's dream of replicating the experience of the newly industrialized economies of Asia via technology-driven and high-tech production patterns may be said to have materialized as it is now being categorized among nations that have significant potential to develop new technologies. This arose directly from the strategic focus on S&T human capital development. Malaysia progressed quite rapidly in terms of human capital development as a way to make up for its erstwhile low Science and Technology graduates. Between 1990 and 2000, the percentage of S&T based graduates increased from 32% to 39% and research scientists and engineers increased from 2 to 10 per 10,000 persons. The biggest portion of the R&D expenditure in Malaysia is used for applied research targeted at solving current and immediate future needs of industry as against the dominant basic research in Nigeria. Malaysia also imitated the Asian Tigers in the establishment of S&T parks having identified its tremendous prospects. For instance, the Multimedia Super Corridor which caters for software and IT services employed 13,000 people and had 53 institutions in its first 4 years of establishment. As part of the dividends of such efforts, Malaysia recorded a technology transfer index of 1.08, higher than China of 0.90 in 2001.

LESSONS AND RECOMMENDED STRATEGIES

From the foregoing, it is apparent that for Nigeria to progress in transforming R&D outputs to tangible products and services, appropriate measures must be taken in the areas of institutional framework, support structure, knowledge generation, education and technological entrepreneurship. It is clear from our review of the selected NICs that education, and in particular, technological education is the power of wise nations. It is generally agreed that the single most important key to development and to poverty alleviation is education. The greatest threat to Nigeria's future is indeed the present wrong perception of the place of education in national development. Thus, it is imperative that massive investments be made in education, especially S&T education, till a critical mass of S&T personnel whose collective activities can imapct the system is achieved.

The observed disconnect between economic needs and scholastic research in Africa may not be unconnected to inadequate funding. In the absence of serious local funding for research and innovation, much of Africa remains industrially underdeveloped. Many African researchers therefore rely on external donors, aid agencies and grant-giving bodies to sponsor their work (Soboyejo, 2006). The direct relevance of these research efforts to local developmental needs is thus not guaranteed. For instance, developing anti-malarial vaccines, creating cheap means of transport in the rural areas and many more issues that still challenge the African populace might not be adequately addressed with extra-African funding.

Additionally, current investment levels in S&T must go up; in fact it has been argued that to achieve meaningful economic and social development, a country must invest at least 1-4 per cent of its gross national product in science and technology research (Soboyejo, 2006) and that it would be a mistake for policy advisers to concentrate too much on the research element of R&D spending without giving adequate attention to spending on development (Oyelaran-Oyeyinka, 2006).

While the academia and industry each have their individual and unique roles to play in development, bringing these actors together is paramount for S&T-led development. Academia and industry could be brought together through consultancies, contracts, and research or network partnerships. Following Kruss (2006), we argue that it is not enough for a country or region to simply increase the total number of consultancies and contracts between academia and industry. This is because such arrangements are generally myopic, focusing narrowly on specific interests. By their very nature, they preclude serendipitous discovery and seldom generate publications or postgraduate work. Though they can supplement individual academic salaries and help retain staff, many are not officially reported, and may happen at the expense of teaching and research.

Network partnerships across academia and industry, that include government and even international development partners, would be significantly more beneficial. Besides serving as instruments to address innovation market failures, these industry-academia-government partnerships could also act as mechanisms for 'informal' knowledge transfer.

In the UK, for instance, innovative solutions are provided by academia for many real-life industrial problems under the Knowledge Transfer Partnership program which supports joint supervision of PhDs by university and industry. As a consequence of the extensive policy support for industry-academia relationship, it is not surprising that the degree of interaction with industry from UK universities had increased in recent years, with precious results (UNICO, 2005)[1].

The following examples from South Africa (Kruss, 2006) go further to illustrate the power of strategic industry-academia network partnerships.

"The Tree Protection Co-operative Programme is a biotechnology research network of large paper firms and small timber producers, working on tree pathogens with academic partners at the University of Pretoria, to the benefit of all.

"The university research unit is building an international scientific reputation by producing a large number of postgraduate students and accredited publications. It has become a sponsored 'centre of excellence' that attracts considerable government research funding.

"The industry partners depend for their competitive edge on the costly research and development and the risk-management strategies the network provides. For example, the university researchers provide DNA technology to produce trees resistant to pests and pathogens.

"A second example of a successful strategic partnership is the remote sensing Multi-Sensor Microsatellite Imager project. In this government-funded research network, university, industry and government partners work together to design micro-satellites that can supply affordable high-resolution imagery to African governments. The images can help monitor, regulate and manage resources, for example, water distribution, crop management and settlement infrastructure.

"A Stellenbosch University laboratory conducts fundamental research for the network. A spin-off company manages the technology development, while application research managed by a government science council informs the design. Finally, a Belgian university and industrial partner develop specific technical components.

"Mutually beneficial network partnerships like these — where university, industry and intermediary partners work towards a shared objective — generate knowledge and technological innovation for all. They help universities harness the innovation potential of their researchers while still maintaining academic integrity. They meet industrial needs for technological progress, and also contribute to national development."

There is also a dire need for institutions to re-structure, refocus and re-organise their R&D activities different from the present approach whereby research planning is not systematically carried out; and research priorities and targets are not usually clearly defined. When proposed projects are not adequately evaluated for desirability and many projects are based on the personal initiatives of the research staff, attracting the support of the industrial sector is highly likely to be difficult. For them to readily attract the attention of the private sector as well as funding support for their R&D projects, EROs need to engage in interdisciplinary research that will cut across departments, always looking to solve specific problems emanating from the industrial sector. A particularly beneficial initiative would be to send Doctoral students out to the industry for them to find out problems which they could study and help to solve. By so doing, institutions would be focusing more on market-driven research rather than basic or curiosity-oriented projects.

Several commercialisable inventions and research results have been generated by educational institutions and research institutes in Nigeria, but regrettably, one of the main reasons given for their low utilisation by industrial firms was that industries were not aware of these inventions and results. The National Office for Technology Acquisition and Promotion (NOTAP) is a facilitator, a catalyst and an information hub in marketing of research results. This, and other cognate agencies should be so structured and strengthened as to be able to adequately forge strong research-industry connections.

As a result of the fear of the unknown, researchers in Nigeria are usually cautious about embarking on activities that would take them out of their institutions/institutes because this might lead to loss of their job. However, the tacit nature of the inventions may require the inventors to work closely with the industrial firms that are exploiting the inventions in some instances. The current conditions of service do not allow the researcher to take up such appointments. Therefore, researchers could be motivated if the conditions of their employment are modified to allow them nurture such projects outside their institutions/institutes for a period, and later return to their jobs. This could be facilitated by designing policy to encourage researchers to nurture spin-off companies within or outside their institutions. The institutions should also map out mutually beneficial guidelines for disbursement and sharing of royalties that emanate from such activities.

The domestic industries' lack of interest in the research outputs of Nigerian R&D organisations has been associated with the strong preference for foreign technologies. The foreign-based multinational companies do not have confidence in the quality of personnel and research outputs of the Nigerian R&D organisations and this affects the rate of adoption of the research outputs. The quality of R&D outputs could be improved through the involvement of the industry in the formulation of research agenda/projects of the R&D organisations, adequate funding, and the provision of shared R&D facilities and equipment. Each institution could use whatever meagre fund is available to provide facilities in areas where they have comparative advantage and which other institutions could share.

CONCLUSION

This chapter has reviewed the concepts of Science, Technology and Innovation. The factors militating against the translation of scientific R&D outputs into innovations have been discussed. We have also reviewed the experiences of selected countries to specifically highlight the importance of institutional framework and education/knowledge pool in the process of national development. Policies, strategies and structures that could facilitate effective management of R&D within the Nigerian NIS are suggested.

Appropriate institutional framework must be put in place to coordinate and disseminate information about commercialisable inventions to interested entrepreneurs and industrial firms. The relevant government agencies should also strongly promote entrepreneurship, business creation, commercialisation of research and business development. Effective links should be encouraged among all stakeholders in the NIS. Conclusively, with a stable political environment, the benefits of the measures recommended above would be channelled to build internal and international competitiveness for growth and development.

REFERENCES

Ahmed, A., & Russel, H. C. (1988). *Science and Technology Policy for National Development – A Window on the Asian Experience*. Foundation for International Training.

Ajoku, K. B. (2002). *Evaluation of the Management of Technology Development*. A thesis submitted to National Centre for Technology Management Ife, Nigeria for award of Post Graduate Diploma on Technology Management.

Bell, M. (1984). Learning and the accumulation of industrial technological capability in developing countries. In Fransman, M., & King, K. (Eds.), *Technological Capability in the Third World* (pp. 187–209). London: Macmillan.

Coe, D., & Helpman, E. (1993). *International R&D spillovers*. NBER Working Paper 4444, National Bureau of Economic Research.

Coe, D., Helpman, E., & Hoffmaister, A. (1997). North-South R&D spillovers. *The Economic Journal, 107*, 134–149. doi:10.1111/1468-0297.00146

Coe, D., & Moghadam, R. (1993). Capital and Trade as Engines of Growth in France. *IMF Staff Papers*, 40.

Commission of the European Communities. (2006). Putting knowledge into practice: A broad-based innovation strategy for the EU. *Communication from the Commission to the Council, the European Parliament, the European Economic and Social Committee and the Committee of the Regions, COM(2006) 502 final*. (available online at http://eur-lex.europa.eu/LexUriServ/site/en/com/2006/com2006_0502en01.pdf)

Department of Science and Technology (DST). (2005). *'About the System' Science and Technology System in India.* (available at http://dst.gov.in/stsysindia/about-sys.htm, accessed January 8, 2006)

Egbetokun, A. A., & Siyanbola, W. O. (2008a). *Reliance on Natural Resources is Africa's Undoing. African Executive Magazine (online)*, April 30 - May 7 (available at http://www.africanexecutive.com/modules/magazine/articles.php?article=3030)

Egbetokun, A. A., & Siyanbola, W. O. (2008b). *Linking Science, Technology and Innovation (STI) For Development. African Executive Magazine (online)*, April 30 - May 7 (available at http://www.africanexecutive.com/modules/magazine/articles.php?article=3043)

Egbetokun, A. A., & Siyanbola, W. O. (2008c). *Making Science, Technology and Innovation Relevant. African Executive Magazine (online)*, April 30 - May 7 (available at http://www.africanexecutive.com/modules/magazine/articles.php?article=3069)

Egbetokun, A. A., Siyanbola, W. O., Sanni, M., Olamade, O. O., Adeniyi, A. A., & Irefin, I. A. (2009). What drives innovation? Inferences from an industrywide survey in Nigeria. *International Journal of Technology Management, 45*(1/2), 123–140. doi:10.1504/IJTM.2009.021524

Freeman, C. (1987). *Technology Policy and Economic Performance: Lessons from Japan*. London: London Printer Publishers.

Hamel, G. Prahalad. C.K. (1994). *Competing for the future*(pp.231-232). Boston:Harvard Business School Press.

Ilori, M. O. (2006). *From Science to Technology and Innovation Management. Inaugural Lecture Series, Faculty of Technology, Obafemi Awolowo University, Ile-Ife, Nigeria*. Oxford, UK: Obafemi Awolowo University Press. Ltd.

Ilori, M. O., Adeniyi, A. A., Oyewale, A. A., Sanni, S. A., & Irefin, I. A. (2002). Developing a manufacturing-based economy in Nigeria through science and technology. *Technovation, 22*(1), 51–60. doi:10.1016/S0166-4972(00)00054-7

Ilori, M. O., Oke, J. S., & Sanni, S. A. (2000). Management of new product development in selected food companies in Nigeria. *Technovation, 20*, 333–342. doi:10.1016/S0166-4972(99)00147-9

Joint Admissions and Matriculation Board (JAMB). (2007).Retrieved from http://www.jamb.org.ng (April 17, 2007).

Jolly, V. K. (1997). *Commercializing New Technologies: Getting from Mind to Market*. Boston: Harvard Business School Press.

Jones, C., & Williams, J. (1999). *Too much of a Good Thing? The Economics of Investment in R&D. I.Cambridge*. MA: National Bureau of Economic Research.

Jones, R. (2005). *The 2005 Economic Survey of Korea. Presentation at the OECD Breakfast Series, Washington, DC, 30 November 2005*. Paris: OECD.

Katz, J. (Ed.). (1987). *Technology generation in Latin American manufacturing industries*. Macmillan.

Kruss, G. (2006). University-industry links: lessons from South Africa. Retrieved from Science and Development Network (SciDev.net) - http://www.scidev.net/dossiers/index.cfm?fuseaction=dossierreaditem&dossier=13&type=3&itemid=542&language=1 on December 3, 2007.

Lall, S. (1987). *Learning to industrialize. The acquisition of technological capability by India*. New York: Macmillan.

Lall, S. (1991). Human Resources, Technology and Industrial Development in Sub-Saharan Africa. In Chhiber, A., & Stanely, F. (Eds.), *Economic Reform in Sub-Saharan Africa*. Washington, DC: World Bank.

Lall, S. (1992). Technological Capabilities and Industrialisation. *World Development, 20*(2). doi:10.1016/0305-750X(92)90097-F

Liao, Z., & Greenfield, P. F. (1997a). The impact of generic competitive strategies on corporate R&D: an empirical study in Japan. *International Journal of Technology Management, 13*(5/6), 542–555. doi:10.1504/IJTM.1997.001687

Liao, Z., & Greenfield, P. F. (1997b). Major considerations in the corporate development of R&D strategies within Australian technology-based firms. *International Journal of Technology Management, 13*(5/6), 542–555. doi:10.1504/IJTM.1997.001687

Marinova, D., & Phillimore, J. (2003). Models of Innovation. In Shavinina, L. V. (Ed.), *The International Handbook on Innovation*. Oxford, UK: Elsevier Science. doi:10.1016/B978-008044198-6/50005-X

National Universities Commission (NUC). (2007). *List Of Nigerian Universities And Years Founded*. Retrieved from http://www.nucnigeria.info/universities.htm (accessed April 17, 2007)

Nonaka, I. (1994). A dynamic theory of organisational knowledge creation. *Organization Science, 5*(1), 14–37. doi:10.1287/orsc.5.1.14

Norris, K., & Vaizey, J. (1973). *The Economics of Research and Technology* (Cater, C., Ed.). London: George Allen and Unwin Ltd.

Norris, K., & Vaizey, J. (1973). The Economic of Research and Technology. In Cater, C. (Ed.), *London: George Allen and Unwin Ltd*. London.

OECD. (1981). *North/South Technology Transfer Adjustment Ahead. Organisation for Economic Co-operation and Development (OECD)* (pp. 17–19). Publication.

Oyebisi, T. O., Ilori, M. O., & Nassar, M. L. (1996). Industry-academic relations: an assessment of the linkages between a university and some enterprises in Nigeria. *Technovation, 16*(4), 203–209. doi:10.1016/0166-4972(95)00049-6

Oyelaran-Oyeyinka, B. (1997). *NNEWI: An Emergent Industrial Cluster in Nigeria*. Ibadan, Nigeria: Technopol Publishers.

Oyelaran-Oyeyinka, B. (2005). *Translating Research into Innovation in Nigeria: A Critical Perspective.* Paper Presented at the Seminar organized by the Nigerian Academy of Engineering, May 17-19, Abuja, Nigeria

Oyelaran-Oyeyinka, B. (2005). *Translating Research into Innovation in Nigeria: A Critical Perspective.* Paper Presented at the Seminar organized by the Nigerian Academy of Engineering, May 17-19, Abuja, Nigeria

Oyelaran-Oyeyinka, B. (2006). *Don't make a fetish out of R&D spending*. Retrieved from Science and Development Network (SciDev.net) - http://www.scidev.net/Opinions/index.cfm?fuseaction=readOpinions&itemid=543&language=1 on December, 12, 2007.

Oyelaran-Oyeyinka, B., Laditan, G. O. A., & Esubiyi, A. O. (1996). Industrial innovation in Sub-Saharan Africa: the manufacturing sector in Nigeria. *Research Policy, 25*(7), 1081–1096. doi:10.1016/S0048-7333(96)00889-X

Oyewale, A. A. (2005). *Addressing the research-industry linkage impasse in Nigeria: The critical issues and implementation strategies*.Paper presented at Globelics-South Africa, 2005. www.globelics2005africa.org.za/papers/p0032/Globelics2005_Adesina%20Ayobami%20Oyewale.pdf (accessed December 2, 2006).

Prusak, L. (1996). *The knowledge advantage*. Strategy and Leadership, March-April.

Prusak, L. (1997). *Knowledge in Organisations*. Woburn, MA: Butterworth-Heinemann.

Prusak, L., & Davenport, T. H. (1998). *Working Knowledge* (p. 5). Boston: Harvard Business School Press.

Romijn, Henny &Albaladejo, Manuel (2000) *Determinants of Innovation Capability in Small UK Firms: An Empirical Analysis.* QEH Working Paper 40

Soboyejo, W. (2006). *A network of excellence for African development*. Retrieved from Science and Development Network (SciDev.net) - http://www.scidev.net/dossiers/index.cfm?fuseaction=dossierreaditem&dossier=13&type=3&itemid=548&language=1&CFID=1495413&CFTOKEN=24816639 on June 29, 2007.

Souitaris, V. (2003). Determinants of technological innovation: Current state of the art, modern research trends and future prospects. In Shavinina, L. V. (Ed.), *The International Handbook on Innovation*. Oxford, UK: Elsevier Science. doi:10.1016/B978-008044198-6/50035-8

Stanton, W. J., Etzel, M. J., & Walker, B. J. (1994). *Fundamentals of Marketing* (10th ed.). New York: McGraw Hill, Inc.

Stern, S., Porter, M., & Furman, J. (2000). *The determinants of national innovative capacity*, NBER Working Paper 7876, National Bureau of Economic Research.

Tiffin, S. (1997). *Building Science Technology Innovation System in Africa*. Paper delivered at the First Regional Workshop on the Restructuring of National Science and Innovation Systen in Africa Regional Programme for Technology Management (REPTEM) Lagos; May 12 – 16, 1997.

United Nations Development Programme (UNDP). (2002). *Human Development Report*, Nigeria 2000/2001 Millennium Edition, 80p

United Nations Development Programme (UNDP). (2005). *Human Development Report 2005*. Geneva: UNDP.

United Nations Educational Scientific and Cultural Organisation (UNESCO). (1982). *An Introduction to Policy Analysis in Science and Technology*. Paris: UNESCO.

University Companies Association (UNICO). (2005). UK University Commercialisation Survey. Financial Year 2004. Cambridge, UK.

Valentas, K. J., Levine, L., & Clark, J. P. (1990). *Food Processing Operations and Scale-up*. New York: Marcel Dekker Inc.

ENDNOTE

1 The University Companies Association's (UNICO) UK University Commercialisation Survey: Financial Year 2004 reports that during that year, 141 patents were granted, 1406 licences executed with total revenue of 65.2m USD as licence income and 229 start-ups were successfully initiated – all as a direct result of interactions between universities and industry.

Chapter 12
Theoretical Approaches to the Marketing of Innovation in Biotechnology Firms:
New Products for New Markets

Tomas Gabriel Bas
University Adolfo Ibañez, Chile

ABSTRACT

The marketing of new products for new markets is uncertain and often leads to confusion. One of the many causes is the wrong use of marketing techniques for mature products whose life cycles are different, although they are very familiar. Biotechnology, with a strong component of technological knowledge and innovative cross-refers not to an industry, but rather a set of applications of intricate characteristics mainly due to genetic manipulation, which represents a particular challenge unlike other more mature or traditional product or technology. For this reason, through an innovation based marketing for new product, new customers and new markets are essential to maximize the success of this technology.

INTRODUCTION

Marketing of innovation is highly complex because the potential customers usually are unaware of the new product. In the same way, the markets as known in traditional marketing are frequently absent. The biotechnology is based on a cross-sectional knowledge, which represents a particular challenge and is different from other more mature and more traditional market products. Customers and biotech markets are different from traditional products as well as their goals and needs.

The studies related to behavior, tactics and influence of marketing in the process of developing a new technology product are relatively poor (Millier, 1997; Di Benedetto, 1997 and Atuahene-Gima, 2000). The most cases of biotechnology refer not to a particular industry, but to crossing a set of applications with multiple characteristics, which can be used in a variety of fields, such as agriculture, pharmaceutical, information technology, environment, etc. This makes both the customers, markets, and the objectives and needs of biotechnology, different from that of any other technology or industry.

DOI: 10.4018/978-1-61692-846-9.ch012

Copyright © 2011, IGI Global. Copying or distributing in print or electronic forms without written permission of IGI Global is prohibited.

The uses of marketing strategies, which increase the likelihood of success of a new product, are crucial for biotechnology firms. In this context, it is important to highlight how these companies are faced with two major problems of marketing. The first is the novelty of the product, a situation that makes it unknown to potential customers; the second is whether this novel product will meet the needs of those customers, a distance by the nature of biotechnology (doctors, biologists, engineers, etc.). The challenge is how to reconcile the two major problems through marketing innovation, and not through the operational marketing (Montalvo, 2005) that is commonly used by companies with more mature technologies.

METHODOLOGY

On the base of a review of literature on marketing innovation in general and biotechnology in particular, the study describes the factors involved in marketing innovation so as to increase understanding of its importance, and assesses the impact on decisions that are commensurate with the product to be marketed, i.e. new products for new markets. We analyze the conditions necessary for insertion of a new product, taking into account the different life cycles that must face.

THE CHARACTERISTICS OF THE MARKETING OF INNOVATION

Many authors claim that the development of a new technological product is characterized by a high degree of uncertainty and risk, while retaining the context of turbulence, segmentation and high volatility exist in biotechnology sector. These are elements that somehow prevent the use of conventional marketing themselves useful for more traditional and stable products (Shanklin and Ryan, 1984; Jolly, 1997; Millier, 1997; Mohr, 2000; Day, 2000, Maltz and Kohli, 2000, Van Birgelen et al., 2001).

Biotechnology companies must operate differently than those serving more mature technologies. Foray (1996) explained that at the time of the birth of a technology, its future is essentially uncertain for several reasons:

- The atmosphere is different and the speed of change too, as biotechnology evolves extremely fast.
- Distribution channels are either absent or embryonic.
- We know badly the reaction of potential customers because they do not know the product.

In biotechnology, what often separates the winners from the losers are the measures that take into account the quality of the R&D organization, product advantage, the pre-development activities through preliminary studies that have account of the trilogy -market, technical, finance, also the specifications of the needs of the market and product, competition and compatibility among staff (R&D, marketing, finance) using channels and fluids compatible communication (Jolly, 1997). The intuitive, proactive attitude is necessary to try to anticipate markets, and in that sense, the marketing staff in conjunction with other staff must have the ability to investigate the environment related to potential customers.

In fact, the literature considers that the context of turbulence and high volatility segmentation gives a prominent role to the intercommunication between the different R&D members of the firm. The cross talk between marketing and R&D sector is one of the key elements in the success or failure of a new biotechnology product (Atuahene-Gima and Evangelista, 2000; Atuahene-Gima and Li, 2000, Sherman et al., 2000; Maltz and Kohli, 2000). Thus, we observe the importance of interaction between marketing, finance, manufacturing sector, customers and definitely R&D. The intercommunication needs to address socio-cultural differences of the departments that can lead to internal conflicts. The companies have structured by departments other than the organizational behavior of such dimensions that lack them. Departments usually have a divergent culture due to their training differently, which eventually ends up influencing their perception. In these cases, the communicative learning is paramount, the content of communication is represented by knowledge, which in turn can be tacit or explicit. In this regard, Nonaka and Takeuchi (1995) define explicit knowledge as something that can be expressed in words and numbers to be easily communicable and transferable. When more explicit is the knowledge, much more easier is for the members of the company to learn and interact more smoothly. At the other extreme is the tacit knowledge, which is highly personal and hard to formalize, making learning more difficult to communicate. The latter often presents serious difficulties that can lead to discourage communication between departments or between members of the company.

CO-EVOLUTION OF MARKETING BUDGET INNOVATION / R&D

The competition for the upgrade or the continued development of knowledge generates a steady increase of investment in R&D resulting in an increase of the marketing budget. Unfortunately, these costs are often poorly managed, however, these should be considered by the managers since the start of the project and evolve as the project develops. It is necessary to consider whether it is the technology that is creating the market- technology push or on the contrary the market that pushes the technology to develop- market pull. Given these factors, the treatment and marketing strategies to consider would be different. From this perspective, coordination of marketing staff with R&D is crucial. Marketing personnel should not only aim at exploiting potential markets, but also check what competitors are doing through technological monitoring chord. We must also consider the uncertainties related to biotechnology in order to understand the limits and possibilities of success of a new product. Mohr (2000) identifies five factors to explain the uncertainty related to technology:

1. Questioning whether the product will meet expectations.
2. Product availability in time and in shape. That is the firm's ability to meet demand
3. The real ability of suppliers to ensure a service if a problem arises.
4. The unanticipated consequences.
5. Ignorance about the viability of the technology and the time at which it will be obsolete.

INCREASING PROBABILITY OF SUCCESS OF NEW BIOTECHNOLOGY PRODUCTS

The successes of a new biotechnology product is dependent (Figure 1) on a set of interrelated actions and immersed in a highly complex and subject to a number of actions in the multiple stages of product development. Part of the marketing literature gives priority to different notions regarding the next steps, from conception of the original idea to the development and introduction into the market. However, these concepts are mostly linear in nature and led to products that more traditional and potential customers already know.

Following this guideline, a bit of simplicity when it comes to developing and marketing a new product technology is concerned, Urban and Hauser (1993) propose five scales of activities in the process of developing a new product, among which are: 1) Identification and selection of opportunities, 2) Design of new product, 3) Test, 4) Marketing, and 5) post-launch control. Shanklin et Ryans (1984) use a similar approach but with nine sections, which are: 1) Conception of the

Figure 1. Set of interrelated actions in the multiple stages of product development (Source: Cooper, 1986)

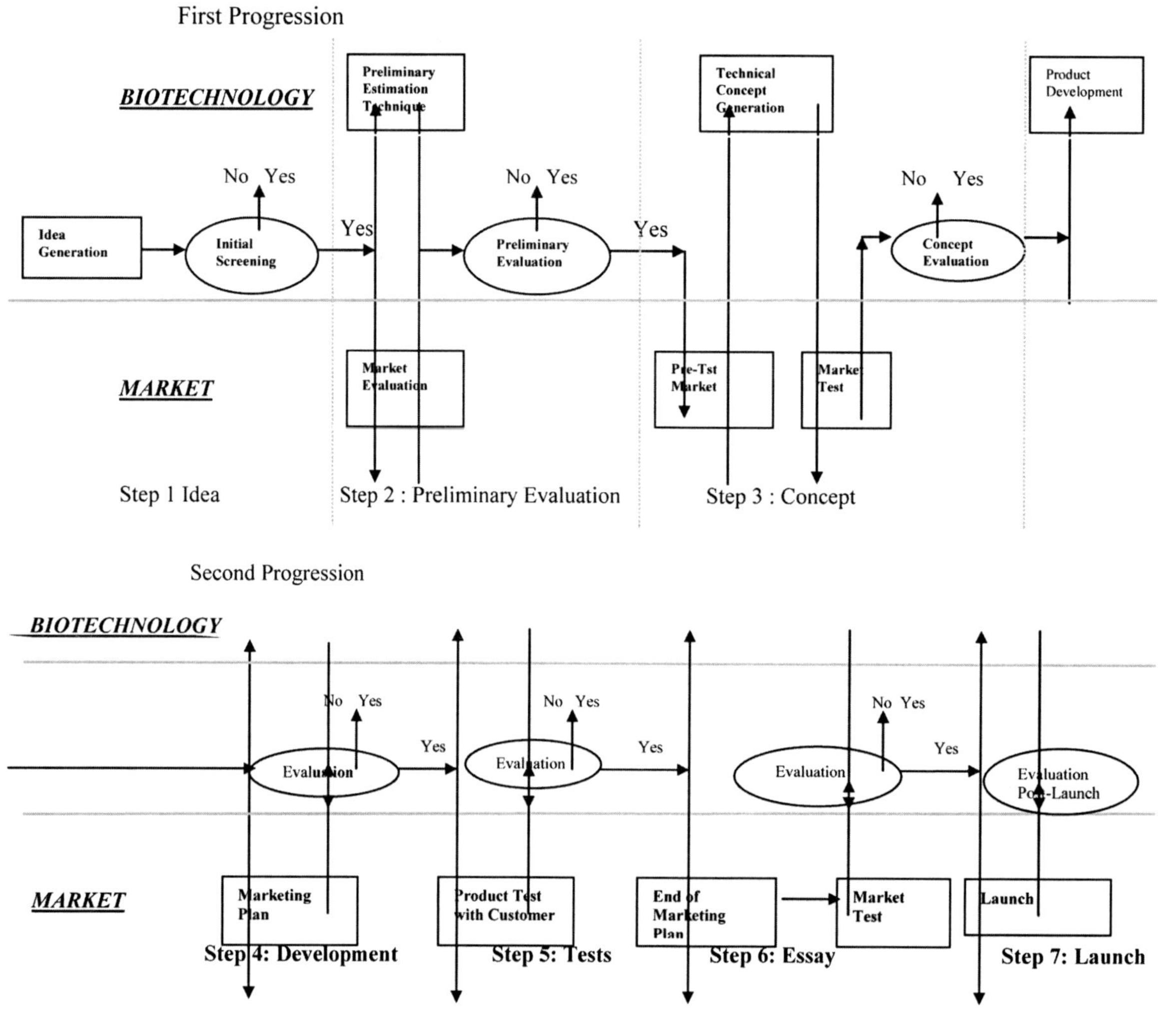

idea, 2) Identification of potential customers, 3) Search for markets, 4) Estimate potential demand, 5) development and test, 6) Decision to introduce the product on the market, 7) Development of the final product, 8) Review of Marketing and 9) Marketing. While, Kotler et al. (2000) do a marketing analysis closest to innovation and identify eight stages starting with: 1) Search for different ideas, 2) Filtering the best or most feasible of these, 3) Development and test technological product concept, 4) Development of marketing strategy, 5) Economic Analysis on the feasibility of the product, 6) Development itself, 7) Market testing that performs extremely complex considering the novelty of the innovator product and 8) Launching the final product.

Thus we see how the literature identifies different stages that are framed to an operational and marketing rather linear, traditional products, instead of a more strategic and targeted marketing innovative products that are unknown by the customer and the market in general. This requires the implementation of specific methods, aimed at new technological products, according to the innovation called marketing.

Figure 2. Process and cost of one molecule derived of biotechnology sector

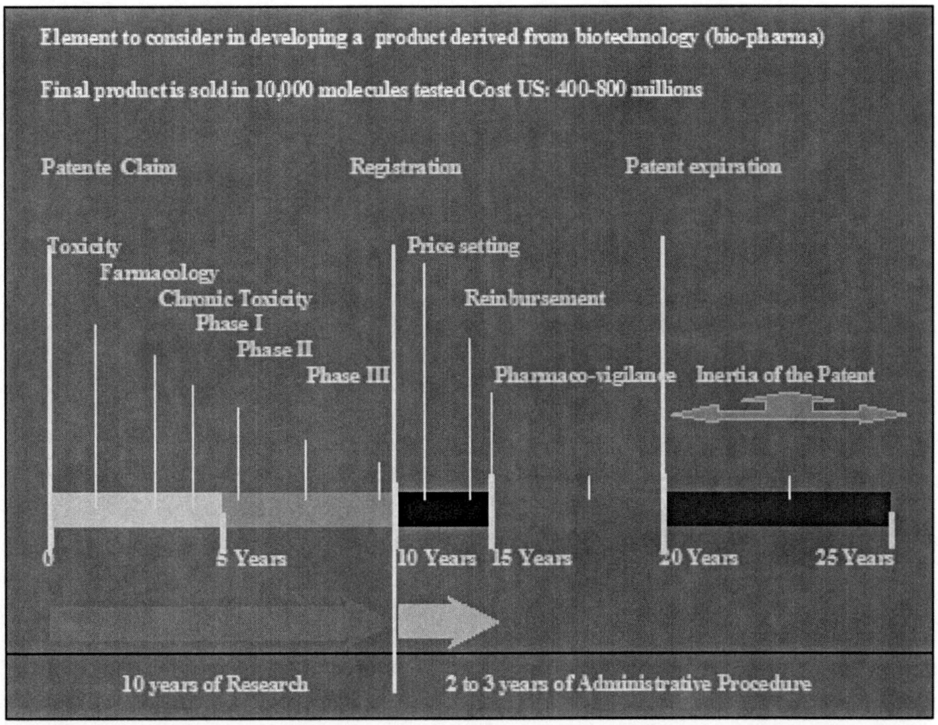

THE SHORTCOMINGS OF THE LINEAR CONCEPT WITH RESPECT TO MARKETING INNOVATION

It is important to establish certain parameters regarding the literature cited above, in order to understand better than the linear concept of the development process of a new technological product and its distribution in this case of a new molecule derived from biotechnology, and is not adequate (Millier, 1997 and Allen, 1985). The same problem occurs if we use the model of the product life cycle as a rule, as is easily seen that this is deterministic and normative. Is deterministic, because everything happens as if the first months of life determine their future product, as is normative, because it gives the rules to respect for their actions. These are precisely the gaps in these tools, when to test a new technological product. Here you do not know the validity of the product life cycle, but their application is considered to be suitable only for products that have passed the crucial stage of release. That theory is hardly applicable to new and innovative products which are unknown or emerging markets, such as those derived from biotechnology, it is not enough to introduce a product to market so that it is disseminated in a sustained manner and regular, as in a curve of "Foster" (Kotler et al., 2000). It is essential to consider that in biotechnology, the probability of success of a new molecule lies in following from the conception of the original idea, which means devoting a lot of time (about ten years) and significant amounts of money (from 400 to 800 million U.S. $) as shown in Figure 2 well in advance of the market, even if the increased spending on marketing will begin with the launch of the product more time later. This means that we should involve not only the product life cycle, but also the notions of life cycle of technological innovation and the project life cycle, which

Figure 3.

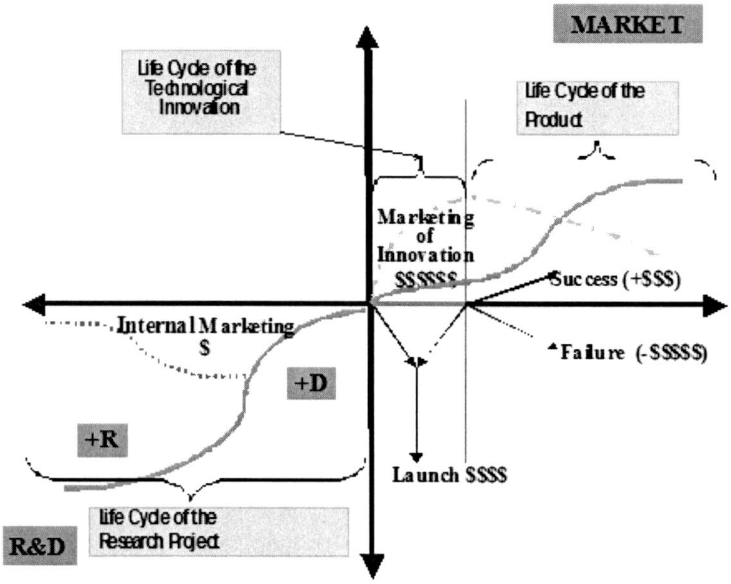

are often neglected or underestimated by other marketing (Millier, 1997).

On the other side, in Figure 3, we observe the impact that each has on the development cycle and the likely success or failure of the final product. The three life cycles should be considered as a whole that is indivisible and interdependent since the inception of the idea to post-launch of the innovative product.

- **The Project Life Cycle Research:** Refers to what happens in the phases of design and project feasibility. In this case, there is increased research spending (I) Developing (D) and internal marketing plays a crucial, since you have to choose the best idea in all the proposals that emerge in the enterprise, which carries an enormous risk, since a poor choice can mean not only the failure of the project but also the death of the company.
- **The Life Cycle of Technological Innovation:** This involves the stage at which the project is based and moves from a purely laboratory object, a product expected (or not) by potential customers. Due to its transient, there is great instability. It is time to product launch, so expenditures on marketing innovation are very important. It is in these early states, where it was found if the product will allow a glimpse of its potential success, or simply a failure.
- **The Product Life Cycle:** It represents the evolution of product sales after being released. This is where play begins to enter the rules of traditional strategic marketing. If the product is accepted, as they evolve their sales, and the same is best known in marketing costs tend to decrease innovation, to focus more on a maintenance marketing.

Experience shows that the highly successful biotechnology products (blockbusters), have been those in which the strategy and marketing investments have started early and continued throughout the entire R&D. That is, long before the product is placed on the market. This strategy is not necessarily very expensive, even if it absorbs some cases, up to 20% of total marketing budget during the four years preceding its release (Radal, 2000). However, this is a crucial strategy to increase the likelihood of success of new product. The key is to concentrate since the birth of the project, market research and meetings with various opinion leaders (professional, industrial, customers, suppliers, distributors, etc.). In order to identify needs, market prospects and even to observe the competition. Increase the chances of success of a new technological product, is an arduous and long-term, where the challenges are enormous and numerous causes. Among the major difficulties, we can cite the property of "novelty", which implies the use of non-traditional communication channels to understand the technology. Strategies must lean communication channels accessible to inter-departmental teams (Le Gall, 1999; Allen, 1985 and 1987). Different makers have to work using the same "language", reducing conflicts and their cultural differences, ensuring that all team members work on the same objectives (Maltz and Coolio, 2000 and van Birgelen et al, 2001), thereby optimizing the competitive positioning of the new product to develop.

THE CHALLENGES OF A NEW PRODUCT

Many times, an innovation often transforms an existing technology obsolete in a relatively short time, so we talk about market volatility. Such instability could doom a small firm to disappear, if they invest all their assets in developing a technology that can be outdated, even long before it is marketed (Jolly, 1997). Another flow instability appears with its own market segmentation. This context complicates the biotech firm and must make a clear decision on marketing strategies to follow, because applications of this technology tend to be numerous, and usually they are not obvious.

Millier (1997) explains that evaluating future product markets that does not exist for clients who do not know, is a challenge with great instability. One of the biggest problems is the parameters used to evaluate the emerging markets because the correlation between development of technology and the degree of acceptance that this will, is very difficult to prevent.

One wonders whether the new technology is able to meet the needs of potential customers, better than the alternative available in the market and applications will be more attractive and in what order these should make an appearance. We must analyze whether the market size justifies the investment and how quickly this could conquer. The answers are not in pre-established parameters, since they will depend on many variables, which often escape the control of the firm, especially those related to the action of competitors (Thomke, 1997 and Day, 2000).

To reduce uncertainty to levels that allow the company, to have an advantage over its rivals, can be achieved considering at least three parameters:

1. **The dissemination and adoption:** Refers to the time the product takes to penetrate the market. Some have a very long gestation (biotechnology), while in other cases; this is relatively short (some software). In this figure, we must consider the risk perceived by prospective customers as possible barriers to the adoption of the product and finally, we must ensure that this is marketed as soon as possible so that potential customers can try it and learn (Roberts, 2000). Kotler et al. (2000) allude to the marketing of the users *"vedette"*, since they have a particular status, positioning itself among the leaders to test the product. Furthermore, the characteristics of each user will differ according to their personal tastes, needs and values, with respect to a name (brand) or the time taken to adopt the new product (van Birgelen et al., 2001). You can sort these customers into five categories: innovators, early adopters, early majority, late majority and the latecomers.

2. **The market exploration and learning:** Many technology products have uncertain developments and impregnated with successive transformations, before emerging from learning a series of mistakes. This leads to the company, a process of successive accumulation of knowledge and learning. This process, which requires active distribution of information regarding potential customers, their problems, their needs, the criteria used to make a decision, contingencies, etc. We must interpret the learning and react accordingly, but paying attention to the quality of information and communication channels used, and in the perception of the various members of the firm.
3. **The value of multiple methods:** As we have said previously, traditional methods to mature markets are best suited for use in biotechnology, due to the uncertainty and the proliferation of differences during the long, successive trial and error, which must pass this technology to marketing the product. Day (2000) suggests however, that certain methods used for mature products could be adapted to the latent needs of new technologies. Nevertheless, there is no foolproof method, all have strengths, weaknesses and limitations, but marketing strategies should always consider the quality of novelty and uncertainty when it comes to biotechnology.

THE PIONEERS-BUSINESS CUSTOMER INTERACTIONS

If we talk about new technology products, it is expected that joint customers will have some difficulty in recognizing their earliest versions (Bettman et al., 2000). To avoid this obstacle, some authors (Kotler et al., 2000; Day, 2000, Roberts, 2000; Atuahene-Gima and Evangelista, 2000; Jassawalla and Sashittal, 1998, Sherman et al., 2000, von Hippel, 1986 and 1995 and Goldenberg et al., 2001) propose the education of pioneering customers who have latent needs an urgent product today and that the market will need tomorrow. These customers need new technology products, and to express their needs, potential problems they may encounter, the different situations of use or application, as well as their needs change as technology evolves. Such interaction (Atuahene-Gima and Evangelista, 2000), may allow the marketing staff, to better understand the needs of new clients, to meet the best possible way.

Von Hippel (1986) estimates that customers must be identified according to their position and trajectory, which can give an index of their loyalty and experience they possess in their sector of operation. Therefore, one solution is to develop the new product in direct interaction with researchers, suppliers and customers in default markets. Firms that can come to have an internal fluid and permeable, between actors from the original idea may (even with some limitations) to anticipate, analyze and exploit favorable market conditions much more easily than competitors that do not consider this situation. However, some solutions are emerging to try to counter these limitations, which are designed through the interrelationship of functions within the firm (R&D, finance, suppliers, marketing, etc.). This can be done through co-development contracts, which have the function, development of the product in which the client has a particular interest. The contract has the power to bind the interests of both parties to commit long term. With trust to such co-development, and given that the client knows more about the product from the embryonic stages, the chances of success go up exponentially.

CONCLUSION

To increase the chances of success of new products for new technology markets, firms must adapt to live with limits and risks, the uncertainty generated by this new technology. Consider that a firm can have an enormous intellectual capacity of R&D, but be unable to make it a commercially viable product. Following the same principle, to have the best technology does not guarantee success without proper management of marketing innovation. In the development of biotech products, which often separates the successful from the rest, are precisely the measures they consider innovative marketing strategies. Those companies that address the pre-development activities, the needs of product-market, the potential advantage of the product, capacity and support staff and customer firm (R&D, finance, etc. and marketing) through fluid communication channels, will increase the chances of success.

The communion between the creators of a new biotechnology product and a strategic marketing plan for innovation is necessary for customers to learn about and future products, evaluate them according to their needs and discover their strengths and limitations. Investing in marketing is "mandatory", but we must invest in good marketing. If the product is unknown and an innovative and technological, marketing will have to use the innovation.

REFERENCES

Allen, T. (1985). *Managing the Flow of Technology*. Cambridge, MA: MIT Press.

Allen, T. (1987). Performance of Information Channels in the Transfer of Technology. In Roberts, E. B. (Ed.), *Generating Technological Innovation* (pp. 89–104). New York: Oxford University Press.

Atuahene-Gima, K., & Evangelista, F. (2000). Cross-Functional Influence in New Product Development: An Exploratory Study of Marketing and R&D Perspectives. *Management Science*, *46*(10), 1269–1284. doi:10.1287/mnsc.46.10.1269.12273

Atuahene-Gima, K., & Li, H. (2000). Marketing's Influence Tactics in New Product Development: A Study of High Technology Firms in China. *Journal of Product Innovation Management*, *17*, 451–470. doi:10.1016/S0737-6782(00)00059-X

Bettman, J., Luce, M. F., & Payne, J. (2000). Processus de choix construit du consommateur. *Recherche et Applications en Marketing*, *15*(2), 187–217.

Day, G. (2000). Assessing Future Markets for New Technologies. In Day, G., Schoemaker, P., & Gunther, R. (Eds.), *Wharton on Managing Emerging Technologies*. New York: John Wiley & Sons.

Di Benedetto, C. (1997). Identifying the Key Success Factors in New Product Launch. *Journal of Product Innovation Management*, *16*, 530–544. doi:10.1016/S0737-6782(99)00014-4

Foray, D. (1996). Diversité, sélection et standardisation: Les nouveaux modes de gestion du changement technique. Revue d'Économie Industrielle, N° 75.

Goldemberg, J., Lehmann, D., & Mazursky, D. (2001). The Idea Itself and the Circumstances of Its Emergence as Predictors of New Product Success. *Management Science*, *47*(1), 69–84. doi:10.1287/mnsc.47.1.69.10670

Jassawalla, A., & Sashittal, H. (1998). An Examination of Collaboration In High-Technology New Product Development Processes. *Journal of Product Innovation Management*, *15*, 237–254. doi:10.1016/S0737-6782(97)00080-5

Jolly, V. (1997). *Commercializing New Technologies Getting from Mind to Market*. Boston: Harvard Business School Press.

Kotler, P., Filiatrault, P., & Turner, R. (2000). Le management du marketing. Gaetan Morin éditeur. Montréal. Canada.

Le Gall, J-M. (1999). L'apprentissage communicationnel» *Revue française de gestion*. 130, 31-44.

Maltz, E., & Kohli, A. (2000). Reducing Marketing's Conflict with other Functions: The Differential Effects of Integrating Mechanisms. *Academy of Marketing Science*, *28*, 479–492. doi:10.1177/0092070300284002

Millier, P. (1989). Le marketing des produits "High Tech". *Les éditions d'organisation*, Paris.

Millier, P. (1997). *Stratégie et marketing de l'innovation technologique*. Paris: Dunod.

Mohr, J. (2000). The Marketing of High-technology Products and Services: Implications for Curriculum Content and Design. *Journal of Marketing Education*, *22*, 246–259. doi:10.1177/0273475300223008

Montalva, M. (2005). El Marketing Estratégico Aplicado en Chile y en Canadá para Empresas de Tecnología de Información y Comunicación. *Trend Mangement*, *7*, 130–137.

Nonaka, I., & Takeuchi, H. (1995). *The Knowledge-creating Company – How Japanese Companies Create the Dynamics of Innovation*. New York: Oxford University Press.

Radal, C. (2000). Le nouveau rôle du Marketing dans l'industrie pharmaceutique. Retrieved from http://www.edoceo.net/articles/le_role_marketing_indu_pharma_me.htm

Roberts, J. (2000). Developing New Rules for New Markets. *Academy of Marketing Science*, *28*, 31–44. doi:10.1177/0092070300281004

Shanklin, W., & Ryans, J. (1984). *Marketing High Technology. Washington, DC: LexingtonBooks*. USA: Heath and Company.

Shanklin, W., & Ryans, J. (1993). Organizing for High-tech Marketing. *Harvard Business Review*, *62*, 164–171.

Sherman, D., Souder, W., & Jenssen, S. (2000). Differentiel Effects of the Primary Forms of Cross Functional Integration on Product Development Cycle Time. *Journal of Product Innovation Management*, *17*, 257–267. doi:10.1016/S0737-6782(00)00046-1

Thomke, S. (1997). The Role of Flexibility in the Development of New Products: An Empirical Study. *Research Policy*, *26*, 105–119. doi:10.1016/S0048-7333(96)00918-3

Urban, G., & Hauser, J. (1993). *Design and Marketing of New Products*. Englewood Cliffs,NJ: Prentice Hall. van Birgelen, M.; De Ruyter, K. & Wetzels, M. (2001). Conceptualizing and Isolating Cultural Differences in Performance Data in International High-Tech Industrial Markets. *Industrial Marketing Management*, *30*, 23–35.

von Hippel, E. (1986). Lead Users: A Source of Novel Product Concepts. *Management Science*, *32*(7), 791–805. doi:10.1287/mnsc.32.7.791

von Hippel, E. (1995). *The sources of Innovation*. Oxford, Uk: Oxford University Press

Chapter 13
Strategic Regional Networks in Higher Education

Mauri Kantola
Turku University of Applied Sciences, Finland

Juha Kettunen
Turku University of Applied Sciences, Finland

Satu Helmi
Turku University of Applied Sciences, Finland

ABSTRACT

The purpose of this study is to present a framework, which can be used to analyze the regional dimension of the institutional performance in higher education. This chapter combines strategic planning and corporate social responsibility reporting. The strategic planning of a higher education institution defines strategic objectives, which take into account educational policy, national and regional networked strategies, and the regional demand for labor. The evaluation of the external impact of an institution means that a general conceptual framework must describe the outreach and engagement of the institution with the region. The framework presented in this study includes the elements of the networked knowledge society, strategic management, global reporting initiative and quality assurance. The empirical part of this study is based on the network study of the Turku University of Applied Sciences. The framework can be used in decision making when an institution is invited to join local, national and international networks.

1. INTRODUCTION

The greater autonomy of higher education institutions has increased the importance of their accountability. Higher education institutions are no longer ivory towers, but active participants in their social networks. In this situation, institutional management is growing in importance, and this growth increases the role of management approaches and frameworks. Institutions must be able to plan their future, but at the same time they must be able to evaluate and report their performance which takes into account social responsibility and quality assurance.

This study utilizes two theoretical approaches to describe the regional dimension of higher education institutions: the Balanced Scorecard introduced by Kaplan and Norton (2001, 2004)

DOI: 10.4018/978-1-61692-846-9.ch013

Copyright © 2011, IGI Global. Copying or distributing in print or electronic forms without written permission of IGI Global is prohibited.

and the Global Reporting Initiative (GRI) (2009). Strategic planning can be extended to consider the networked and potential external partners in addition to internal stakeholders. The GRI framework provides the widely communicated protocol for sustainability reporting, along with guidelines explaining how social responsibility can be promoted and communicated using the set of core indicators.

Networks are important elements of human capital when it is necessary to increase the institutional impact on a region (Coleman, 1988). Most of the best higher education institutions, like companies, are serious regional actors, that are sensitive both to local and global circumstances. The creation of human capital creation can factor into the strategic planning of a higher education institution just as it does in business.

The empirical part of the chapter is based on the study of the Turku University of Applied Sciences (TUAS). The institution collaborated with partners in social networks with the help of the corporate social responsibility report in order to support regional development. Collaboration with the partners and stakeholders is one of the themes of the strategic plan and of the regional development of the institution (Kettunen, 2004a, b, 2006, 2008). Regional development has been embedded in education and applied research and development and it is linked with the institution's other strategic objectives.

In keeping with its theme of social responsibility, the institution is focusing its activities on the needs of the region. The strategic plans are implemented using the Balanced Scorecard and the management information system that has been tailored to the institution (Kettunen & Kantola, 2005). The institution collaborates with many outstanding global and local partners. The external impact is described in the region and customer perspectives of the Balanced Scorecard and the measures with target values are annually set to maximize the external impact (Kettunen & Luoto, 2008, 278).

This chapter is organized as follows. First, the study introduces strategic planning and the Balanced Scorecard. Then the Global Responsibility Reporting (GRI) framework is introduced in the context of higher education. Thereafter the general framework of networked collaboration and regional development is presented. The empirical part of the study presents the strategic networks of the TUAS combined with quality assurance. Finally, the results of the study are summarized and discussed in the concluding section.

2. STRATEGIC PLANNING

Strategic planning entails a holistic view of the organization and its future with a clearly articulated mission, vision, strategic choices and values. The purpose of the strategy process is to produce a strategic plan and objectives. The overall strategic or the functional strategic plans may describe how an institution will support regional development in the future. The purpose of an institution is to adapt the financial and human resources to the changing environment, markets and customers to meet the expectation of stakeholders (Johnson & Scholes, 2002; Steiss, 2003). Higher education institutions are moving from the traditional public orientation to the more market oriented approach which emphasizes the engagement in its environment, outreach, and regional development (Kettunen & Kantola, 2006).

The TUAS uses the Balanced Scorecard, introduced by Kaplan and Norton (2001, 2004) for strategic planning. The Balanced Scorecard is also used to communicate and implement the strategic plan. The perspectives of the Balanced Scorecard and the strategic objectives of the TUAS are following:

- **Region and customer perspective:** regional development and customer satisfaction

- **Financial perspective:** funding from central government, external funding, cost-efficiency
- **Internal processes perspective:** teaching and learning, research and development and internalization
- **Learning and growth perspective:** teacher, research and development, and international capabilities

The local networks of experts and scientists transmit information, knowledge, attitudes and norms; through these networks individuals are exposed to each others' opinions and functional modes. Within these very same interactive and linked processes, the networks mediate both individual and public attitudes. The influence of networks extends to experts in different social appearances (Goffman, 1963). Thus, the experts learn through the networks and these specialists' own individual expertise and learning are developed at the same time (Wenger, 1998).

Individual motives and interests determine the functioning of networks. The networks link the learning of individuals to that of organizations and the region. The sharing of knowledge and partnerships between individuals are the main activities of higher education institutions. In order to change the behavior of a system, enough individuals in the system have to change their behavior. The learning organization is the result of the learning and the changed behavior of the organization (Sandaker & Andersen, 2008).

The core idea of network theories is that networks of experts have value and that the value added is hidden in the social relations, which are characterized by trust and reciprocity, thus raising the ability of community and members to participate and interact when resolving their shared problems (Cross & Baker, 2004). By deliberately creating regional networks, the institution may strengthen its ability to design its outreach and engagement in regional development.

3. EXTERNAL IMPACT OF THE INSTITUTION DESCRIBED BY THE GRI

The Global Reporting Initiative is a network-based body, a non-profit organization whose mission is to develop globally applicable sustainability reporting guidelines (GRI, 2009). The GRI has pioneered the development of the one of the most widely used sustainability reporting frameworks. The GRI network collaborates by continuously improving and applying the framework. The reporting framework of the GRI is a generally accepted framework for reporting on an organization's economic, environmental, and social performance. The purpose of the GRI is to take into account the practical considerations faced by organizations, ranging from small enterprises to extensive and geographically dispersed operations. The GRI reporting framework contains general and sector-specific content that has been accepted by stakeholders around the world in order to report on organization's sustainability performance.

It has been stipulated in Finland that the profession-oriented universities of applied sciences should support the regional development and consider the regional characteristics of business and working life. The regional development strategies reflect the aim of the institutions to meet the needs of the region in which they operate. The universities of applied sciences should also collaborate with other institutions, especially region ones. It is preferable to use the term "regional development" instead of "outreach and engagement" because this language is stipulated in the law. In addition, the selected term is widely used in the European Higher Education Area (Kettunen & Luoto, 2008).

The science-oriented universities have traditionally served international and national academic communities with few locally embedded activities (Chatterton & Goddard, 2000). Due to the expansion of higher education there has been

Table 1. The main goals of corporate social responsibility of the TUAS

Economic performance	Environmental performance	Social performance
Effective in degree production	Protection of natural resources	Supporting the well-being and know-how of students and personnel
Regional employment of graduates	Production of the positive environmental impacts	Developing good practices in working life

a recent trend to expect Western universities to serve their communities directly in return for public funding (Schuttle & van der Sijde, 2000). The responsibility of universities to contribute to the social and economic development of their environments has been called their third mission (Kettunen & Luoto, 2008, 279). The new mission of the science universities emphasizes the need to serve their society, compared with the earlier task to serve the whole mankind. Universities can contribute to the attainment of regional development goals in all their activities by establishing new institutional management structures to meet the demands of regional stakeholders.

The sustainability reporting of the GRI is a useful way to measure, disclose, and be accountable to internal and external stakeholders for organizational performance in the interest sustainable development. Sustainability reporting is synonymous with other terms used to describe economic, environmental, and social impacts (GRI, 2009).

According to the GRI, Sustainability Reporting Guidelines (2009) the corporate social responsibility (CSR) report can be used for the following purposes:

- Benchmarking and assessing sustainability performance with respect to laws, norms, codes, performance standards, and voluntary initiatives
- Demonstrating how the organization influences and is influenced by expectations about sustainable development
- Comparing performance within an organization and between organizations over time.

The CSR report has established its role as an important communication channel to express the external impacts of organizations.

The TUAS has reported its outcomes in its CSR report to its stakeholders since 2003 (Helmi, 2008; Luoto & Helmi, 2008). The reports include the indicators of the external impact of the institution and the thematic map of the coverage the collaboration with the local community. The CSR reporting follows the guidelines of the GRI. The TUAS has also analyzed and evaluated its collaboration with working life (Kettunen & Luoto 2008, 290).

Table 1 describes the main objectives of the corporate social responsibility of the TUAS. The annual CSR report of the TUAS, "Working for the Region" covers all three areas of sustainable development: economic, environmental and social performance. Since the CSR report is covers all the necessary activities there is no need for the TUAS to publish a traditional annual report.

The main goal of the economic perspective is to produce a skilled regional labor force. The institution has met this goal: 75% of the graduates of the TUAS remain in Southwest Finland. This figure is higher than that of comparable with other institutions and regions in Finland. That is based on the strategic plan, where the institution has made a commitment to serve the region. The active communication and collaboration in the local networks have been instrumental in reaching this desired outcome.

Environmental performance includes the protection of the local archipelago, rivers and reservoirs. The positive environmental impact means environmentally conscious graduates, the promoting of sustainable development and environment-oriented development projects. The development of regional production technology also has an important role. For example, the TUAS has a research laboratory that is designing cleaner motors for vehicles.

Social performance at the TUAS is divided into sections for students, personnel and working life, each of which is essential. The institution should be a reliable partner in the development of working life in the region, promoting stakeholder collaboration, networking and innovation. According to the feedback from companies and other organizations the TUAS is an expert community between higher education and working life in the region. The TUAS also brings new international innovative elements to the local level through the national and global networks typical of the higher education knowledge transfer and expert and student mobility.

The TUAS has identified its stakeholders. Its report explains how the TUAS has responded to stakeholders' expectations and interests. The GRI reporting framework defines stakeholders as entities or individuals that can reasonably be expected to be significantly affected by the organization's products and services. Stakeholders are also entities or individuals whose rights under law or international conventions provide them with legitimate claims to the organization (GRI, 2009). The organization needs stakeholders to implement its strategic plan and achieve its strategic objectives.

4. THE CONCEPTUAL FRAMEWORK

CSR reporting is developed in this study to utilize the concept of the strategy map introduced by Kaplan and Norton (2004) to create a conceptual framework for networked collaboration and regional development. The strategy map is a graphical representation of the functioning of the organization. The purpose of the strategy map is to help managers, personnel and stakeholders understand why the strategic objectives of the organization have been set and how to achieve them. The strategy map can be used to describe the strategic themes and objectives in the different perspectives and the causal relationships between those objectives (Kettunen & Kantola, 2006). The strategy map is conceptually extended in this study to a framework describing networking and regional development.

The description and communication of intangible activities, such as regional development, require a transparent framework which considers the essential elements but without the details. A conceptual framework can help the management of the institution translate the strategic objective of regional development into action. The measurement system can be developed so that the indicators describe the achievement of the strategic objective. The indicators express the development through a sequence of relationships between performance drivers or leading indicators and outcome measures or lagging indicators.

With the conceptual framework of networked collaboration and regional development, it is possible to communicate the objectives of an organization and to describe why they have been set, thus opening a dialogue among stakeholders. The outlines of regional development should be understood before the managers, personnel and partners in the network can make a significant contribution. The elements of the framework are aligned with education policy, regional strategies, and the annual budget, and action and human resources plans.

Figure 1 depicts the conceptual framework of networked collaboration and regional development in higher education, according to the perspectives of the Balanced Scorecard. The region and customer perspective includes the outcomes

Figure 1. The conceptual framework of networked collaboration and regional development

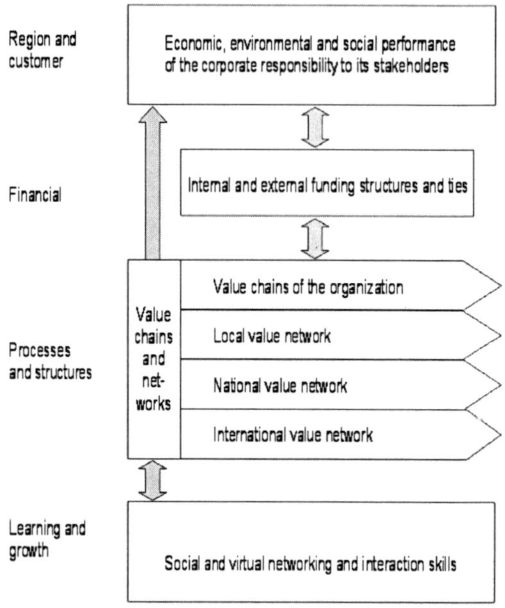

described in the corporate social responsibility report. In budgeting, the financial perspective is aligned with the processes and structures, and with the important networks of the institution to obtain external funding for the collaborative projects. The processes include the value chains of the organization and extramural local, national and international value networks. The structures include organizational units collaborating with each other and in the networks. The learning and growth perspective consists of the capabilities, abilities, work experience and organizational skills related to the value networks.

The value networks generate economic, environmental and social value added through complex dynamic exchange processes among organizations, customers, suppliers, strategic partners and the community. Strategic partners are those who are the most valuable to the future of higher education institutions (Allee, 2000). Important elements in these processes are the exchange for services, goods and revenues including knowledge that creates revenue. In addition, the exchange of strategic information, planning knowledge, process knowledge, technical know-how, collaborative design and policy development play an important role which flows around the core production processes. The third important element is the intangible benefit: the exchange of value that goes beyond the actual services and which is not accounted for traditional measures. These values and benefits include, among others, customer loyalty, image enhancement, and collaborative branding.

5. STRATEGIC NETWORKS

Since 2003, the TUAS has systematically collected quantitative data on working life collaboration using the GRI approach. The information about the collaboration with working life is used to locate geographical or branch-specific coverage in the customer network and identify potential customers. The institution developed a framework to evaluate its networked collaboration and regional development, because it was essential to

Table 2. The quality levels of networked collaboration

Quality levels	The description of the levels
1.	The collaboration with partners is random.
2.	The personnel are active in some regional networks. Collaborations with the working life exists in some local districts
3.	Regional networking is extensive in some industries and developing in others. In addition, some national and international networking is observed. Working life collaboration is strong in most of the local districts
4.	The administrative units collaborate in the essential networks. The collaboration covers most of the degree programs and the region. Some of the networks are national and others are international.
5.	All the essential networks are covered. Networks are mainly national and international. Working life collaboration covers the whole province.

know who the important partners are and where they are located.

In recent years, the faculty has reported more than 8000 collaboration contacts each year. The contacts were related, for instance, to work placements (54%), bachelor's theses (11%), research and development (5%), service (9%) and education (9%). The institution collaborated with nearly 3000 organizations. Collaboration was mostly connected with the practical training of students, supervision of their bachelor's theses, and other projects. Fifty-one percent of collaboration partners operate in the public sector, 36% in business and 13% in the tertiary sector.

The partners of the TUAS are located in the more than 170 Finnish municipalities or local regions, which cover more than one-third of the country's municipalities. The majority of the partners are located in Southwest Finland, mainly in cities in which the TUAS has a campus or branch. In addition, the TUAS has more than 250 international collaboration partners in 40 countries. In many cases, these partnerships resulted from networked projects funded by the European Union or other bodies.

The administrative units at the TUAS can use quantitative data to evaluate and develop their networked collaboration. The regional development unit of the institution has collected data and analyzed the networked collaboration. The stakeholder classification of the GRI approach has been used as a basis for the analysis. Because the GRI scheme was designed for private organizations, the TUAS had to make some interpretations and modifications before applying it.

Table 2 depicts the quality of networked collaboration at the TUAS. The purpose of the quality assurance is to ensure that the strategic objective of regional development can met. Each year, the organizational units evaluate their networking activities using the quality levels of networked collaboration. The evaluations are finally agreed in the internal target discussions between the top management of the institution and each organizational unit and then saved in the management information system. If necessary, the action plans are updated to make sure that the objectives can be achieved.

Table 3 describes the institution's strategic regional networks. The networks have been classified according to the perspectives of the Balanced Scorecard and the strategic objectives. The quality levels have been evaluated using GRI's stakeholder classification. The evaluation shows that the networks related to the different perspectives are balanced, and that the collaboration with other universities is strong. In addition, the institution is able to take advantage of the financial opportunities provided by the European Union. Nevertheless, there is still plenty of room for improvement especially in the recruitment of potential applicants, collaboration with the supply

Table 3. The strategic regional networks of the TUAS

Perspectives of the Balanced Scorecard	Networks	Quality Levels
Region and Customer	Potential Applicants	2
	Students and Student Unions	3
	Alumni	4
	Companies and Industry Clusters	4
	Networks of Universities	5
	Networks of the City of Turku	4
	Regional Development Bodies	3
	County Administrative Board	3
	Supply Chains of Products and Services	2
Financial	Ministry of Education and Central Government	4
	Elected Officials of the Owner of the Institution	3
	The Finnish Funding Agency for Technology and Innovation	4
	The Programs of the European Union	5
	Companies and Organizations	3
Processes and Structures	Board, Delegation and Advisory Boards of the Institution	3
	Local Development Projects and Clusters	3
	National University Networks and Learning Environments	4
	Regional Development Network of Local Universities	4
	Educational Development Programs of Ministries and European Union	4
	Personnel and Students	4
	Global Network of the Universities and Multinational Companies	4
Learning and Growth	Personnel and Personnel Training Organizations	3
	International Development Networks of Higher Education	4
	Scientific Further Education Programs and Networks	4
	Development Networks of Administration	3
	Occupational Networks of Personnel	2

chain of products and services, and the occupational networks of personnel.

Institutions of higher education typically receive many invitations to join networks. In ever case, however, it is important to determine whether or not the networks have strategic importance. On the one hand, some of the networks are necessary to survive, even though they do not have the strategic importance. On the other hand, some networks make it possible for an institution to achieve its strategic objectives. It is important to assess the value added of the networked collaboration using a systematic framework whenever the institution receives a new proposal. Otherwise the networking may waste the institution's scarce resources.

6. CONCLUSION

One of the main issues of regional impact of the higher education is networking and the strategic attitude toward the networks. Collaboration with the potential and actual partners and other stakeholders is stimulated through the networks. Ideally, the networks strengthen the ability of the higher educational institution to design its external engagement, communicate the results, and prepare for external evaluations.

Each higher education institution and its environment is unique. By using collected quantitative data, each institution and its organizational units can benchmark and analyze the strategic importance of its networks. The new framework and evaluation scheme is presented in this study based on the Balanced Scorecard and the classification of the GRI. These instruments assist the management of the higher education institution and its organizational units evaluate and develop its networked collaboration in the region.

The networks are influenced not only by the interests of organization but also by individuals. Individuals and organizations may have different motives for joining a network. Individuals have professional or career reasons for seeking collaboration; an organization will be interested in achieving its strategic objectives. Working in a regional network may also have personal value, which supports the position of individuals in the labor market. In the successful case the interests of the organization coincide, but in general it is the organization that should evaluate the importance of networks.

Networking should be analyzed and supported when it can help an organization to fulfill its mission or achieve its strategic objectives. The evaluation of networks is important for the strengthening of human capital, which is often a neglected management task even in higher education institutions. It is very important to notice that the higher education institutions are not in a position to accept all networking proposals. It is necessary for an institution to adopt a systematic approach to networking in order to support strategic and human resources management.

Although the members of networks are locally active, the network itself can be national or international. A national network promoting internationalization may enhance national competitive advantage and indirectly support the regional development. On the other hand, some effects become apparent only in the long run. On the other hand, higher education institutions are often forced to rather compete with each other and to with other organizations in some regional activities. This competition may interfere with or even prevent fruitful collaboration between organizations.

REFERENCES

Allee, V. (2000). Reconfiguring the value network. *The Journal of Business Strategy*, *21*(4), 36–39. doi:10.1108/eb040103

Chatterton, P., & Goddard, J. P. (2000). The response of higher education institutions to regional needs. *European Journal of Education*, *35*(4), 475–496. doi:10.1111/1467-3435.00041

Coleman, J. S. (1988). Social capital in the creation of human capital. *American Journal of Sociology*, *94*, S95–S120. doi:10.1086/228943

Cross, R., & Parker, A. (2004). *The Hidden Power of Social Networks. Understanding how work really gets done in organizations*. Boston, MA: Harvard Business School Press.

Goffman, E. (1963). *Stigma*. Englewood Cliffs, NJ: Prentice-Hall.

GRI. (2009) *Sustainability Reporting Guidelines*. Retrieved 20.07.2009 fromhttp://www.globalreporting.org.

Helmi, S. (2008). *Turun ammattikorkeakoulun yhteiskuntavastuuraportti – Käytännön tekoja Varsinais-Suomen työelämälle*. Turku: Turun ammattikorkeakoulu.

Johnson, G., & Scholes, K. (2002). *Exploring Corporate Strategy*. Englewood Cliffs, NJ: Prentice-Hall.

Kaplan, R. S., & Norton, D. P. (2001). *The Strategy-Focused Organization, How Balanced Scorecard Companies Thrive in the New Business Environment*. Boston, MA: Harvard Business School Press.

Kaplan, R. S., & Norton, D. P. (2004). *Strategy Maps, Converting the Intangible Assets into Tangible Outcomes*. Boston, MA: Harvard Business School Press.

Kettunen, J. (2004a). The strategic evaluation of regional development in higher education. *Assessment & Evaluation in Higher Education, 29*(3), 357–368. doi:10.1080/0260293042000197591

Kettunen, J. (2004b). Bridge building to the future of Finnish polytechnics. *Journal of Higher Education Outreach and Engagement, 9*(2), 43–57.

Kettunen, J. (2006). Strategic planning of regional development in higher education. *Baltic Journal of Management, 1*(3), 259–269. doi:10.1108/17465260610690917

Kettunen, J. (2008). A conceptual framework to help evaluate the quality of institutional performance. *Quality Assurance in Education, 16*(4), 322–332. doi:10.1108/09684880810906472

Kettunen, J., & Kantola, I. (2005). Management information system based on the Balanced Scorecard. *Campus-Wide Information Systems, 22*(5), 263–274. doi:10.1108/10650740510632181

Kettunen, J., & Kantola, M. (2006). The implementation of the Bologna Process. *Tertiary Education and Management, 12*(3), 257–267.

Kettunen, J., & Luoto, L. (2008). Cooperation between universities and ICT enterprises. In Zhao, F. (Ed.), *Information Technology Entrepreneurship and Innovation* (pp. 277–292). Hershey, PA: IGI Global.

Luoto, L., & Helmi, S. (2008). *Corporate social responsibility report*. Turku: Summary in English. Turun AMK.

Sandager, I., & Andersen, B. 2008. *Complexity and learning regions: learning regions as complex adaptive systems*. In Doyle, Adams, Tibbitt E Welsh (eds.), *Building stronger communities: Connecting research, policy and practice*. Plymouth, England: Latimer Trend.

Schutte, F., & van der Sijde, P. (2000). *The university and its region*. Enschede: Twente University Press.

Steiss, A. W. (2003). *Strategic management for public and nonprofit organizations*. Boca Raton, Fla: CRC Press and Taylor and Francis.

Wenger, E. (1998). *Communities of practice. Learning, meaning and identity*. Cambridge, UK: Cambridge University Press.

Chapter 14
Innovation Policy, Competitiveness, and Growth:
Towards Convergence or Divergence?

Aikaterini Kokkinou
University of Glasgow, UK

ABSTRACT

This chapter focuses on the actors of knowledge, innovation and entrepreneurship, as comparative advantages towards regional sustainable development. The chapter describes the analytical framework in which technical change, innovation and entrepreneurship activities are considered as among the most important factors influencing sustainable economic development, both in national and regional level. Within this analytical framework, the chapter aims to provide a clear understudying of the main related questions summarized in the following: To which extent economic development depends on the presence of a network favorable to knowledge, innovation and entrepreneurship, based on the endogenous development capabilities? How could the level of regional innovation activities be enhanced? How could the effect of innovation activities on regional development be estimated? What is the relative position of a region as far as development and innovation competitiveness are concerned?

INTRODUCTION

During past decades, major improvements in productivity and development have been interpreted as a movement towards a knowledge-based economy. In today's globalised markets new businesses and corporations have emerged, trying intensively to find new investment opportunities and new channels for their products.

DOI: 10.4018/978-1-61692-846-9.ch014

Significant changes have taken place and firms all over the world are described taking part in a race seeking the most appropriate and effective ways that could provide them with the strengths and opportunities necessary to obtain and sustain a competitive advantage over their rivals. Currently, output and employment are expanding fast in high-technology industries such as computers and electronics, as well as in knowledge-based services such as financial and other business services and more resources are spent on the production and

Copyright © 2011, IGI Global. Copying or distributing in print or electronic forms without written permission of IGI Global is prohibited.

development of new technologies, in particular on information and communication technology. Consequently, globalization and worldwide competition has shifted the comparative advantage of countries and regions towards the factor of knowledge and innovation, where productivity based on endogenous development capabilities plays a rather important role, as far as growth and competitiveness enhancement are concerned.

Regarding this development race, productivity enhancement is a major element towards economic growth and development. Economies increase their productivity in two ways—micro and macro. Microeconomic gains take place within an enterprise as it invests, trains workers, innovate and compete. Macroeconomic gains occur when the overall economy reorganizes and shifts resources so they produce more than before. Within this micro and macro framework, productivity has always played a leading role in raising economic growth and development, by boosting output, improving quality, and saving time and other resources. As companies and workers become more efficient, the economy reallocates resources to more productive uses, either in existing companies or new ones. As the market recycles workers and other resources, the economy grows. The payoff from productivity growth could be summarised in higher GDP, more leisure time, better working conditions, new and improved products, more variety, greater safety and security, as well as cleaner environment.

Figure 1 represents these linkages among the economic and entrepreneurial practices aiming to achieve and sustain positive results towards development and growth. The combination of the three practices target to the enhancement and convergence as far as productivity, competitiveness and development process are concerned.

Within this framework, development and innovation consist two of the core subjects both in economic and political debates and analyses. Respectively, in the modern economy era, there is an increasing interest in the contribution of knowledge in the sustainable long-term economic growth, taking into consideration the need that the modern capitalistic competition forces the enterprises to import technological innovations, that increase the productivity, or to adopt other strategies of organization for the growth of their activities and their purchases.

INNOVATION AND SUSTAINABLE GROWTH: AN INSEPARABLE RELATIONSHIP

The developments in the theory of economic growth have renewed the interest for the role that the innovation plays in the development process, underlining the interaction between the investment in innovative activities, the technological change and sustainable economic growth. The base for this analysis was provided by the main sources of economic growth:

- Increase in the productive base in order to increases the productive possibilities of the economy within a time period (as, for example though increases in total work force or Gross Fixed Capital formation)
- economies of scale that are related with increase in the factors of production
- technological progress

From an economic analysis point of view, the theoretical framework of the effects of innovation on the economic efficiency, productivity and growth is based on endogenous growth theory developed by Solow, 1957, Arrow, 1962, Romer 1986 and 1990, Lucas, 1990 and 1993. Endogenous growth theory claimed that not only the accumulation of capital, but mainly the development and accumulation of knowledge and technological change leads to increased and sustainable growth.

Endogenous growth theory, as represented by Romer (1986), takes innovation as an endogenous variable which can explain the different national growth rates and why economies, even with dif-

Figure 1. Linkages among economic and entrepreneurial practices

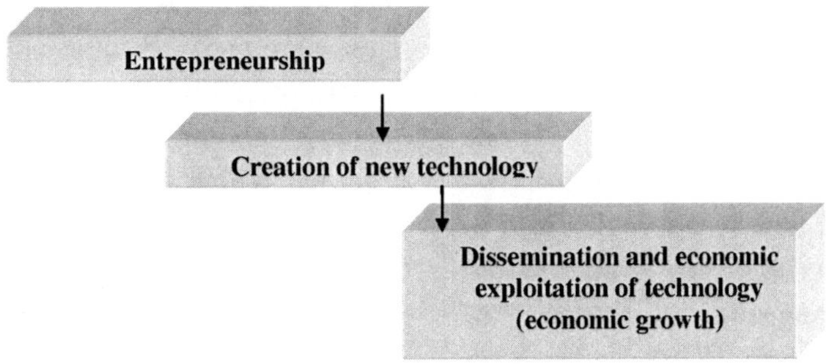

ferent rates, do not converge to long-run steady state equilibrium. The reason is that the long-run productivity decrease is avoided, due to capital accumulation through the qualitative-technological improvements of natural and human capital. According to Romer (1986, 1990), knowledge and technological progress are the main engines of economic dynamism and the economy grows endogenously through the accumulation and spillover of knowledge. Growth rate depends on the amount of technological activity within the economy and on the ability of the economy to exploit external technological achievements (Martin and Ottaviano, 1999, Grossman and Helpman, 1994, Coe and Helpman, 1995). Increasing returns and technical change are incorporated within the production function as determinants of the endogenous growth rate (Romer 1986, Lucas 1988, Grossman and Helpman 1994, Barro and Sala-i-Martin, 1997) and economic growth is sustained because of the continuous creation and diffusion of knowledge.

An important contribution of the endogenous growth theory (Romer, 1987 and 1990) has been to identify the central role that knowledge and knowledge spillovers play in creating and sustaining growth. Pavitt and Soete (1982) examined growth as a result of the development of new knowledge in a country and the diffusion of knowledge between countries. According to Fagerberg (1987) there is a close relation between a country's economic and technological level of development. The rate of economic growth of a country is positively influenced by technological level of the country and its ability to increase it through imitation and exploitation of the possibilities offered by technological achievements elsewhere. Krugman (1991) identified the major role that knowledge spillovers play in generating increasing returns and higher growth. Geroski and Machin (1993) asserted that innovations positively affect the development of enterprises and economies. Moreover, according to Silverberg and Verspagen (1995), technological change and diffusion constitute important factors in long-run macroeconomic growth and development. Moreover, Barro and Sala-i-Martin (1995 and 1997) asserted that growth rate may increase in correlation with technological growth. Furthermore, Freeman and Soete (1997) focused on the importance of technology and innovation claiming that lack of innovation leads to economic death. At the same point of view, Sternberg (2000) said that in industrialized economies the rate of long-term macroeconomic growth depends on the ability of constant development of innovative products and processes.

The economic processes that create and diffuse the new knowledge are critical in the development process and there are powerful contacts between the investment in the human capital, the technological change and finally, economic growth (Acs, Anselin and Varga, 2002). As a motive force, it prompts the enterprises to long-term development objectives and the advancement of productive structures, so that they maintain the elements of growth, competitiveness and employment. Investments in new technologies aim to the modernisation of productive process and the qualitative upgrade of products, which is one from the basic factors of increase of enterprises. The reason is that the new technologies lead to increase of productivity of factors of production, contributing in the long-term improvement of competitiveness (Griliches, 1980). Technology, also, contributes in the growth of economy, on the one hand because the new or improved products that result from innovations improve the level of existence, and on the other hand, because, with regard to the international trade, the record of open economy depends also from the propensity to innovativeness (Fagerberg, 1988). One additional reason is that through innovation the individual and collective needs are satisfied better which constitutes fundamental element of entrepreneurial spirit. The same holds also for countries and economies, which in order to maintain the elements of growth, competitiveness and employment, owe to change fast the new ideas in technical and commercial successes.

INNOVATION AND ENTREPRENEURSHIP: ANOTHER INSEPARABLE RELATIONSHIP

As far as the characteristics of innovation are concerned, it is influenced by many factors, both internal and external to the firm (Dosi 1988). Innovation refers to the collective learning process between several departments within a company, as well as to external collaborations with external bodies (Cooke et al. 2000) and it is characterized by two features: uncertainty (risks) and accumulation (Camagni 1991). What is more, as far as the innovation process is concerned, the necessary information is asymmetrically available, which makes it rather difficult and costly to collect and exploit it. Moreover, the necessary inputs, as well as their attributes are difficult to be defined and, as a consequence, it becomes, also difficult to evaluate their potential effects and results. Furthermore, innovation process requires cooperation and collaboration of a great number of different actors, which, to a large extent, incorporates high transaction cost and high uncertainty level. Because of these qualities of knowledge, namely uncertainty, asymmetries and high transactions cost – entrepreneurship becomes more important in a knowledge-based economy, since entrepreneurship activities are closely related with uncertainty, risk, investment, return and profits and of course with innovative actions.

Since early, differences in economic success have been related to the presence or lack of entrepreneurial activities, as processes of exploiting opportunities that exist in the environment or that are created through innovation in order to create value (Wennekers and Thurik, 1999, Drucker, 1985, Mueller and Thomas, 2000, Ulijn and Weggeman, 2001). Entrepreneurship refers to activities undertaken in order to convert ideas into economic opportunities. These activities include factors such as opportunity seeking, risk and uncertainty bearing, innovativeness, coordination, capital supply, decision making, ownership and resource allocation. Then, entrepreneurship focuses on creating the adequate economic opportunities in order to introduce new ideas in the market. In accordance to these characteristics, entrepreneurship could be mainly considered to be the exploitation of technological opportunities by profit seeking agents, process which actually leads to economic growth and development.

More specifically, according to Penrose (1959), entrepreneurs are important for the growth of firms since they provide the vision and imagination necessary to carry out opportunistic expansion. Acs – Audretsch (1989) claimed that entrepreneurship generates innovations. Shane (1992, 1993) has related cultural norms to levels of innovation, which he assumes would precede economic development. In generally, entrepreneurship and innovation activity can be seen as key factors to promoting growth and increasing productivity. According to this view, economic success and competitiveness result from the combination of favorable entrepreneurial environment, network systems and innovative behavior and the establishment of new combinations of factors of production is a process that will become the engine that drives economic development (Schumpeter 1934, Schumpeter 1942, Thurik and Wennekers 1999).

On the other hand, as mentioned before, due to information asymmetries, uncertainty and high cost features of innovation, entrepreneurship becomes more important in a modern economy, since it may provide one of the mechanisms by which new economic knowledge is disseminated into different networks. Entrepreneurship generates growth because it serves as a link between innovation and change. Thus, by serving as a vehicle for knowledge transmission and spillover, entrepreneurship plays a key role in the link between knowledge and growth.

Moreover, economic success depends a great deal on the quality of the internal innovation network within an economy and the collective learning process is seen as being extremely important for the quality of the innovation network (Harmaakorpi and Pekkarinen, 2002). The relationship between entrepreneurial culture and economic growth is considered to be rather strong and entrepreneurial economies are more innovative and subsequently grow faster (Beugelsdijk, 2004).

Technology and innovation play an important role in the creation of wealth and economic growth and technology has become one of the most important factors in the models of growth (Geroski and Machin, 1993, Barro and Sala-i-Martin, 1995, 1997, Freeman and Soete, 1997, and Sternberg, 2000). The role of innovation is multiple: as motive force it directs the enterprises to ambitious and long-term objectives, it leads to the renewal of methods of production, supply and distribution, and management and marketing, as well as industrial structures and the appearance of new sectors of economic activity, achieving a wider spectrum of products and services, as well as relative markets. Inputs affect the intermediate inputs, which consequently affect and define the productivity and competitiveness level. Technological change, innovation and technology creation and diffusion are an important factor to economic progress.

Furthermore, innovative actions stimulate investments which introduce new commodities and processes, which improve the living standards of the society. Moreover, they lead to new developments, which increase the comparative advantage of an economy and affect positively the trade performance and competitiveness of a country worldwide. These effects result in a greater level of economic growth. On the other hand, innovation is rather important to an individual firm for two main elements, namely a double role in the incentives of the companies to pursuit and invest on it. Firstly, a corporation, which undertakes R&D programmes, acquires new information and knowledge to embody in the new commodities, as well as new production and marketing processes, ready to be employed in product and process innovation. As a result, through innovation, a company is able to develop directly new products and processes and bring them to the market acquiring an advantage over its competitors. Furthermore, it can enhance the ability of the firm to develop and maintain capabilities to absorb and expand technology information available by external sources, and identify, assimilate and exploit new knowledge and technology produced elsewhere (Cohen and Levinthal, 1989).

Innovative actions are considered to be rather important to economic growth, development and welfare. Firstly, they stimulate investments which introduce new commodities and processes, which improve the living standards of the society. Moreover, they lead to new developments, which increase the comparative advantage of an economy and affect positively the trade performance and competitiveness of a country worldwide. These effects result in a greater level of economic growth. While innovation may lead to divergence between firms or nations, imitation through diffusion and dissemination tends to erode differences in technological competencies, and hence lead to convergence (Fagerberg and Verspagen, 2002). On the other hand, combining the production functions in order to create and disseminate innovations leads to improvements in productivity and economic development (Malecki and Varaia 1986; Malecki 1991, Fagerberg and Verspagen, 2002). The economic processes that create and diffuse the new knowledge are critical in the development process and there are powerful contacts between the investment in the human capital, the technological change and finally economic growth (Acs, Anselin and Varga, 2002). The reason is that the new technologies lead to increase of productivity of factors of production, contributing in the long-term improvement of competitiveness (Griliches, 1980). Technology, also, contributes in the growth of economy, on the one hand because the new or improved products that result from innovations improve the level of existence, and on the other hand, because, with regard to the international trade, the record of open economy depends also from the propensity to innovativeness (Fagerberg, 1988).

INNOVATION SYSTEMS: A MAJOR TOOL FOR REGIONAL CONVERGENCE... OR DIVERGENCE

As it has been strongly declared in the Third Report on Economic and Social Cohesion (2004), strengthening national competitiveness throughout the Union will boost the growth potential of the European Union economy as a whole. And, by securing a more balanced spread of economic activity across the Union, it will reduce the risk of imbalances and divergence, making it easier to sustain the European model of economy and society. In policy terms, the objective is to help achieve a more balanced development by reducing existing disparities, avoiding regional imbalances, by making policies more coherent, improving integration and encouraging cooperation between states and regions.

However, in the modern knowledge economy, growth depends extensively on the presence or the formation of a network and environment favorable to innovation, which is based on the endogenous development capabilities. Even though the firm-specific factors are important determinants of innovation activity, technological opportunities and favorable entrepreneurial environment have a positive effect on innovation activity, as well. As a consequence, among other development factors, countries or regions facing lags in innovative activities face also a lag in growth.

Regional development and convergence of the lagging regions in European Union has been one of the main objectives of the European development strategy. Development problems are more intense in lagging regions which present major differences in level of prosperity, economic performance, output, productivity and employment, compared to other European Union regions. These disparities arise due to structural deficiencies in factors, which restrain economic activities and overall development. The imbalances in the European Union, threatening the convergence path, are summarised in Table 1.

These territorial disparities cannot be ignored, since they affect the overall competitiveness of the European Union economy. Covering costs of congestion or treating the social consequences of disparities implies a sub-optimal allocation of resources, as well as a lower level of efficiency and economic competitiveness than could potentially be attained in the regions affected. To combat territorial disparities and achieve a more spatially balanced pattern of economic development requires some coordination of development policies if they are to be coherent and consistent with each other.

Table 1. Threats to E.U. regional convergence

Regional level	Threatens
at EU level	high concentration of economic activity and population in the central metropolitan areas, which account for the major percentage of population, GDP and R&D expenditure.
at national level	persistence of pronounced imbalances between the main metropolitan areas and the rest of the country in terms of economic development.
at regional level	persistence of territorial disparities beyond those measured by GDP or unemployment, such as, social exclusion, inadequate economic links and falling population.
within regions and cities	development of poverty and social exclusion in areas with often only limited availability of essential services.
in areas constrained by geographical features (islands, sparsely populated areas)	declining population and ageing, while accessibility continues to be a problem and the environment remains fragile and threatened.
in outermost areas, with natural and geographical handicaps	continuation of severe social and economic problems which are difficult to tackle because of their remoteness, isolation, topological features, climate, small size of market and dependence on a small number of products.

Figure 2. Economic and social regional E.U. policy

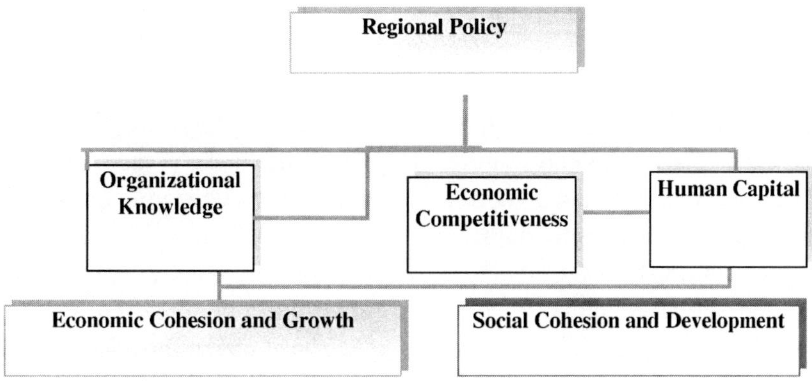

Within this framework, the European Union has defined a comprehensive strategy aimed at long term economic growth, full employment, social cohesion and sustainable development in a knowledge based society. Enhancement and convergence of growth and productivity are a major topic in the economic and social policy agenda. The framework of these policy objectives could be illustrated in Figure 2.

European cohesion policy makes a major contribution to these objectives, especially in those regions where there is unused economic and employment potential which can be realized through targeted cohesion policy measures, so adding to the growth of the EU economy's a whole, implementing an overall scheme, as presented in Table 2.

From a policy perspective, for regional development to be sustained requires favourable conditions being established at the national level, in particular a macroeconomic environment conducive to growth, employment and stability and a tax and regulatory system which encourages business and job creation. At the regional level, two complimentary sets of conditions need to be satisfied. The first is the existence of suitable

Table 2. Productivity and competitiveness enhancement framework

1st phase - Inputs (Productivity enhancement)	**Macroeconomic, entrepreneurial and work environment**
	Economic and technological infrastructure
	Education and skills
	Entrepreneurship and business development
	Innovativeness and creativity
2nd phase - Intermediate output (Productivity enhancement)	Productivity
	Production factors cost
	Prices and wages
3rd phase - Final output (Competitiveness enhancement)	Development
	Employment
	Living standards
	Quality of life
	Competitiveness

endowment of both basic infrastructure (in the form of efficient transport, telecommunications and energy networks, good water supplies and environmental facilities and so on) and a labour force with appropriate levels of skills and training, strengthening of both physical and human capital, together with improvements in institutional support facilities and the administrative framework in place. The second set of conditions, which directly relates to the factors of regional competitiveness which are important in the knowledge-based economy, is that innovation should be accorded high priority, that information and communication technologies (ICT) should be widely accessible and used effectively and that development should be sustainable in environmental terms.; a business culture which encourages entrepreneurship; and the existence of co-operation networks and clusters of particular activities.

EUROPEAN UNION REGIONAL INNOVATION POLICY OBJECTIVES

European Union, through Lisbon Summit (2000), set itself the goal of becoming the most competitive and dynamic knowledge-based economy in the world, capable of sustainable economic growth and closer regional, as well as social cohesion. At the Lisbon European Council, European Union defined a comprehensive strategy aiming at long term economic growth, full employment, social cohesion and sustainable development in a knowledge-based society. Into doing, it has identified a number of priorities. (see Table 3)

These changing conditions have imposed a great number of challenges to economies. In this perception, science, technology and innovation are major elements towards economic growth and development.

At the national level, two complimentary sets of conditions need to be satisfied. The first is the existence of suitable endowment of both basic infrastructure (in the form of efficient transport, telecommunications and energy networks, good water supplies and environmental facilities and so on) and a labor force with appropriate levels of skills and training, strengthening of both physical and human capital, together with improvements in institutional support facilities and the administrative framework in place. The second set of conditions, which directly relates to the factors of regional competitiveness which are important in

Table 3. Economic development priorities

Priority	Means and actions
give priority to innovation and entrepreneurship	creating closer links between research institutes and industry, developing conditions favourable to R&D, improving access to finance and know-how and encouraging new business ventures;
ensure full employment	by emphasizing the need to open up employment opportunities, to increase productivity and quality at work and to promote lifelong learning;
ensure an inclusive labour market	unemployment is reduced and social and regional disparities in access to employment are narrowed;
'connect' Europe	closer integration and by improving transport, telecommunications and energy networks;
protect the environment	stimulating innovation and introducing new technologies, for example, in energy and transport.

the knowledge-based economy, is that innovation should be accorded high priority, that information and communication technologies (ICT) should be widely accessible and used effectively and that development should be sustainable in environmental terms.; a business culture which encourages entrepreneurship; and the existence of cooperation networks and clusters of particular activities.

European cohesion policy makes a major contribution to these objectives, especially in those countries where there is unused economic and employment potential which can be realized through targeted cohesion policy measures. From a policy perspective, for national development to be sustained requires favorable conditions being established at the national level, in particular a macroeconomic environment conducive to growth, employment and stability and a tax and regulatory system which encourages business and job creation.

Community enterprise, industrial and innovation policy is aimed at strengthening the competitiveness of European Union producers by promoting competition, ensuring access to markets and establishing an environment which is conducive to R&D across the Union. As is recognized, a lack of innovative capacity at regional level stems not only from deficiencies in the research base and low levels of R&D expenditure but also from weaknesses in the links between research centres and businesses, and slow take-up of information and communication technologies. Knowledge and access to it has become the driving force for growth in advanced economies like the European Union known-how and intellectual capital, much more than natural resources or the ability to exploit abundant low-cost labor, have become the major determinants of economic competitiveness since it is through these that economies can not only increase their productive efficiency but also develop new products. Innovation, therefore, holds the key to maintaining and strengthening competitiveness which in turn inessential for achieving sustained economic development. To achieve both sets of conditions requires an effective institutional and administrative framework to support development. The cost of not pursuing a vigorous cohesion policy to tackle disparities is, therefore, measured in economic terms, as a loss of the potential real income and higher living standards. Given the interdependencies inherent in an integrated economy, these losses are not confined to the less competitive states but affect every state in the Union.

Information and communications technology (ICT) has brought both new opportunities and challenges for businesses and represents a new factor of regional competitiveness. For regions, ICT has increased the pace of change with potentially profound effects on living and working conditions and on the territorial distribution of economic activity. From a cohesion perspective, ICT seems to offer a major opportunity for reducing the 'friction of distance and the problems of remoteness which many peripheral regions — and even more, outermost areas —suffer from. At the same time, however, there is growing concern over the territorial dimension of the so called 'digital divide' and a fear that restrictions on access to ICT networks or limitations in the ability of enterprises and households to use the new technology could serve to widen rather then narrow disparities in regional performance. Although the pattern of development of different aspects of ICT varies, a number of regional disparities are already evident, taking into account that there is a north-south divide in the present EU in the development of most of the new technologies, which is broadly to a divide between cohesion and non-Cohesion countries.

Summarising, development and innovation consist two of the core subjects both in economic and political analyses. In European Union there is an increasing interest in the contribution of knowledge in the sustainable long-term economic growth, taking into consideration the need that competition forces technological innovations, that increase productivity, renewed the interest for the role of innovation in the development process, underlining the interaction between investment in innovative activities, technological change and sustainable economic growth.

PROSPECTS

As it has been asserted above, globalization and worldwide competition has shifted the comparative advantage of corporations and economies towards the factor of knowledge and innovation, where entrepreneurship based on the endogenous development capabilities plays a rather important role, as far as the growth, productivity and competitiveness enhancement are concerned. In order to promote innovation activities and technological opportunities entrepreneurship enhancement seems to have a significant importance not only to business success, but also to the long run performance of the economy as a whole. Under this perspective, growth policies should focus on creating favorable environment for the co-operation between firms and institutions that support the development and exploitation of knowledge and innovation. Furthermore, policies should promote the entrepreneurial relations between firms and institutions, fostering the development and dissemination of the expertise, the mobility of human and physical capital and the enhancement of the relationships between business and research entities. Specifically, they should encourage actions such as, promoting innovation, technology transfer and interactions between firms and higher education and research institutes, networking and industrial co-operation and support for research and technology supply infrastructure.

As it has already been mentioned, innovation and technology is an important source of regional competitiveness through facilitating cooperation between the various parties involved in both the public and private sectors. In particular, they can improve collective processes of learning and the creation, transfer and diffusion of knowledge and transfer, which are critical for innovation. Such cooperation and the networks that are formed help to translate knowledge into economic opportunity, while at the same time building the relationships between people and organizations which can act as a catalyst for innovation. Such actions should extend to all the policy areas relevant for economic, scientific and social development and should ideally establish a long-term policy horizon.

This, however, needs to happen not just in central parts where productivity and employment are highest and innovative capacity most developed but throughout the Union. Countries and regions need assistance in overcoming their structural deficiencies and in developing their comparative advantages. This means, among others, that encouraging the development of knowledge-based economic activities and innovation and that particular attention needs to be given to:

- developing new innovation promotion policies which focus much more on the provi-

sion of collective business and technology services to groups of firms which can affect their innovative behaviour, rather than direct grants to individual firms which tend only to reduce costs temporarily.
- developing new policies to strengthen the capacity of SMEs to innovate through business networks and clusters and improving their links with the knowledge base, including with universities and research centres.
- encouraging the development of the indigenous R&D potential of weaker regions and their capacity to adapt technological advances made elsewhere to local circumstances and needs.
- facilitating access of researchers, businesses and others in less favoured regions to international networks of excellence, sources of new technology and potential R&D partners.

These conditions are largely related to economic competitiveness and include, among others, the capacity of a regional economy to generate, diffuse and utilize knowledge and so maintain an effective regional innovation system.

Furthermore, policies should promote the entrepreneurial relations between firms and institutions, fostering the development and dissemination of the expertise, the mobility of human and physical capital and the enhancement of the relationships between business and research entities. Specifically, they should encourage actions such as, promoting innovation, technology transfer and interactions between firms and higher education and research institutes, networking and industrial co-operation and support for research and technology supply infrastructure. Such cooperation and the networks that are formed help to translate knowledge into economic opportunity, while at the same time building the relationships between people and organizations which can act as a catalyst for innovation. Such actions should extend to all the policy areas relevant for economic, scientific and social development and should ideally establish a long-term policy horizon. Under this perspective, growth policies should focus on creating favorable environment for the co-operation between firms and institutions that support the development and exploitation of knowledge and innovation. Furthermore, policies should promote the entrepreneurial relations between firms and institutions, fostering the development and dissemination of the expertise, the mobility of human and physical capital and the enhancement of the relationships between business and research entities. Specifically, they should encourage actions such as, promoting innovation, technology transfer and interactions between firms and higher education and research institutes, networking and industrial co-operation and support for research and technology supply infrastructure.

These conditions are largely related to economic competitiveness and include, among others, the capacity of a regional economy to generate, diffuse and utilize knowledge and so maintain an effective regional innovation system, contributing into a sustainable economic growth path.

REFERENCES

Abramovitz, M. (1986). Catching-up, foreign ahead and falling behind. *The Journal of Economic History*, 46.

Acs, Z. J., Anselin, L., & Varga, A. (2002). Patents and innovation counts as measures of regional production of new knowledge. *Research Policy*, *31*, 1069–1085. doi:10.1016/S0048-7333(01)00184-6

Acs, Z. J., & Audretsch, D. B. (1989). Patents as a measure of innovative activity. *Kyklos*, *42*(2), 171–180. doi:10.1111/j.1467-6435.1989.tb00186.x

Aghion, P., & Howitt, P. (1992). A model of growth through creative destruction. *Econometrica, 60*(2), 323–351. doi:10.2307/2951599

Arrow, K. J. (1962). The economic implications of learning by doing. *The Review of Economic Studies, 29*(3), 155–173. doi:10.2307/2295952

Audretsch, D. B. &Thurik, R. (2001). Linking entrepreneurship to growth. *OECD, STI Working Papers* 2001/2, DSTI/DOC (2001).

Barro, R., & Sala-i-Martin, X. (1997). Technological diffusion, convergence and growth. *Journal of Economic Growth, 2*, 1–26. doi:10.1023/A:1009746629269

Barro, R. J., & Sala-i-Martin, X. (1995). *Economic Growth*. New York: McGraw-Hill.

Beugelsdijk, S. (2004). Entrepreneurial Culture, Regional Innovativeness and Economic Growth, *Proceedings of ERSA 2004 Conference*, Porto

Camagni, R. (Ed.). (1991). *Innovation Networks—Spatial Perspectives*. London: Belhaven Press.

Coe, D., & Helpman, E. (1995). International R&D spillovers. *European Economic Review, 39*, 859–887. doi:10.1016/0014-2921(94)00100-E

Cohen, W. M., & Levinthal, D. A. (1989). Innovation and learning: the two faces of R&D. *The Economic Journal, 99*, 569–596. doi:10.2307/2233763

Cooke, P., Boekholt, P., & Todtling, F. (2000). *The Governance of Innovation in Europe: Regional Perspectives on Global Competitiveness*. London: Pinter.

Dosi, G. (1988). Sources, procedures, and microeconomic effects of innovation. *Journal of Economic Literature, XXVI*, 1120–1171.

Drucker, P. (1985). *Innovation and Entrepreneurship*. New York: Harper and Row.

Englander, S., & Gurney, A. (1994). Medium-term determinants of OECD productivity. *OECD Economic Studies, 22*, 49–109.

European Commission. *Union, Regional Policy*. Inforegio, Retrieved from http://europa.eu.int European Union.Retrieved from http://www.europa.eu.int

European Union (1992). *Treaty of the European Union.*

European Union. (2006*). Regional Innovation Scoreboard 2006*.Retrieved from http://www.proinno-europe.eu/inno-metrics.html

Eurostat.Retrieved from http://www.europa.eu.int/eurostat

Fagerberg, J. (1987). A technology gap approach to why growth rates differ. *Research Policy, 16*, 87–99. doi:10.1016/0048-7333(87)90025-4

Fagerberg, J. (1988). a. International competitiveness. *The Economic Journal, 98*, 355–374. doi:10.2307/2233372

Fagerberg, J. (1988) b. Why growth rates differ. In Dosi, G., Freeman, C., Nelson, R.R., Silverberg, G., Soete, L. (Eds.), *Technical Change and Economic Theory (pp.432-457),*London: Pinter, London.

Fagerberg, J., & Verspagen, B. (2002). Technology-gaps, innovation-diffusion and transformation: an evolutionary interpretation. *Research Policy, 31*, 1291–1304. doi:10.1016/S0048-7333(02)00064-1

Freeman, C., & Soete, L. (1997). *The Economics of Industrial Innovation*, 3rd EditionLondon:Pinter.

Geroski, P., Machin, S., Van, R., & Geroski, J. (1993). Innovation and profitability. *The Rand Journal of Economics, 24*(2), 198–211. doi:10.2307/2555757

Gordon, R. (2000). Does the New Economy measure up to the great inventions of the past? *The Journal of Economic Perspectives*, *14*, 49–74. doi:10.1257/jep.14.4.49

Gordon, R. (2003). *Hi - tech innovation and future productivity growth: Does supply create its own demands?* Manuscript, Northwestern University.

Griliches, Z. (1980). R&D and the productivity slow down. *The American Economic Review*, *70*, 2.

Grossman, G. M., & Helpman, E. (Eds.). (1991). *Innovation and Growth in the Global Economy*. Cambridge, MA: MIT Press.

Grossman, G. M., & Helpman, E. (1994). Endogenous innovation in the theory of growth. *The Journal of Economic Perspectives*, *8*, 23–41.

Harmaakorpi, V., & Pekkarinen, S. (2002). Regional Development Platform Analysis as a tool for regional innovation policy. *Proceedings of the ERSA 2002 Conference*, Dortmund, Germany

Hollanders, H. (2006). *European Regional Innovation Scoreboard, Past, Present and Future, 2nd MLP Regional Benchmarking Workshop*. MERIT – Maastricht University.

Jones, L. E., & Manuelli, R. (1990). A convex model of equilibrium growth: Theory and policy implications. *The Journal of Political Economy*, *98*, 1008–1038. doi:10.1086/261717

Jorde, T. M., & Teece, D. J. (1990). Innovation and cooperation: Implications for competition and antitrust. *The Journal of Economic Perspectives*, *4*(3), 75–96.

Jorgenson, D. W., Gollop, F. M., & Fraumeni, B. M. (1987). *Productivity and US Economic Growth*. Cambridge, MA: Harvard University Press.

Jorgenson, D. W., & Stiroh, K. (2000). Raising the Speed Limit: US Economic Growth in the Information Age. *Brookings Papers on Economic Activity*, (1): 125–211. doi:10.1353/eca.2000.0008

Kedrova, J. (2004). *Measuring Productivity*, Federal Reserve Bank of Dallas Expand Your Insight, June 2004, Free Enterprise, Measures of Productivity, Krugman, P. (1991). Increasing returns and economic geography. *The Journal of Political Economy*, *99*, 483–499.

Lucas, R. E. (1988). On the Mechanics of Economic Development. *Journal of Monetary Economics*, *22*, 3–42. doi:10.1016/0304-3932(88)90168-7

Lucas, R. E. (1990). Why doesn't capital flow from rich to poor countries. *The American Economic Review*, *80*, 92–96.

Lucas, R. E. (1993). Making a miracle. *Econometrica*, *61*, 251–272. doi:10.2307/2951551

Maddison, A. (2001). *The World Economy: A Millennial Perspective. Development Centre Studies*. Paris: OECD.

Malecki, E. J. (1991). *Technology and economic development: the dynamics of local regional and national change*, (eds.) Longman Scientific and Technical.

Malecki, E. J., & Varaia, P. (1986). Innovation and Changes in Regional Structure. In Nijkamp, P.Elsevier Science Publishers (Ed.), *Handbook of Regional and Urban Economics* (*Vol. I*).

Mansfield, E. (1968). *Industrial Research and Technological Innovation: An Econometric Analysis*. New York: W.W. Norton and Co.

Martin, P., & Ottaviano, G. I. P. (1999). Growing locations: Industry location in a model of endogenous growth. *European Economic Review*, *43*, 281–302. doi:10.1016/S0014-2921(98)00031-2

OECD. (2003). *The Sources of Economic Growth in OECD Countries*. Paris: OECD.

OECD. (2005). [Paris.]. *Economic Outlook*, 76.

OECD. (2005). *Economy - wide Indicators of Productivity*, Retrieved from www.oecd.org

OECD. (2005). *STAN Indicators Database*, Retrieved from www.oecd.org

OECD. (2005). *Productivity Database*, Retrieved from www.oecd.org

Oliner, S. D., & Sichel, D. E. (2001). The resurgence of growth in late 1990s: Is information technology the story? *The Journal of Economic Perspectives*, *14*, 3–22. doi:10.1257/jep.14.4.3

Pavitt, K., & Soete, L. (1982). International differences in economic growth and the international location of innovation. In Giersch, H. (Ed.), *Emerging Technologies: The Consequences for Economic Growth, Structural Change and Employment* (pp. 105–133). Tübingen: Mohr.

Penrose, E. (1959). *The theory of the growth of the firm*. Oxford: Blackwell.

Rebelo, S. (1991). Long run policy analysis and long run growth. *The Journal of Political Economy*, *99*, 500–521. doi:10.1086/261764

Romer, P. M. (1986). Increasing Returns and Long-Run Growth. *The Journal of Political Economy*, *94*, 1002–1037. doi:10.1086/261420

Romer, P. M. (1987). Growth based on increasing returns due to specialization. *The American Economic Review*, *77*(2), 56–62.

Romer, P. M. (1990). Human Capital and Growth: Theory and Evidence, *Carnegie-Rochester Conference Series on Public Policy, 32*, 251-86.

Romer, P. M. (1990). Endogenous Technological Change. *The Journal of Political Economy*, *98*, 71–102. doi:10.1086/261725

Schumpeter, J. A. (1934). *The theory of economic development*. Cambridge, MA: Harvard Economic Studies.

Schumpeter, J. A. (1942). *Capitalism, socialism and democracy*. New York: Harper.

Shane, S. (1992). Why do some societies invent more than others? *Journal of Business Venturing*, *7*, 29–46. doi:10.1016/0883-9026(92)90033-N

Shane, S. (1993). Cultural influences on national rates of innovation. *Journal of Business Venturing*, *8*, 59–73. doi:10.1016/0883-9026(93)90011-S

Sheehan, J., & Wyckoff, A. (2003). Targeting R&D: Economic and Policy Implications of Increasing R&D Spending. *STI Working Paper 2003/8*, OECD, Paris.

Silverberg, G., & Verspagen, B. (1995). Long term cyclical variations of catching up and falling behind. An evolutionary model. *Journal of Evolutionary Economics*, *5*, 209–227. doi:10.1007/BF01198304

Smith, A. (1937). *The wealth of nations*. New York: Random House.

Solow, R. (1956). A Contribution to the Theory of Economic Growth, Quarterly Journal of Economic Growth. *The Quarterly Journal of Economics*, *50*, 65–94. doi:10.2307/1884513

Solow, R. (1957). Technical change and the aggregate production function. *The Review of Economics and Statistics*, *39*, 312–320. doi:10.2307/1926047

Sternberg, R. (2000). Innovation Networks and Regional Development – Evidence from the European Regional Innovation Survey (ERIS): Theoretical Concepts, Methodological Approach, Empirical Basis and Introduction to the Theme Issue. *European Planning Studies*, *8*, 389–407. doi:10.1080/713666420

Thomas, A., & Mueller, S. (2000). A case for comparative entrepreneurship: assessing the relevance of culture. *Journal of International Business Studies*, *31*, 287–301. doi:10.1057/palgrave.jibs.8490906

Ulijn, J., & Weggeman, M. (2001). Towards an innovation culture: what are its national, corporate. marketing and engineering aspects, some experimental evidence. In Cooper, C., Cartwright, S., & Early, C. (Eds.), *Handbook of Organizational Culture and Climate* (pp. 487–517). London: Wiley.

Wennekers, S., & Thurik, R. (1999). Linking entrepreneurship and economic growth. *Small Business Economics*, *13*, 27–55. doi:10.1023/A:1008063200484

Chapter 15
Mapping the Innovation Activities in Europe

George M. Korres
University of Aegean, Greece

ABSTRACT

This chapter analyses the innovation activities and the way they contribute essentially to the regional dimension and growth. The chapter aims to investigate how technological infrastructure and innovation capabilities affect not only regional growth, but also the whole periphery and economy, as well as how technological innovation is essential if companies - and countries - are to maintain their competitiveness in global markets. Furthermore, this chapter describes the relative measures and indexes introduced by EU and OECD during the last years, concerning Research and Development Expenditures, patents etc., namely measuring innovation activities. However, there are a lot of underlying problems and questions regarding the measurement of innovation activities at a regional level. This chapter attempts to analyze the whole framework of innovation statistics and in particular to examine the measurement and also the statistical estimation of innovation activities in EU. Within this context, it is also aiming to emphasize and review the appropriate techniques, the most common methods and the particular related problems.

INTRODUCTION

The growing importance of technological change in world production and employment is one of the characteristics of the last four decades. Technological innovation is essential to maintain their competitiveness in global markets. The effectiveness of measures to stimulate or facilitate innovation depends to a very large extent on the ability of policy-makers to answer these kinds of question. Without a detailed understanding of innovation's complex processes, resources may not be directed where they will do most good. There is a huge literature suggesting and demonstrating that research and scientific indicators make an important contribution to the growth at the firm, industry and national levels. Most of these studies

DOI: 10.4018/978-1-61692-846-9.ch015

Copyright © 2011, IGI Global. Copying or distributing in print or electronic forms without written permission of IGI Global is prohibited.

have investigated the relation between productivity, employment, growth and R&D.

The first important point is to distinguish between diffusion and the adoption of technology. In the analysis of adoption one considers the decisions taken by agents to incorporate a new technology into their activities. This paper investigates the mechanisms of knowledge economy. In addition, it examines the national system of innovations within the framework of the European technological policy and its effects to the member states.

LEADING INDICATORS OF SCIENTIFIC AND RESEARCH MEASURES

The main expenditure aggregate used for international comparison is gross domestic expenditure on R&D (GERD), which covers all expenditures for R&D performed on national territory in a given year. It thus includes domestically performed R&D which is financed from abroad but excludes R&D funds paid abroad, notably to international agencies. The corresponding personnel measure does not have a special name. It covers total personnel working on R&D (in FTE, full-time-equivalent) on national territory during a given year. International comparisons are sometimes restricted to researchers (or university graduates) because it is considered that they are the true core of the R&D system, (Bienaymé, A. 1986).

The use of research and technological data implied a lot of problems with the collection and measurement. The problems of data quality and comparability are characteristic for the whole range of data on dynamic socio-economic activities. However, most of the research and technological indicators capture technological investment in small industries and in small firms only imperfectly. Usually only, the manufacturing firms with more than 10,000 employees have established some research and technological laboratories, while industrial units with less than 1,000 employees usually do not have any particular research activities. Finally, the research and technological statistics concentrate mostly on the manufacturing sectors, while usually neglecting some service activities, (Dosi G., et. al. 1988).

The collection of R&D data of regional statistics implied a lot of problems in comparison to data of national statistics. For the collection of regional statistics, we should take into the local differences and the difficulties. R&D units can operate in more than one regions and we should allocate these activities between regions. Innovation indicators measure aspects of the industrial innovation process and the resources devoted to innovation activities. They also provide qualitative and quantitative information on the factors that enhance or hinder innovation, on the impact of innovation, on the performance of the enterprise and on the diffusion of innovation, (Korres G. 1996 and 2008).

The variables common used variables for S-R&T activities are:

- R&D expenditures
- R&D personnel
- Patents of New Technologies.

Technological product and process (TPP) innovations comprise implemented technologically new products and processes and significant technological improvements in products and processes. The term product innovation on the one hand includes technologically new products whose technological characteristics or intended uses differ significantly from those of previously produced products and on the other hand technologically improved products which mean that the performance of an existing product is significantly enhanced or upgraded.

Figures 1 to 3 illustrate the European Patent Applications (EPO), Gross Domestic Expenditures in R&D (GERD), and the Patents granted by United States & Trademark Office (USPTO), respectively.

Figure 1. Patent applications (EPO): number of applications per million inhabitants

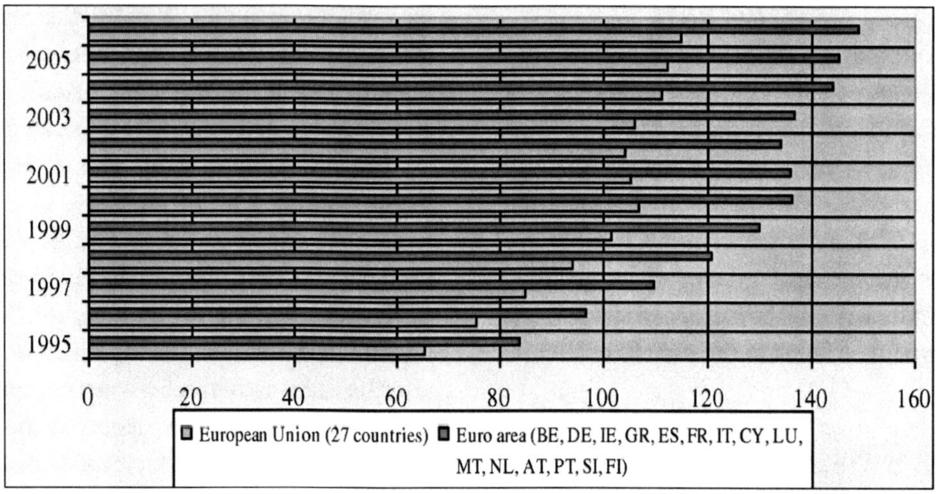

Figure 2. Gross domestic Exp.R&D (GERD): as percentage of GDP

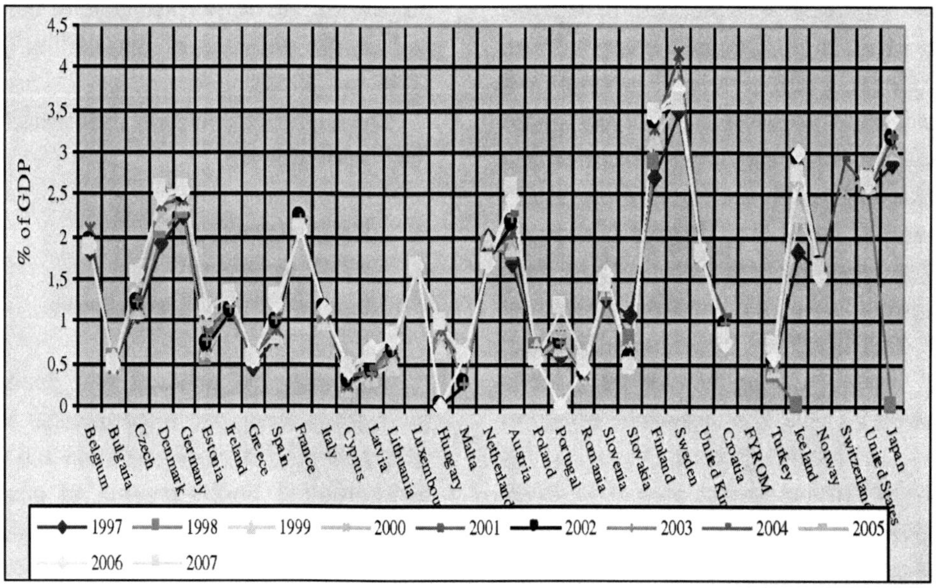

The term R&D covers three activities: basic research, applied research and experimental development. Basic research is "experimental or theoretical work undertaken primarily to acquire new knowledge of the underlying foundation of phenomena and observable facts, without any particular application or use in view". Applied research is also "original investigation undertaken in order to acquire new knowledge". However, it is directed primarily towards a specific practical aim or objective. Experimental development is "systematic work, drawing on existing knowledge gained from research and/or practical experience, which is directed to producing new materials, products or devices, to installing new processes, systems and services, or to improving substantially those already produced or installed". R&D covers both formal R&D in R&D units and

Figure 3. Patents granted by United States Patent & Trademark Office (USPTO)

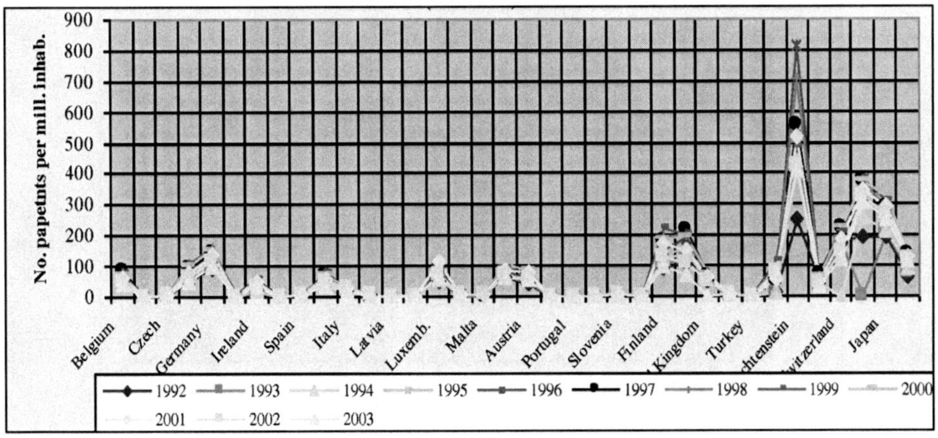

informal or occasional R&D in other units (Dosi Giovanni 1984).

MAPPING THE INNOVATION AND CATCHING UP PROCESS

A higher level of innovation activities tends to have a higher level of value added per worker (or a higher GDP per head) and a higher level of innovation activities than others. Following the technological-gap arguments, it would be expected that the more technologically advanced countries would be the most economically advanced (in terms of a high level of innovation activities and in terms of GDP per capita). Innovation requires, first and foremost, a state of mind combining creativity, entrepreneurship, willingness to take calculated risks and an acceptance of social, geographical or professional mobility (Fageberg J. 1988).

At national level, continue reviewing courses and teaching methods, above all for their ability to stimulate creativity and a spirit of enterprise from the earliest age, and think about any changes which may be necessary to the training of trainers. Member States should also continue to develop lifelong training. The Commission's contribution will be to set up a permanent "training and innovation" forum to stimulate the exchange of experience and best practice in this area. Member States are invited to adopt similar measures and to set up the conditions for making this mobility a reality. Greater priority should be given at both national and Community level to disseminating organisational innovations and using information and communication technologies in this field. Figures 4 and 5 illustrate the European Innovation Gap towards E.U., United States and Japan, respectively. The European Green Paper on Innovation opened up a number of pathways. For the sake of efficiency, this "First Action Plan" refers to a limited number of priority initiatives to be launched very soon at Community level and includes a number of schemes put into action or announced since the launch of the Green Paper, identified as essential to the innovation process, (Freeman C. and Lundvall B. 1988). The Action Plan for Innovation identifies three areas for action:

- to foster an innovation culture;
- to establish a framework conducive to innovation;
- to better articulate research and innovation

Figures 6 and 7 illustrate the convergence in innovation performance and the country-group in innovation performance per dimension, respectively. For the calculation of the average annual

Figure 4. EU Innovation gap towards EU & US

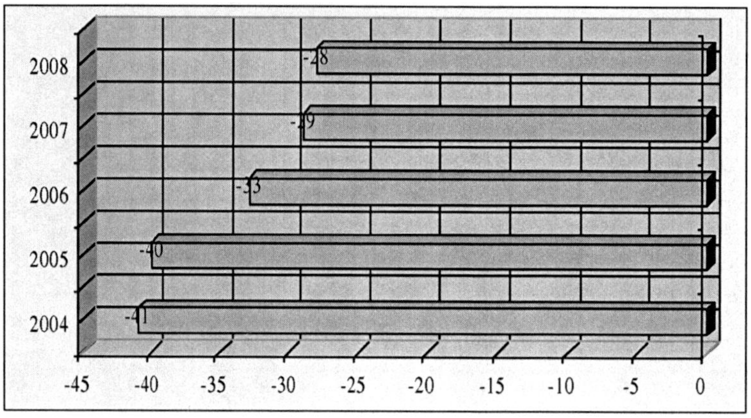

Figure 5. EU Innovation gap towards EU & Japan

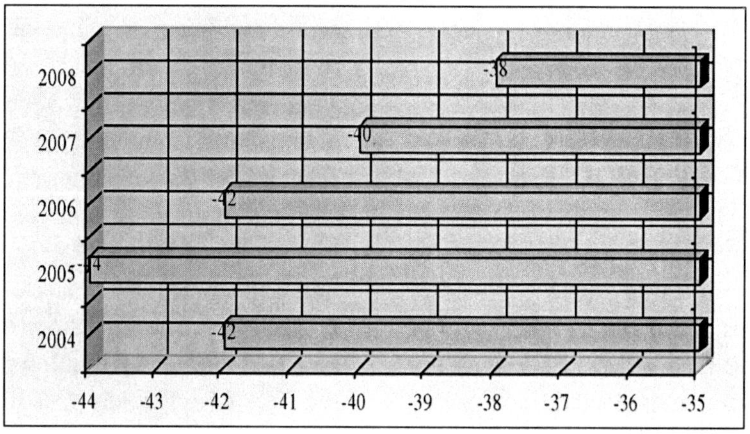

growth rate in innovation performance we have adopted a generalized approach. Using data from European Innovation Scoreboard, we first define growth for each country *c* per indicator *i* as $y^{t}_{ic} / y^{t-1}_{ic}$, as for instance is the ratio between the non-normalised values for year t and year t-1. In order to minimize the effect of growth outliers on the overall growth rate, these ratios are restricted to a maximum of 2 (such that growth in an individual indicator is restricted to 100%) and 0.5 (such that a decrease in an individual indicator is limited to -50%).

We aggregate these indicator growth rates between year t and year t-1 using a geometric average to calculate the average yearly growth rate τ^{t}_{c}:

$$1 + \tau^{t}_{c} = \prod_{i=1} \left(\frac{y^{t}_{ic}}{y^{t-1}_{ic}} \right)^{w_i}$$

We then calculate for each country c the average annual growth rate in innovation performance as the geometric average of all yearly growth rates:

$$1 + \text{Innovation Growth Rate}_{c} = \prod_{t} \left(1 + \tau^{t}_{c}\right)^{w_i}$$

Figure 6. Convergence in innovation performance

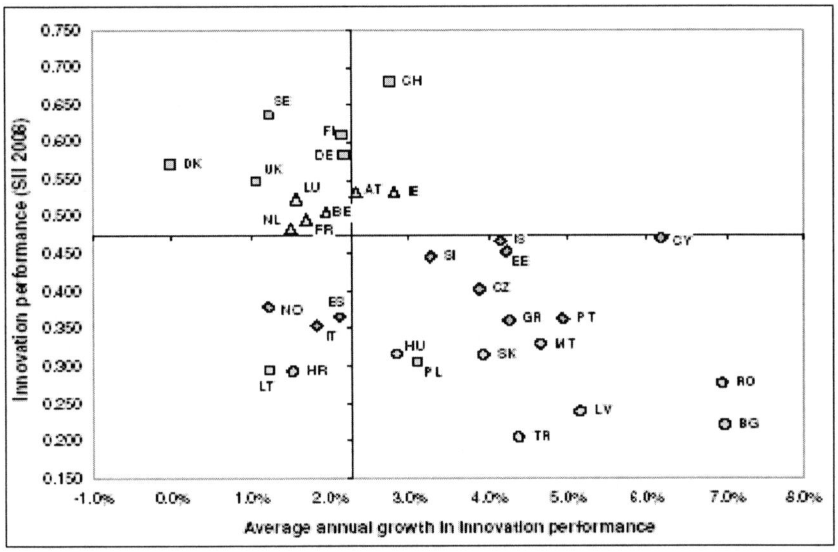

Figure 7. Country groups: Innovation performance per dimension

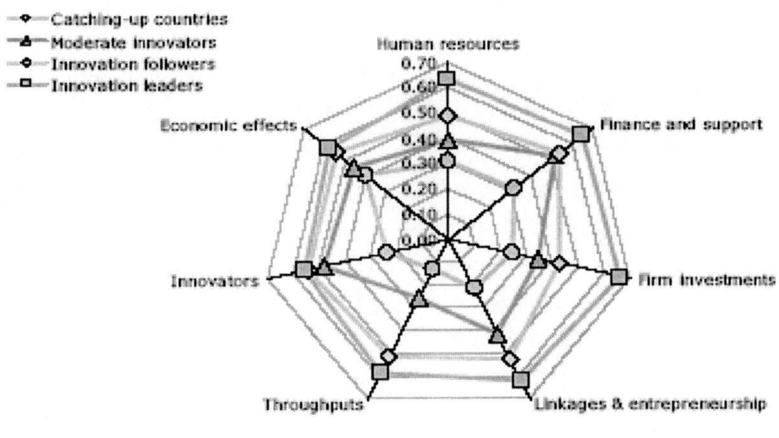

where $t\varepsilon[2004, 2008]$ and each average yearly growth rate receives the same weight w_t. The average annual growth rate in innovation performance does not measure the change in the SII but the average change in the 29 innovation indicators. Table 1 illustrates the innovation growth leader-groups. The average growth rates calculated over a five-year period.

POLICY IMPLICATIONS AND CONCLUSIONS

Technological progress has become virtually synonymous with long run economic growth. It raises a basic question about the capacity of both industrial and newly industrialized countries to translate their seemingly greater technological capacity into productivity and economic growth. Usually, there are difficulties in the estimation the relation between technical change and productiv-

Table 1. Innovation growth leaders

Group	Growth Rate	Growth Leaders	Moderate Growers	Slow Growers
Innovation leaders	1,6%	Switzerland	Germany, Finland	Denmark, Sweden, United-Kingdom
Innovation followers	2,0%	Ireland, Austria	Belgium	France, Luxembourg, Netherlands
Moderate Innovators	3,6%	Cyprus, Portugal	Czech Rep. Estonia, Greece, Iceland, Slovenia	Italy, Norway, Spain
Catching-Up countries	4,1%	Bulgaria, Romania	Latvia, Hungary, Malta, Poland, Slovakia, Turkey	Croatia, Lithuania

ity. Technological change may have accelerated, but in some cases there is a failure to capture the effects of recent technological advances in productivity growth or a failure to account for the quality changes of previously introduced technologies (Freeman C. 1994).

This article attempts to identify the R&D activities and also to investigate the estimation-methods, the techniques of scientific and technological activities and the measurement problems. Series of R&D statistics are only a summary of quantitative reflection of very complex patterns of activities and institutions. In the case of international comparisons, the size aspirations and institutional arrangements of the countries concerned should be taken into consideration. One way of constructing reliable indicators for international comparisons is to compare R&D inputs with a corresponding economic series, for example, by taking GERD as a percentage of the Gross Domestic Product. However, its quite difficult to make detailed comparisons between R&D data and those of non-R&D series both because of the residual differences in methodology and because of defects in the non-R&D data.

The collection of R&D data of regional statistics implied a lot of problems in comparison to data of national statistics. For the collection of regional statistics, we should take into the local differences and the difficulties. In addition, we can use either the "local-units" or the "local-economic-units". The first method «top-to-the-bottom method» focused on the collection of aggregate R&D data (for the whole country) and after that on the distribution of these figures into a regional-level; the disadvantage of this method is that there is not a direct collection of data from the regions or the second method «bottom-to-the-top method» for the collection dissaggregate R&D data (for the whole regions) based on the direct-collection at a regional-level and after that on the summation of these figures in order to obtain the aggregate-total R&D data (for the whole country). Technological gap models represent two conflicting forces: innovation which tends to increase the productivity differences between countries and diffusion which tends to reduce them. In the Schumpeterian theory, growth differences are seen as the combined results of these forces. Research on why growth rates differ has a long history which goes well beyond growth accounting exercises.

REFERENCES

Abramovitz, M. (1986). Catching-up, foreign ahead and falling behind. *The Journal of Economic History*, 46.

Berndt, E. R., & Triplett, J. (1990). Productivity and economic growth. In Jorgenson, D. (Ed.), *Fifty years of economic measurement studies in income and wealth* (pp. 19–119). NBER.

Bienaymé, A. (1986). The Dynamics of Innovation. *International Journal of Technology Management, 1*, 133–159.

Dosi, G. (1984). *Technical change and Industrial Transformation: the theory and the application to the semiconductor industry*. New York: MacMillan Press LTD.

Dosi, G., Freeman, C., Nelson, R., Silverberg, G., & Soete, L. (1988). *Technical change and economic theory*. London: Pinter Publishers.

European Commission. (2009). *European Innovation Scoreboard: Comparative Analysis of Innovative Performance*. Enterprise and Industry, Innometrics.

Fageberg, J. (1987). A technology gap approach to why growth rates differ, in *Research Policy,16*, 87-99.

Fageberg, J. (1988). Why growth rates differ, chapter 20. In Dosi G., Freeman Ch., Nelson, R., Soete, L. et (eds.) *Technical change and Economic theory*.

Fageberg, J.(1994). Technology and international differences in growth rates, *Journal of Economic Literature,* September, XXXII (3).

Fageberg, J. (1995). User-producer interaction, learning and comparative advantage. *Cambridge Journal of Economics, 19*, 243–256.

Freeman, C. (1994). The economics of technical change: a critical survey. *Cambridge Journal of Economics, 18*, 463–514.

Freeman, C., & Lundvall, B. (1988). *Small countries facing the technological revolution*. London: Pinter.

Korres, G. (1996). *Technical change and economic growth: an empirical analysis of EEC countries*. London: Avebury Press.

Korres, G. (2008). *Technical change and economic growth: Inside to the Knowledge Based Economy*. London: Avebury Press.

UNESCO (1969). *The measurement of scientific and technological activities*. United Nations publications.

Chapter 16
Biopharma Drugs Innovation in India and Foreign Investment and Technology Transfer in the Changed Patent Regime

Rakhi Rashmi
University of Nice, France

ABSTRACT

This chapter addresses the relationship between the intellectual property rights and Foreign Direct Investment in the context of Indian biopharma Industry to assess the impact of the TRIPs agreement of WTO on the biopharmaceutical industry of developing countries. The central issue in this study is the extent to which patent reform (after the imposition of the TRIPs agreement in 1995) affects India's ability to attract technology transfer for biopharmaceutical drugs innovation. The study analyzes FDI flow in the biopharma industry in Pre TRIPs (before the imposition of product patent protection 1991-1999) and post TRIPs (after product patent protection, 1999-2005). It does a comparative analysis of the relationship between the amount of foreign investment in different Indian states and the investment climate ranking of those states that are the part of Indian bio cluster.

1 INTRODUCTION

Trade related aspects of Intellectual Property Rights (TRIPs) agreement of the World Trade Organization (WTO) are considered to be one of the milestones in the international harmonization of patent protection. Product patent protection has been imposed on all its member countries by its Article-27.1.[1] Article-27.1 places a strict obligation on member to provide patents for all patentable pharma and biopharma products. After the start of enforcement of the TRIPs agreement, there has been a significant change in the patent regime in the developing countries. Before TRIPs, a vast majority of developing countries such as Argentina, China, India, Korea and Mexico protected process but not product.

After the enforcement of the TRIPs agreement, these countries are no longer allowed to carry out reverse engineering (Lanjouw 2000). In reverse engineering the new chemical entity or molecule is manufactured with a new process whose de-

DOI: 10.4018/978-1-61692-846-9.ch016

Copyright © 2011, IGI Global. Copying or distributing in print or electronic forms without written permission of IGI Global is prohibited.

velopment cost is minimal and clinical testing is not required. It has cost the developers several hundred million dollars to discover, develop and gain regulatory approval for a new medicine, even before incurring costs for development of the new drug molecule (A Maria et al; 2004).

After the imposition of product patent protection, generic production of new biopharmaceutical drugs (i.e. those patented post 2005) are not allowed and Indian firms must innovate novel biopharma drugs. The Indian biopharmaceutical industry is basically a biogeneric (Therapeutic products based on genetically engineered or recombinant technologies that are already in the market at least in some industrialized countries - Maria and Ramani; 2004) industry. India is the supplier of cheap generic drugs to the world. Main innovator products which have been produced as biogeneric drugs by the Indian biogeneric industry are Erythropoietin, G-CSF, recombinant Hepatitis B vaccine, Insulin, Interferon Alfa, streptokinase, among others.

The Indian generic drug industry grew to meet the drug needs of the country, and the government permitted it to ignore international patent protection law. The heavy burden of diseases in India comprises 18% of worldwide mortality and 20% of worldwide morbidity (WHO; 2003). However, it has limited resources to innovate novel drugs, since India's resources are 2% of world GDP and it invests just 1% the world healthcare investment (Sheena Reddy; 2006). As of 2006, the generic biopharmaceutical industry of a developing nation like India does not have the monetary resources to innovate novel drugs on its own. FDI and foreign technology transfer is essential to India's entry into the chain of novel drug innovation.

This gives an important situation to study the impact of the TRIPs agreement on the biopharma drug innovation in India by foreign technology transfer. So, the study is concentrated on the Indian biopharmaceutical industry and attempts to analyze the effect of the product patent protection on the foreign technology transfer in biopharmaceutical (Biopharmaceuticals are the large molecule drugs produced using rDNA technology e.g. Interferon, human growth hormone, human insulin - Arundel and Mintz; 2004) drugs innovation.

Also, it has been argued by the developed countries that strengthening of patent protection will bring innovation through increased foreign direct investment and greater transference of technology in developing countries. The objective of the TRIPs agreement, stated in Article 7 says; Protection and enforcement of Intellectual Property Rights should contribute to the promotion of:

Technological innovation and transfer and dissemination of technology to the mutual advantage of producer and user of the technological knowledge.

Thus, it can be interpreted that implementation of Art 27.1 will lead to greater foreign technology transfer in Pharma and biopharma sector. In Doha declaration too (Art 37)[2] it was agreed that the WTO would set up a working group to examine the relationship between trade and transfer of technology and to report findings to the fifth session of the Ministerial Conference (Art 38, 41).[3]

In this way the relationship of two very Articles (Article 7 and 27) is interesting to analyze as the same will measure the relationship between the intellectual property rights and FDI in the context of Indian biopharma Industry in order to know the impact of TRIPs agreement on the biopharmaceutical industry of developing countries.

These have been investigated in the first part of the chapter by analyzing the trend of foreign investment and transfer of technology in Indian biopharma sector as an effect of product patent regime.

The central issue in this study is the extent to which patent reform (after the imposition of the TRIPs agreement in 1995) affects India's ability to attract technology transfer for the biopharmaceutical drugs innovation. This is in order to see the impact of changed intellectual property protection on FDI and technology transfer.

Further, in the second part of the chapter in order to be more sure in the finding about the role of IP in pulling FDI and technology transfer, this study has made a comparative analysis of the amount of foreign investment and the investment climate ranking of the different states of India which are the part of Indian bioclusters. By investment climate we mean a menu of policy, regulatory and institutional factors that provide incentives sufficiently robust to induce private sector to invest in socially desirable projects (Weingast;1992). According to Stern Investment climate is policy, institutional and behavioral environment both present and expected that influences the returns and risks associated with investment.1 Also, as per John Dunning's (Dunning;1992) ownership location and internalization (OLI) approach ownership specific location, location specific and internalization specific ownership advantage create monopolistic advantages which can be used to prevail in market abroad. The OLI paradigm of John Dunning shows that under certain conditions it becomes profitable for the investor to produce in foreign market. In the specific case of biopharmaceuticals for partnering or collaborating or transferring technology, the multinational companies generally consider factors like emerging therapeutic areas beneficial markets for products, long term commitment to absorb the cost of failure and partnering with similar entities for sharing cost of failure and partnering with similar entities for sharing cost of establishment facilities(Maskus K; 1998). Organization decides to outsource business process or collaborate for various reasons. These reasons can be divided into the following categories:

1. Improve competitiveness
2. Reduce Costs
3. Access to higher level human resources at lower cost
4. Obtain advanced technologies without significant upfront costs
5. Acquire innovative concepts and products
6. Capitalize on low cost structure of the provider.

Apart from these, the location theory is also linked to the cluster effect which provides fertile ground for learning, experimentation and innovation due to short distances, short information time lag and relatively inexpensive communications (Von Zedwitz & Heimann, 2006). Therefore the chapter intends to add to this strand of research by analyzing and anatomizing the actual relationship between the FDI and higher patent protection by measuring the strength of factors of investment climate (including higher patent protection for biopharma) in attracting the foreign investment and technology transfer overall on country level as well as state level (in different bioclusters of India).

2 RESEARCH QUESTIONS

The basic question which has been addressed in this chapter is the role of product patent protection in stimulating technology transfer for the biopharma drugs innovation in Indian biopharma sector and to what extent differences in investment climate of different states of India affects inflow of foreign direct investment and technology transfer in the same patent regime which is applicable to all the Indian states equally?

3 METHODOLOGY

The chapter has addressed these questions using an industry level FDI and foreign technology transfer data collected from FIPB, Ministry of commerce and Industry, Government of India which was procured by self from the ministry. Also the firm-level database derived by the personal interviews of the biopharma MNC heads based in India. The survey has covered 10 biopharma MNS firms based in India. The study has selected 5 Indian states: Maharashtra, Gujarat, Andhra Pradesh, Karnataka, and Delhi for the research which forms the biocluster of India. In these five states out of total 142 biopharma firms in India more than 130 (90%) firms are located in just 5 states, in Andhra Pradesh 38 (27%), Maharashtra 38 (27%), Karnataka 17 (12%), Delhi 16 (11%) (Arora;2005) and Gujarat which form the Indian biocluster whose Investment climate grading or ranking has done by the World Bank it its survey of 2000 and 2003. (Table 1) (Figure 1)

On the basis of collected data from Ministry and IC data of World Bank the relationship between investment climate and FDI have been investigated by comparing both.

Study Design

For measuring the impact of product patent protection on foreign investment and technology transfer, data has been collected from Foreign Investment Promotion Board, Ministry of Commerce and Industry, Government of India between 1991 and 2005. 1991-1999 to see post liberalization effect and pre TRIPs effect and 1999-2005 (post TRIPs). TRIPs agreement came into effect on January 1st 1995, it set out transitional period of 10 years (until 2005) for the implementation of the TRIPs agreement for the developing countries like India, which had not granted patents for pharmaceutical products. India had however, accepted applications (the mail box provision)[4] and, beginning in 2000, offered "exclusive marketing rights (EMR) by the Patent amendment Act 1999. EMR are very similar to patents in offering monopoly marketing rights to the inventor, so effectively product patent protection was available since 1999.Thus, for analyzing impact of product patent protection on the biopharmaceutical innovation, the time period of year 2000 to 2005 has been considered.

Table 1. Number of biopharma firms in different states of India (Source: Based on the information of Arora Praveen (2005); Healthcare biotechnology firms in India: Evaluation, structure and growth; current science. vol. 89, no 3)

State	No of Firms	Percentage
Andhra Pradesh	38	(27%)
Maharashtra	38	(27%)
Karnataka	17	(12%)
Delhi	16	(11%)
Gujarat	12	(8%)
Tamil Nadu	10	(10%)
Utter Pradesh	4	(3%)
Haryana	2	(1%)
West Bengal	2	(1%)
Jharkhand	1	(1%)
Kerala	1	(1%)
Orissa	1	(1%)
Total	**142**	**(100%)**

Two types of data have been collected for the survey, first industry level and second firm level. Industry level data on the flow of foreign technology transfer has been collected majorly from foreign investment promotion board (FIPB), Ministry of Industry and commerce. Firm level data has been collected from the biopharma firms (MNC which are in India). This has been done by visiting individual biopharma MNCs established in India. Visits have been made to the biopharma MNCs which are working in India such as Novo Nordisk, Aventis pharma, Glaxo smith cline, Eli Lily and Co, LG Chemicals, and AstraZeneca etc in order to know the influence of the factors which influenced their decision to invest in India. Survey has been done by interviewing CEOs of 10 firms personally which are located in these states mainly in Bangalore, Hyderabad, Mumbai, Pune, Delhi, Gurgaon etc from July 2006 to January 2007. Personal visit has been made to these firms located in the abovementioned states in order to get the results.

Figure 1. Number of biopharma firms in different Indian clusters (Source: Foreign Investment Promotion Board, Ministry of Commerce and Industry, Government of India)

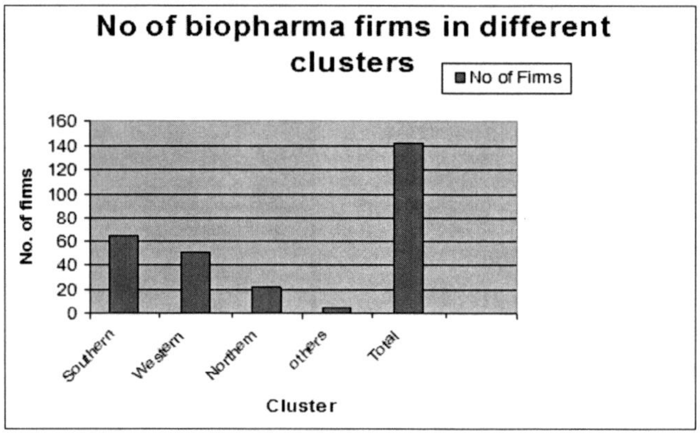

Figure 2. Total number of technology transfer cases in Indian biopharmaceutical sector from January 1991 to December 2004 (Source: Foreign Investment Promotion Board, Ministry of Commerce and Industry, Government of India, 2005)

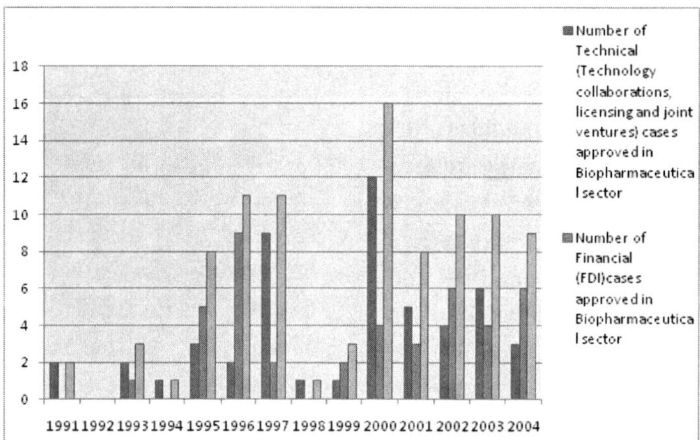

The study has employed graphical analysis as well as interview analysis to answer the research question.

4 CASE STUDY

Part A

If we see the number of foreign collaborations in Indian biopharma industry in the initial years of liberalization then we find that the change in the liberalization policy did not yield dramatic results immediately (Figure: 18). The economic reforms in India started in the early 1990s, or to be more precise, in July 1991. This reform process and the liberalization of the economy continued throughout the decade and into the 21st century. In brief, the reforms include flexible exchange rates regime, full convertibility of current account transactions and removal of many international trade restrictions, transfer in India by the announcement of the new industrial policy in July 1991.

Figure 3. FDI in Indian biopharmaceutical and pharma sector (January 1991–March 2005) (Source: Foreign Investment Promotion Board, Ministry of Commerce and Industry, Government of India, 2005)

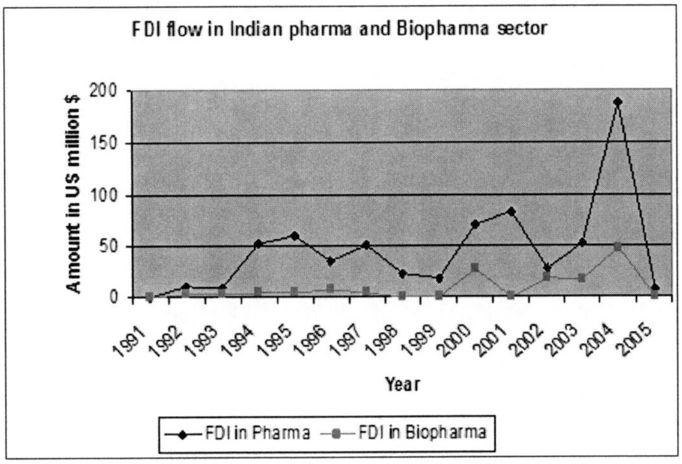

After 1991, with the liberalization[5] of the Indian economy, foreign investment started flowing into various sectors of the India ranging from the manufacture of cereals to the IT industry (Figure 4). However, by analyzing the technology transfer data in biopharma sector ranging from 1991 to 2005, it can be inferred that the Indian biopharmaceutical industry too did not attract much foreign investment in the initial years of liberalization (Figure 2).

As per Figure 2 there is increase in the technology transfer trend though the increase is not sharp and continuous after 1995 but the trend is better than the initial years of liberalization therefore it can be referred as the combined effect of liberalization and globalization. The Figure shown above also shows that the number of technology transfer cases has increased in post TRIPs era although not sharply as there are ups and downs in the flow. So, it cannot be said for sure that the strong IPR leads to an increase in foreign investment alone because if it is product patent protection only then the trend would have been sharp and continuous. Therefore it can be inferred that there are other factors as well which affect the foreign technology transfer. At the same time it can also be said for sure that product patent is one factor which influence technology transfer in the Indian biopharmaceutical sector.

By examining the trend of technology transfer in Indian biopharma sector carefully, it can be seen that the flow has increased more sharply in the post TRIPs era (1999-2005) than in pre TRIPs period (From 1991 to 1999) (see Figures 2 and 3).

Similar trend has been found in FDI flow in Indian pharma industry (see Figure 3). The trend of FDI flow is more or less the same in both the sectors and the same has increased more in post TRIPs period (1999- 2005). The foreign investment flow in Indian biopharmaceutical sector is sharper in 2000- 2005 period (total FDI during this period is 40.12 US $ million) than 1995 - 2000 period (total FDI during this period is 83.12 US $ million). The total amount of investment undertaken by foreign firms during 2000–2005 stood at $83.12 million, more than double the investment they had undertaken over 1995–2000 $40.12 million) (Table 2). The total flow of FDI into biopharma sector was $16.22 million between 1991 and 1995 which increased five times during 2000 and 2005. Such trend can be seen as a result of increasing interest in the Indian biopharmaceutical industry through direct investment. As a consequence, the share of biopharmaceuticals in

Figure 4. FDI approved in all sectors in India from January 1991 to March 2005 (Source: Foreign Investment Promotion Board, Ministry of Commerce and Industry, Government of India 2005)

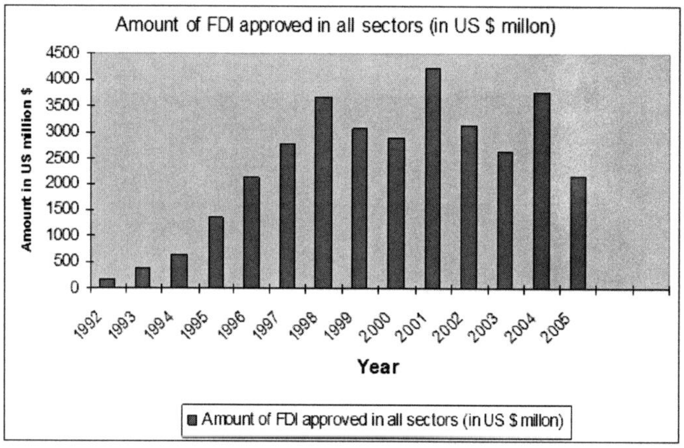

Table 2. Comparison of the statement of year wise breakup for foreign direct investment (FDI) and foreign technology cases approved by government during August 1991 to December 2005 in drugs & pharma and biopharma sectors (Source: Foreign Investment Promotion Board, Ministry of Commerce and Industry, Government of India)

Year	Amount of FDI approved in all sectors (in US $ million)	Amount of FDI approved in Drugs and pharma sector (in US $ million)	Amount of FDI approved in Biopharmaceutical sector (in US $ million)
1991	165	0.32	nil
1992	383	11.13	3.76
1993	654	9.76	3.76
1994	1374	51.91	4.35
1995	2141	59.54	4.35
1996	2770	34.37	8.36
1997	3682	50.99	4.77
1998	3083	23.07	Nil
1999	2908	18.55	0.14
2000	4222	70.85	26.85
2001	3134	82.82	0.33
2002	2634	28.03	18.35
2003	3754	51.92	16.87
2004	2171	188.01	47.57
2005		7.23	.71
TOTAL		688.51	131.57

the total FDI inflows has increased from 2.4 per cent from 1995 to1999 which further increased to 4.9 per cent from 2000 to 2005. The result shows that India's shifting from a weak patent regime to a stronger patent regime has taken place during this period and this final shift in the patent regime might have promoted hesitant foreign enterprises to enter into Indian market. The actual FDI inflow into Indian biopharmaceutical industry is estimated to be only $ 47.57 million in 2005 which is 2.4% of the total amount $2171 million received by the economy and the same has increased significantly from 1991, which was around 1% of the total FDI (Table 2).

Thus, the trend of FDI flow shows that the foreign technology flow is increasing but not very smooth and sharp which suggests that this trend has some positive influence from stronger patent protection which were combined with liberalization but still there are few other factors which are acting as entry barrier for the global players for foraying into the industry therefore the flow is not sharp and continuous. One reason can be given that as the product patent protection has been imposed in India in 2005 and this is the effect of the transition period from EMR[6] protection to product patent protection. So, the actual effect of product patent protection may not be visible as yet. On such trend, it can be interpreted that though, the advent of product patent did not bring dramatic increase in FDI flow in IP driven industry such as drugs and pharma but trend may be sharper with the years to come as product patent protection has just imposed in the biopharma sector.

Part B

Along with the above analysis and discussions, to be more accurate in finding about the actual relation between the product patent protection and foreign technology flow in biopharma sector, the same relationship has been further examined by interview data to validate the outcome stated above. For the same, field research has been done and in the same field research, 10 CEOs and director level officials of Biopharma MNCs were interviewed and asked about the most influencing factor in making decision to invest in India. These MNCs are also located in different bioclusters of the country- Bangalore, Hyderabad, Mumbai – Pune, Delhi and Gurgaon. They were given 7 factors to grade them according to their preference. Factors were Intellectual property Rights, quality and implementation, stability of Indian partner, management skills, bureaucracy, market, geographical distance. On 15 scale value 12 scale values was given to stronger patent protection or higher intellectual property rights. (See Figure 5).

So, by looking at this trend, it can be said that certainly stronger patent protection is one of the strongest factor which has influenced the foreign MNCs to transfer their technology to India which seems to increase with time. At the same time other factors also influence the decision of making the foreign investment such as regulatory policies, fiscal policy, infrastructure, bureaucracy etc along with patent policy.

From the above finding it is clear that changed IP protection is one of the components of the investment climate of India which is influencing the decision of the foreign investors.

Part C

As the change in patent protection imposed equally all over the India and if IP is the only factor which influences the FDI flow in sector like biopharma then the flow of FDI should be more or less the same in all the biocluster states. To be surer about the role of IP in pulling FDI into Indian biopharmaceutical sector the foreign technology transfer data collected from FIPB has been further divided state wise and then relationship of the same had been measured by the investment climate of the state. For the data on the investment climate of different states of India the chapter has taken the data provided by the World Bank in its joint sur-

Figure 5. Comparative importance of factors in decision of technology transfer (Source: based on own research)

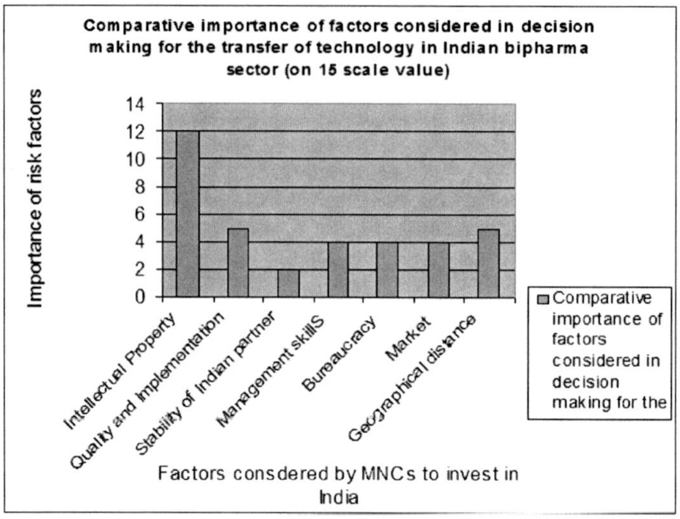

vey with Confederation of Indian Industries (CII) in 2000 and 2003 (Dunning and Narula; 1996).

The World Bank has conducted surveys FACS in 2000 and in 2003 by jointly with Confederation of Indian Industries in which 12 states and large number of industries were covered which include: (i) Food processing, (ii) Textiles, (iii) Garments, (iv) Leather goods, (v) Pharmaceuticals, (vi) Electronic consumer goods, (vii) Electrical white goods, (viii) Auto components, (ix) Metal and metal products, (x) Plastics, and (xi) Machine Tools. Since the pharmaceutical industry is covered so the survey is relevant base our findings on the same. In this survey the business managers were asked to identify the states that they thought had a better or worse IC than the state in which they were currently based. They were also asked to say which of the states in their opinion had the best IC and which had the worst. The subjective ranking of the states according to IC, as expressed by the business managers, is presented in Table 3.

The chapter has tried to find the match or relationship of FDI data collected from FIPB with the World Bank data of IC of the Indian states for the results. For the same the complete FDI data of 1991 to 2005 has been rearranged state wise (see Table 5 and Figure 6).

The table shows that maximum number of foreign collaboration took place in western cluster which is formed by Maharashtra and Gujarat. Now if we see the investment climate of these states then we will find that these two states have the best IC in country as in the 2000 survey Maharashtra stood at the first place whereas Gujarat stood second. Although in 2003 survey Gujarat lost its position from best IC state to Good IC state but remained at the top among the good IC states. In 2003 too Maharashtra maintained its position as the best IC state therefore it can be interpreted that the FDI investment happened the most in this region because of the best investment climate.

As per the FIPB FDI data the second highest foreign collaborations happened in Southern cluster which is formed by Andhra Pradesh and Karnataka, if we see the investment climate of these states the we will find that these two states are in Good IC states and are just behind Maharashtra and Gujarat in 2000 survey whereas in 2003 survey these states are at 4th and 5th position so overall from 1991 to 2005 this cluster main-

Table 3. Subjective ranking of best to worst IC (FACS 2000) (Source: Estimates provided by the World Bank 2000 in CII - World Bank (2002), 'Competitiveness of Indian Manufacturing: Results form a Firm Level Survey' Confederation of Indian Industry)

Category	States	Sectors Covered in this Survey
Best investment climate	1. Maharashtra 2. Gujarat	*1. Auto components* *2. Drugs and pharmaceuticals* *3. Electrical white goods* *4. Electronic consumer goods* *5. Garments* *6. Textiles* *7. Machine tools* *8. Software*
Good investment climate	3. Andhra Pradesh 4. Karnataka 5. Tamil Nadu	
Medium investment climate	6. Delhi 7. Punjab	
Poor investment climate	8. Kerala 9. West Bengal 10. Uttar Pradesh	

Table 4. Subjective ranking of best to worst IC (FACS 2003) (Source: Estimates provided by the World Bank 2003 in World Bank and Confederation of India Industry, 2003b, "Investment Climate in India.)

Category	States	% say best minus % say worst
Best Investment climate	1. Maharashtra	29.1
	2. Delhi	16.7
Good Investment Climate	3. Gujarat	9.6
	4. Andhra Pradesh	8.6
	5. Karnataka	6.8
	6. Punjab	4.9
	7. Tamil Nadu	3.7
	8. Haryana	1.1
Poor Investment Climate	9. Madhya Pradesh	-6.8
	10. Kerala	-15.0
	11. West Bengal	-30.6
	12. Uttar Pradesh	-30.6

Table 5. Foreign technology transfer in different bioclusters of India from 1991 to March 2005 (Source: Foreign Investment Promotion Board, Ministry of Commerce and Industry, Government of India, 2005)

Biopharmaceutical Cluster	Foreign Technology transfer cases	Percentage
Southern	26	26.532
Western	38	38.775
Northern	14	14.285
States not indicated	20	20.408
Total	***98***	***100***

Figure 6. Foreign technology transfer in different bioclusters of India from 1991 to March 2005 (Source: Foreign Investment Promotion Board, Ministry of Commerce and Industry, Government of India, 2005)

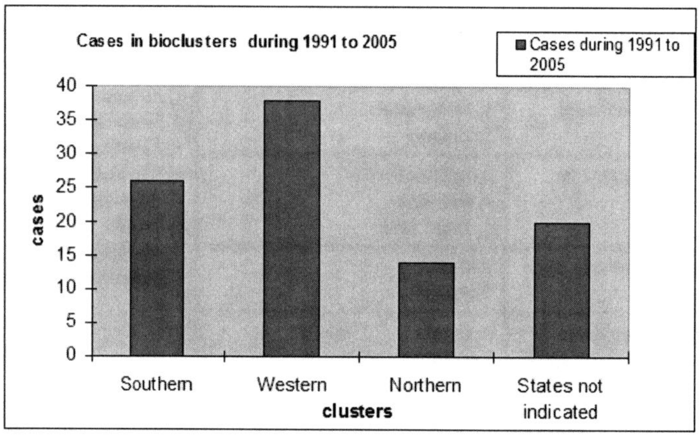

Figure 7. Foreign technology transfer in different Indian States from 1991 to March 2005 (Source: Foreign Investment Promotion Board, Ministry of Commerce and Industry, Government of India, 2005)

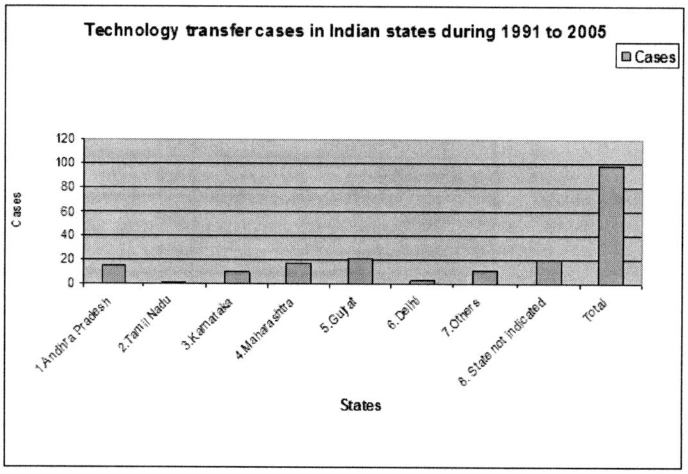

tained the second best IC states position which is directly correlated with the second highest FDI flow which flowed around 26.532% of the total FDI flow in the country after the western cluster of 38.775% (Table 5).

Now if we see the IC state of the northern cluster which comprise of Delhi and Utter Pradesh then we will find that in 2000 survey Delhi was in the category of medium IC states whereas UP was placed in the poor IC state. Although Delhi's IC improved and it occupied second position in 2003 after Maharashtra among the best IC states but still the FDI flow in this region flowed less than western and southern cluster may be because the calculation of FDI is from 1991 to 2005 and till 2000 the IC of Delhi was at 6th position and still Maharashtra occupies the best states position and Gujarat also remained at the 3rd position. Among the southern cluster states also these states maintained their position in the good IC states

Figure 8. Foreign technology transfer in different Indian biocluster states in pre TRIPs and post TRIPs period (Source: Foreign Investment Promotion Board, Ministry of Commerce and Industry, Government of India, 2005)

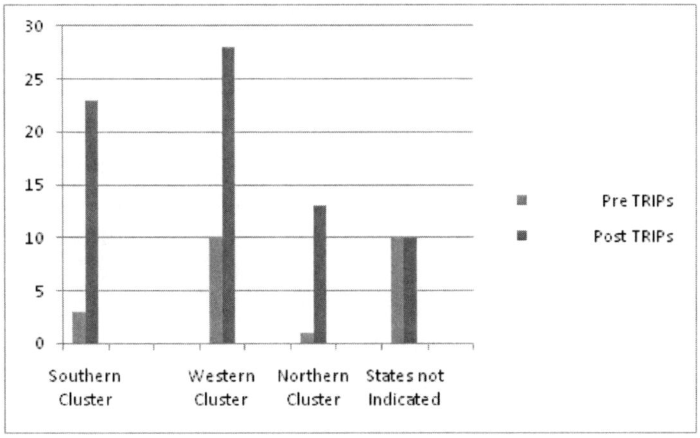

Table 6. Foreign Technology transfer in different Indian biocluster states in pre TRIPs and post TRIPs period (Source: Foreign Investment Promotion Board, Ministry of Commerce and Industry, Government of India, 2005)

		PRE TRIPs			POST TRIPs			
Cluster	**States**	**Technology Collaborations**	**Financial Collaborations**	**Total**	**Technology Collaboration**	**Financial Collaborations**	**Total**	**Grand Total**
Southern Cluster	Andhra Pradesh	1	1	2	7	11	18	
	Karnataka	1		1	5	-	5	
				3			23	3+23=26
Western Cluster	Maharashtra	2	2	4	8	6	14	
	Gujarat	6	0	6	7	7	14	
				10			28	10+28=38
Northern Cluster	Delhi	1		1	8	5	13	
	Himachal Pradesh							
	Punjab							
				1			13	1+13=14
States not Indicated		7	3	10	6	4	10	10+10 =20

throughout therefore it can be said that it remained second choice for the foreign investors to invest.

Now if we further divide the data on FDI and technology transfer which flowed into different Indian biocluster states in pre TRIPs and post TRIPs period and compare them with the investment climate of those states in two periods in order to be surer about the role of product patent in stimulating investment and technology transfer. As per Table 6 in pre TRIPs period highest FDI and technology collaboration took place in the western cluster with 10 such cases (Maharashtra with 4 cases and Gujarat with 6). Then second highest FDI and technology collaboration took place in southern cluster with 3 such cases (Andhra Pradesh with 2 such cases and Karnataka with 1).

Northern cluster remained at the third place with one case only which took place in Delhi. Now if we relate the FDI and technology collaboration in these states with the investment climate of them in these two periods then we will find that investment climate of the western cluster (Maharashtra and Gujarat) was the best in pre TRIPs era. Second position with good investment climate was of Southern cluster (Andhra Pradesh at 3rd position, Karnataka with 4th position and Tamil Nadu with 5th position). Northern cluster (Delhi at 6th, Punjab at 7th and UP at 9th position) remained at 3rd position in terms of investment climate ranking in the pre TRIPs period. In the abovementioned result of FDI flow in the states and their ranking in terms of investment climate show that there is direct relation among them. As the best IC cluster (western) received highest FDI, the second best cluster (Southern) in terms of IC received second highest FDI collaborations and the cluster (Northern) which ranked 3rd received 3rd highest FDI collaborations.

Now we will see how it worked in the post TRIPs period (1999-2005) and for the same comparison has been made between Post TRIPs FDI and technology flow in different biocluster states with the investment climate of those states during that period. We see that the best IC cluster consist of Maharashtra (Western cluster) at first place and Delhi (Northern cluster state) at 2nd position which remarkably rose from 6th position in pre TRIPs investment climate survey.

Second best investment climate position in post TRIPs era was taken by Gujarat (western cluster) and Andhra Pradesh (Southern cluster) with 3rd and fourth position respectively. Karnataka (southern cluster) remained at 5th position and Punjab (Northern cluster state) at 6th.

Now, if we see FDI and technology flow in these states during post TRIPs period then we will find that highest inflow of FDI and technology collaboration happened in the Western cluster with 28 such cases among which 14 cases took place in Maharashtra and 14 in Gujarat. Second best cluster was Southern with 23 such cases in which Andhra Pradesh had 18 and Karnataka bagged 5. Northern cluster remained at 3rd place again in post TRIPs with 13 such cases in Delhi solely. Now looking at the above data it can be said FDI and technology collaboration have a significant influence of the investment climate of the states. Highest collaboration happened in Maharashtra with 14 such cases which ranked 1st among the investment climate of the states. Second highest such case took place in Gujarat with 14 which ranked 3rd in investment climate grading. Delhi remained at 3rd place in attracting FDI whereas investment climate wise it ranked 2nd. Here in this case it can be said FDI and technology collaboration increased significantly in Delhi from pre TRIPs period to post TRIPs period from 1 to 13 which signifies that as the investment climate improved it started attracting more and more investors.

Despite of significant improvement in Delhi's investment climate and its position as 2nd in attracting FDI in post TRIPs period Northern cluster remained at the 3rd position because Delhi did not get support from other Northern cluster states as Punjab and UP still remained backward. At the same time states like Maharashtra and Gujarat maintained their position as best and good IC states. Although Gujarat fell to 3rd position from second in terms in post TRIPs period but its ally Maharashtra maintained 1st position consistently therefore it can be said that this western biocluster remained the favorite hotspot for the foreign investors in both pre and post TRIPs period. Southern cluster states (Andhra Pradesh and Karnataka) remained at 4th and 5th position in terms of investment climate and also ranked 4th and 5th position in terms of getting FDI and technology collaboration.

Therefore it can be said that factors of investment climate play a significant role in attracting foreign investment. Here one more significant

Table 7. Foreign technology transfer in different Indian bioclusters in pre TRIPs and post TRIPs period (Source: Foreign Investment Promotion Board, Ministry of Commerce and Industry, Government of India, 2005)

	Pre TRIPs	Post TRIPs
Southern Cluster	3	23
Western Cluster	10	28
Northern Cluster	1	13
States not Indicated	10	10

point needs to be noted that investment flow has increased in post TRIPs and in no case has decreased during this period which signifies that product patent as a factor of investment climate has positively affected the inflow of FDI. In case the product patent had affected FDI inflow negatively then the flow would have been declining. This is not the case in this case study. Therefore it can be said for sure that patent protection or higher IPR is one factor of investment climate which influence the foreign investment and technology inflow.

5 CONCLUSION

Thus, this research founds that the factors of location advantage is important for the foreign investment destination in the local market, which includes factors like resource endowment and also economic and social factors such as market size and structure, prospect for market growth and degree of development, the cultural, legal, political and institutional environment and Government legislation and policies of the different states of India. In terms of the demand side factors, the host's overall attractiveness to FDI is determined by location advantages it possesses or the investment climate of the country. As resource endowments are not evenly distributed among countries and social and economic factors as well as Government policies are different among countries, even it is different among the states of the same country, the attractiveness different states in India for FDI is different. In biopharma sector attractiveness of the location in country like India depends on the cost advantage in clinical trials and labor as these are largely the factors which affect the drug development cost significantly. Biotechnology is a knowledge-driven sector because it consists of knowledge working on knowledge to create value, decoding in genomics and proteomics being paradigmatic knowledge-based economic activity. In the case of biotechnology, foreign investments are key magnets for the Indian biopharma industry to shift to novel innovation in change patent regime. Since transfer science from the laboratory bench to the market involves complex, interactive chains of transactions among scientists, entrepreneurs, and various intermediaries therefore proximity to services such as investors and lawyers and, in biotechnology, research hospitals for clinical trials are important which also create an innovation system. These factors form the factors of investment climate. Since patent policy has changed and applied uniformly to whole of the India but other factors of the investment climate differs state to state therefore foreign investment has also made according to the investment climate of the states.

Looking at the results it can be said that the four southern and western states, Gujarat, Karnataka, and Maharashtra who are stronger in terms of real GDP and per capita growth than other major Indian states with better investment climate have fetched greater FDI and foreign technology transfer. So, it can be said that these high performing states have excelled due to their better infrastructure, education, business policies and better economic reforms. Therefore the foreign companies could foresee better and higher growing incomes and opportunities from their investment in these states.

Thus, form the above analysis it is clear that the factors of investment climate play their role in influencing the decision of the investment and Intellectual property is the one factor among the different factors of investment climate which influence the decision of foreign investment. Therefore the chapter concludes that product patent protection is certainly worked as stimulant to pull foreign investment in Indian biopharma industry but it is not the only factor as there are other factors of investment climate which affect the foreign technology transfer and investment.

REFERENCES

Arundel, A. and Mintzes, B. (2004). *The benefits of biopharmaceuticals*. Innogen Working Chapter No. 14. University of Edinburgh.

Dunning, J. (1979). Explaining changing pattern of international production: In defence of eclectic theory. *Oxford Bulletin of econometric statistics*, 41, 269-275.

Dunning, & Narula. (1996). *Foreign direct investment and government: Catalysts for economic restructuring*. London.

Lanjouw, J. (2000). *Do patent matter? Evidence after GATT.* NBER working chapter no. 7495, National bureau of Economic Research, Cambridge.

Maria, A., Ruet, J., Zerah, M. H. (2004). *Final report on Biotechnology in India.* Commissioned by French embassy.

Maria, A., & Ramani, S. (2005, February 12-18) TRIPS and its possible impact on the biotech based segment of the Indian Pharmaceutical industry. *Economic and political weekly,* Pages 675-683.

Maskus, K. (1998). The role of intellectual property rights in encouraging foreign direct investment and technology transfer. *Journal of competition and International Law.*

Praveen, A. (2005). Healthcare biotechnology firms in India: Evaluation, structure and growth; *current science, 89*(3).

Reddy, S. (2006, May). *The cost of India complying with world intellectual property rights: effect on the and access to drugs*. PhD Thesis: Berkeley University.

Von Zedtwitz, M., & Heimann, P. (2006). Innovation in Clusters and the Liability of Foreignness of International R&D. In Carayannis, E., & Campbell, D. (Eds.), *Knowledge Creation, Diffusion, and Use in Innovation Networks and Clusters: A Comparative Systems Approach across the United States, Europe, and Asia* (pp. 101–122). New York: Praeger.

Weingast, B. R. (1992) *The Economic Role of Political Institutions*. Working Chapter 46. Institute for Policy Reform.

World Health Organization (2003). *Shaping the future.* World Health Report.

ENDNOTES

1 Patents shall be available for any inventions, whether products or processes in all fields of technology, provided that they are new, involve an inventive step and are capable of industrial application. Patents shall be available and patent rights enjoyable without discrimination as to the place of invention, the field of technology and whether products are imported or locally produced.

2 Art 37 We agree to an examination, in a Working Group under the auspices of the General Council of the relationship between trade and transfer of technology, and of any possible recommendations on steps that might be taken within the mandate of the WTO to increase flows of technology to developing countries. The General Council shall report to the Fifth Session of the Ministerial Conference on progress the examination.

3 Art 38. We confirm that technical cooperation and capacity building are core elements of the

development dimension of the multilateral trading system, and we welcome and endorse the New Strategy for WTO Technical Cooperation for Capacity Building, Growth and Integration. We instruct the Secretariat, in coordination with other relevant agencies, to support domestic efforts for mainstreaming trade into national plans for economic development and strategies for poverty reduction. The delivery of WTO technical assistance shall be designed to assist developing and least-developed countries and low-income countries in transition to adjust to WTO rules and disciplines, implement obligations and exercise the rights of membership, including drawing on the benefits of an open, rules-based multilateral trading system. Priority shall also be accorded to small, vulnerable, and transition economies, as well as to members and observers without representation in Geneva. We reaffirm our support for the valuable work of the International Trade Centre, which should be enhanced. Art 41. We have established firm commitments on technical cooperation and capacity building in various paragraphs in this Ministerial Declaration. We reaffirm these specific commitments contained in paragraphs 16, 21, 24, 26, 27, 33, 38-40, 42 and 43, and also reaffirm the understanding in paragraph 2 on the important role of sustainably financed technical assistance and capacity-building programmes. We instruct the Director-General to report to the Fifth Session of the Ministerial Conference, with an interim report to the General Council in December 2002 on the implementation and adequacy of these commitments in the identified paragraphs.

4 Mailbox system was a TRIPs imposed obligation on developing countries like India that wished to benefit from the TRIPs transitional period by delaying granting of patents for pharmaceutical products until 2005. In exchange for not granting patents, these countries had to established a mailbox system for receiving and filing patents applications from the beginning of the transitional period in 1995 in order to establish filing date.

5 Following independence, India pursued a development policy based on centralized planning, regulation and control of private enterprise, state ownership, trade protection and limits on the penetration of foreign capital and technology. This regime determined India's economic development until the mid-1980s when there began some movement towards economic liberalization and market orientation. The Government implemented a program of structural reforms, aimed at stabilizing the economy and promoting reliance on market mechanisms, broadly referred to as 'liberalization'. The main components of the structural reforms program were exchange and trade liberalization; financial sector reforms and control of the budget deficit; inflation and money supply. A great deal of significance was placed on promotion of foreign technology transfers and foreign investment in key areas, as well as, the further development of the private sector.

6 EMR can be applied in the case the product 'waiting for a patent in the mail box' obtained marketing approval before the mailbox is opened and a decision is made on whether or not to grant the patent. In such a case the manufacturer could request exclusive marketing rights for up to five years or until a decision is rendered on the patentability of the product, whichever is shorter. The grant of EMR stifles with the generic competition. In theory, EMR could reduce the benefit of transition period

Compilation of References

Abramovitz, M. (1986). Catching-up, foreign ahead and falling behind. *The Journal of Economic History*, 46.

Ache, P. (1994). *Wirtschaft im Ruhrgebiet*. Working Paper 1994/123. University of Dortmund: Institute for Spatial Planning.

Acs, Z. J., Anselin, L., & Varga, A. (2002). Patents and innovation counts as measures of regional production of new knowledge. *Research Policy*, *31*, 1069–1085. doi:10.1016/S0048-7333(01)00184-6

Acs, Z. J., & Audretsch, D. B. (1989). Patents as a measure of innovative activity. *Kyklos*, *42*(2), 171–180. doi:10.1111/j.1467-6435.1989.tb00186.x

Affaires Extérieures et Commerce Extérieur Canada (AECEC). (1992). *Les alliances stratégiques: passeport pour l'Europe*. Ottawa: Prospectus Publications Limited.

Aghion, P., & Howitt, P. (1992). A model of growth through creative destruction. *Econometrica*, *60*(2), 323–351. doi:10.2307/2951599

Ahmed, A., & Russel, H. C. (1988). *Science and Technology Policy for National Development – A Window on the Asian Experience*. Foundation for International Training.

Aidemark & Sterner. (2003). A Framework for Strategic Balancing of Knowledge Management Initiatives. *International Conference on System Sciences 36th Proceeding.* Hawaii

Ajoku, K. B. (2002). *Evaluation of the Management of Technology Development*. A thesis submitted to National Centre for Technology Management Ife, Nigeria for award of Post Graduate Diploma on Technology Management.

Albino, V., Carbonara, N., & Schiuma, G. (1998). Le strategie di sviluppo dei network locali. *Sviluppo & Organizzazione*, *170*, 33–46.

Albino, V., & Schiuma, G. (2003). New Forms of Knowledge Creation and Diffusion in the Industrial District of Matera-Bari. In Belussi, F., Gottardi, G., & Rullani, E. (Eds.), *The Net Evolution of Local Systems – Knowledge Creation, Collective Learning and Variety of Institutional Arrangements*. Dordrecht, the Netherlands: Kluwer.

Albino. A.; Garavelli, A. & Schiuma, G. (1999). Knowledge Transfer and Inter-firm Relationships in Industrial Districts: The Role of the Leader Firm. *Technovation*, 9, 53-63.

Aliouat, B. (1996). Les stratégies de coopération industrielle. Collection Gestion. Série: Politique générale, finance et marketing. Ed. Economica. Paris.

Allee, V. (2000). Reconfiguring the value network. *The Journal of Business Strategy*, *21*(4), 36–39. doi:10.1108/eb040103

Allen, T. (1985). *Managing the Flow of Technology*. Cambridge, MA: MIT Press.

Allen, T. (1987). Performance of Information Channels in the Transfer of Technology. In Roberts, E. B. (Ed.), *Generating Technological Innovation* (pp. 89–104). New York: Oxford University Press.

Angang, H., & Wenzhi, X. (2008). Quantitative evaluation on science in China(1980-2004). *Journal of Tsinghua University(. Philosophy of the Social Sciences*, *23*(2), 104–119.

Copyright © 2011, IGI Global. Copying or distributing in print or electronic forms without written permission of IGI Global is prohibited.

Aoki, R., Kubo, K., & Yamane, H. (2006). Patent and public health under countries lessons from Japan. *Bulletin of the World Health Organization*, *84*(5).

Archibugi, D., & Michie, J. (1995). Technology and innovation: an introduction. *Cambridge Journal of Economics*, *19*, 1–4.

Arora, A., & Gambardella, A. (2004). The globalization of the software industry: perspectives and opportunities for developed and developing countries. *NBER Working paper, series 10538*.

Arora, A., & Gambardella, A. (2005, eds.). *From Underdogs to Tigers: The Rise and Growth of the Software Industry in Brazil, China, India, Ireland and Israel*. New York: Oxford University Press

Arrow, K. J. (1962). The economic implications of learning by doing. *The Review of Economic Studies*, *29*(3), 155–173. doi:10.2307/2295952

Arundel, A., & Mintzes, B. (2004). *The benefits of biopharmaceuticals. Innogen Working Chapter No. 14*. University of Edinburgh.

Asheim, B. T., & Isaksen, A. (1997). Location, Agglomeration and Innovation: Towards Regional Innovation Systems in Norway? *European Planning Studies*, *5*(3), 299–330.

Asheim, B. T. (1996). Industrial Districts as 'learning regions': a condition for prosperity? *European Planning Studies*, *4*, 379–400.

Asheim, B. T. (2002). Temporary organizations and spatial embeddedness of learning and knowledge creation. *Geografiska Annaler. Series B, Human Geography*, *84*, 111–124. doi:10.1111/j.0435-3684.2002.00117.x

Asheim, B. T., & Coenen, L. (2005). Knowledge bases and regional innovation systems: Comparing Nordic clusters. *Research Policy*, *34*, 1173–1190. doi:10.1016/j.respol.2005.03.013

Asheim, B. T., & Isaksen, A. (2002). Regional innovation systems: the integration of local sticky and global ubiquitous knowledge. *The Journal of Technology Transfer*, *27*, 77–86. doi:10.1023/A:1013100704794

Asia Pacific Research and Positioning. (2007). *The Chinese Video Game Market: Market Entry for Canadian Firms*. Retrieved from www.asiapacificresearch.com

Atuahene-Gima, K., & Evangelista, F. (2000). Cross-Functional Influence in New Product Development: An Exploratory Study of Marketing and R&D Perspectives. *Management Science*, *46*(10), 1269–1284. doi:10.1287/mnsc.46.10.1269.12273

Atuahene-Gima, K., & Li, H. (2000). Marketing's Influence Tactics in New Product Development: A Study of High Technology Firms in China. *Journal of Product Innovation Management*, *17*, 451–470. doi:10.1016/S0737-6782(00)00059-X

Audretsch, D. B., & Feldman, M. (1996). R&D spillovers and the geography of innovation and production. *The American Economic Review*, *86*, 630–640.

Audretsch, D. (2001). The Role of Small Firms in U.S. Biotechnology Clusters. *Small Business Economics*, *17*, 3–15. doi:10.1023/A:1011140014334

Audretsch, D. B. &Thurik, R. (2001). Linking entrepreneurship to growth. *OECD, STI Working Papers* 2001/2, DSTI/DOC (2001).

Axelrod, R. (1984). *The Evolution of Cooperation*. New York: Basic Books.

Aydalot, P., & Keeble, D. (1998). *High Technology, Industry and Innovation Environments: The European Experience*. London: Routledge.

Azoulay, N., & Weinstein, O. (2000). Les compétences de la firme. *Revue d'Économie Industrielle*, *93*, 117–154. doi:10.3406/rei.2000.3007

Badaracco, J. (1991). *The Knowledge Link. How Firms Compete through Strategic Alliances*. Boston: Harvard Business School Press.

Bade, F. J., & Niebuhr, A. (1998). *Zur Stabilität des räumlichen Strukturwandels*. Working Paper 1998/158. University of Dortmund: Institute for Spatial Planning.

Barley, S., Freeman, J., & Hybels, R. (1992). Strategic Alliances in Commercial Biotechnology. In Nohria, N., & Eccles, R. (Eds.), *Networks and Organizations: Structure, Form, and Action*. Boston: Harvard Business School Press.

Barro, R., & Sala-i-Martin, X. (1997). Technological diffusion, convergence and growth. *Journal of Economic Growth*, *2*, 1–26. doi:10.1023/A:1009746629269

Barro, R. J., & Sala-i-Martin, X. (1995). *Economic Growth*. New York: McGraw-Hill.

Bas, T. G., & Niosi, J. (2007). The Issue of Asymmetrical Growth in Specialized Biotechnology Firms in the United States and the United Kingdom. *International Journal of Biotechnology*, *9*(1), 87–100. doi:10.1504/IJBT.2007.012170

Bashir, S. (2005). India's tryst with TRIPs: The Patents (Amendment) Act, 2005. The Indian Journal of Law and Technology, volume 1.

Basu, I. (2004; Jul 23). *India's clinical trials and tribulations*. Asia Times Online; Accessed 2008; January 16 at www.atimes.com/atimes/South_Asia/FG23Df03.html

Baum, J., Calabrese, T., & Silverman, B. (2000). Don't Go it Alone: Alliance Network Composition and Startup' Performance in Canadian Biotechnology. *Strategic Management Journal*, *21*, 267–294. doi:10.1002/(SICI)1097-0266(200003)21:3<267::AID-SMJ89>3.0.CO;2-8

Baum, J. A. C. (1996). Organizational Ecology. In Clegg, S., Hardy, C., & Nord, W. (Eds.), *Handbook of Organization Studies*. London: Sage.

Becattini, G. (1990). The Marshallian industrial district as a socio-economic notion. In Pyke, F., Becattini, G., & Sengenberger, E. (Eds.), *Industrial* (pp. 37–51). Geneva: Districts and Interfirm Cooperation in Italy.

Behrens, K., & Thisse, J. F. (2007). Regional economics. A new economic geography perspective. *Regional Science and Urban Economics*, *37*, 457–465. doi:10.1016/j.regsciurbeco.2006.10.001

Bell, M. (1984). Learning and the accumulation of industrial technological capability in developing countries. In Fransman, M., & King, K. (Eds.), *Technological Capability in the Third World* (pp. 187–209). London: Macmillan.

Bengt-Åke Lundvall. 2004. *Why the New Economy is Learning Economy*. DRUID Working Papers 04-01, DRUID, Copenhagen Business School, Department of Industrial Economics and Strategy/Aalborg University, Department of Business Studies.

Bennett, R. (2003). *Services marketing: a managerial approach* (1st ed.). Milton: John Willey & Sons Australia.

Berndt, E. R., & Triplett, J. (1990). Productivity and economic growth. In Jorgenson, D. (Ed.), *Fifty years of economic measurement studies in income and wealth* (pp. 19–119). NBER.

Bettman, J., Luce, M. F., & Payne, J. (2000). Processus de choix construit du consommateur. *Recherche et Applications en Marketing*, *15*(2), 187–217.

Beugelsdijk, S. (2004). Entrepreneurial Culture, Regional Innovativeness and Economic Growth, *Proceedings of ERSA 2004 Conference*, Porto

Bhattasali D., (2001). *Sustaining China's development: Some issues*. Presentation to Tsinghua University 90th Aniversary Celebrations Seminar Series.Beijing, People's Republic of China, April 24.

Bienaymé, A. (1986). The Dynamics of Innovation. *International Journal of Technology Management*, *1*, 133–159.

Bils, M., & Klenow, P. (2000). Does schooling cause growth? *The American Economic Review*, *90*(5), 1160–1183. doi:10.1257/aer.90.5.1160

Bing, W., & Zheng, W. (2004). A study on measurement of human capital across provinces of China. *Science Research Management*, *25*(4), 60–65.

Birch, K., Mackinnon, D., & Cumbers, A. (2008). Old industrial regions in Europe: a comparative assessment of economic performance. [iFirst article]. *Regional Studies, 42*, 1–19.

BIS (British Information Services). (2003). *Biotechnology in the UK – The Government View*. Retrieved from www.usembassy.org.uk

Blomstrom, M., & Wolff, E. (1989). [Starr Center for Applied Economics, New York University.]. *Multinational Corporations and Productivity Convergence in Mexico, C*, V.

Blythe, J. (2006). *Principles & Practice of Marketing* (1st ed.). London: Thomson Learning.

Bontis, N. (2004). National Intellectual Capital Index – A United Nations initiative for the Arab region. *Journal of Intellectual Capital, 5*(1), 13–39. doi:10.1108/14691930410512905

Borensztein, E., Gregorio, J. D., & Lee, J.-W. (1998). How does foreign direct investment affect economic growth? *Journal of International Economics, 45*, 115–1351. doi:10.1016/S0022-1996(97)00033-0

Bostyn, S. J. R. (2004/2005). Narrow trousers and narrow patents, a health risk? Product protection or purpose bound protection for biotechnological inventions. *Bio-Science Law Review*; 2 BSLR 95.

Botkin, J. W., & Matthews, J. B. (1992). *Winning Combinations. The Coming Wave of Entrepreneurial Partnerships between Large and Small Companies*. New York: Wiley.

Bouayad, A. (1996). *Les alliances stratégiques*. Paris: Dunod.

Bounfour, A., & Edvinsson, L. (2005, Eds). *Intellectual Capital for Communities*. Boston, MA: Elsevier Butterworth-Heinemann.

Bradley, J. (1997). Intellectual capital and the new wealth of nations. *Business Strategy Review, 8*, 53–62. doi:10.1111/1467-8616.00007

Brooksbank, D. J., & Pickernell, D. G. (1999). Regional competitiveness indicators – a reassessment of method. *Local Economy, 13*, 310–326. doi:10.1080/02690949908726459

Budd, L., & Hirmis, S. A. (2004). Conceptual Framework for regional competitiveness. *Regional Studies, 38*(9). doi:10.1080/0034340042000292610

Burt, R. S. (1992). *Structural Holes: The Social Structure of Competition*. Cambridge, MA: Harvard University Press.

Caballeros, R. J., & Jaffe, A. B. (1993). How high are the giants' shoulders? In Blanchard, O., & Fischer, S. (Eds.), *NBER Macroeconomics Annual*. Cambridge, MA: MIT Press.

Camagni, R. (Ed.). (1991). *Innovation Networks—Spatial Perspectives*. London: Belhaven Press.

Camagni, R. (1991). Local 'milieu', uncertainty and innovation networks: towards a new dynamic theory of economic space. In Camagni, R. (Ed.), *Innovation Networks: Spatial Perspectives*. London: Belhaven.

Caniëls, M. C., & Verspagen, B. (2001). Barriers to knowledge spillovers, and regional convergence in an evolutionary model. *Evolutionary Economics, 11*, 307–329. doi:10.1007/s001910100085

Carlsson, B., Jacobsson, S., Holmen, M., & Rickne, A. (2002). Innovation Systems: Analytical and Methodological Issues. *Research Policy, 2*, 233–245. doi:10.1016/S0048-7333(01)00138-X

Carlucci, D., & Schiuma, G. (2007). Exploring intellectual capital concept in strategic management research. In Joia, L. (Ed.), *Strategies for Information Technology and Intellectual Capital: Challenges and Opportunities*. Hershey, PA: Idea Group Inc.

Carrillo, F. J. (2006). *Knowledge Cities* (1st ed.). Boston, MA: Elsevier Butterworth-Heinemann.

Carroll, G. (1994). A Sociological View on Why Firms Differ. In Rumelt, R., Schendel, D., & Teece, D. (Eds.), *Fundamental Issues in Strategy*. Boston: Harvard Business School Press.

Cassiolato, J. E., & Lastres, H. M. (1999). *National and Regional Systems of Innovation in the Mercosur*. Prepared for the DRUID Summer Conferences on National Innovation.

Chambers, J. (2002). Patent Eligibility of Biotechnological Inventions in the United States, Europe and Japan: How Much Patent Policy is Public Policy? 34 GEO. WASH INT'L L. Rev. 223.

Chaminade, C., & Vang, J. (2008). Globalisation of knowledge production and regional innovation policy: Supporting specialized hubs in the Bangalore software industry. *Research Policy*, *37*, 1684–1696. doi:10.1016/j.respol.2008.08.014

Chan, H. C. (1971). *Nation-building in Southeast Asia: The Singapore case*. Singapore: Institute of Southeast Asian Studies.

Chatterjee (2002). *Knowledge Management-Theories, Issues and Challenges, IIT*.Kanpur, India.

Chatterton, P., & Goddard, J. P. (2000). The response of higher education institutions to regional needs. *European Journal of Education*, *35*(4), 475–496. doi:10.1111/1467-3435.00041

Chen, X. (2005) R&D Internationalization and Two-Shore's High-tech Industry Competition and Cooperation. *Report of China Economy Institute*. Retrieved from http://www.docin.com/p-13289093.html

Chiesa, V. (1996). Managing the Internationalization of R&D Activities. *IEEE Transactions on Engineering Management*, (5): 7–23. doi:10.1109/17.491264

Chohung Bank vs. Deputy Director of Income-tax (2006) 102 ITD.

Chua, B. H. (1995). *Communitarian ideology and democracy in Singapore*. London: Routledge.

CNNIC. (2008). *21st Statistics Survey Report on the Internet Development in China*. Beijing: CNNIC.

Coe, D., Helpman, E., & Hoffmaister, A. (1997). North-South R&D spillovers. *The Economic Journal*, *107*, 134–149. doi:10.1111/1468-0297.00146

Coe, D., & Moghadam, R. (1993). Capital and Trade as Engines of Growth in France. *IMF Staff Papers*, 40.

Coe, D., & Helpman, E. (1995). International R&D spillovers. *European Economic Review*, *39*, 859–887. doi:10.1016/0014-2921(94)00100-E

Coe, D., & Helpman, E. (1993). *International R&D spillovers*. NBER Working Paper 4444, National Bureau of Economic Research.

Cohen, W. M., & Levinthal, D. A. (1989). Innovation and learning: the two faces of R&D. *The Economic Journal*, *99*, 569–596. doi:10.2307/2233763

Coleman, J. S. (1988). Social capital in the creation of human capital. *American Journal of Sociology*, *94*, S95–S120. doi:10.1086/228943

Combe, E. (1995). *Alliances entre firmes et course technologique*. Paris: Économique.

Combe, E. (1996). Alliances en R&D, course à l'innovation et gain stratégique; éléments théoriques et application au segment des DRAM. *Revue d'économie industrielle*. 78, 4e trimestre.

Commission of the European Communities. (2006). Putting knowledge into practice: A broad-based innovation strategy for the EU. *Communication from the Commission to the Council, the European Parliament, the European Economic and Social Committee and the Committee of the Regions, COM(2006) 502 final*. (available online at http://eur-lex.europa.eu/LexUriServ/site/en/com/2006/com2006_0502en01.pdf)

Commission on Intellectual Property Rights. *Innovation and Public Health Report, World Health Organisation* (2003). London.Correa, C. (1999). *Intellectual property rights and the use of compulsory licenses: options for developing countries*. Trade-Related Agenda, Development and Equity, Working Papers, South Centre, Geneva.

Cooke, P. (2001). Regional innovation systems, clusters, and the knowledge economy. *Industrial and Corporate Change*, *10*, 945–974. doi:10.1093/icc/10.4.945

Cooke, P. (2008). Regional Innovation Systems, Clean Technology & Jacobian Cluster-Platform Policies. *Regional Science Policy and Practice*, *1*(1), 23–45. doi:10.1111/j.1757-7802.2008.00002.x

Cooke, P., Gomez Uranga, M., & Extebarria, G. (1997). Regional innovation systems: Institutional and organizational dimensions. *Research Policy*, *26*, 475–491. doi:10.1016/S0048-7333(97)00025-5

Cooke, P., & Memedovic, O. (2003). *Strategies for Regional Innovation Systems: Learning Transfer and Application Policy Papers*. Wien: UNIDO.

Cooke, P., & Schienstock, G. (2000). Structural competitiveness and learning regions. *Enterprise & Innovation Management Studies*, *1*, 265–280. doi:10.1080/14632440010023217

Cooke, P., & Wills, D. (1999). Small firms, social capital and enhancement of business performance through innovation programmes. *Small Business Economics*, *13*, 219–234. doi:10.1023/A:1008178808631

Cooke, P. (1995). *The Rise of the Rustbelt*. London: UCL Press.

Cooke, P., Boekholt, P., & Todtling, F. (2000). *The Governance of Innovation in Europe: Regional Perspectives on Global Competitiveness*. London: Pinter.

Cooke, P. (2004). Evolution of regional innovation systems – emergence, theory, challenge for action. In Cooke, P. (Ed.), *Regional Innovation Systems*. London: Routledge.

Cooke, P., & Braczyk, H. J. (1996). *Regional Innovation Systems: the Role of Government in the Globalized World* (Heidenreich, M., Ed.). London: UCL Press.

Cooke, P., & Morgan, K. (1994) The creative Milieu: A regional Perspective on Innovation. In M. Dodgson & R. Rothwell(Eds.).*The Handbook of Innovation*, 25-32.

Correa.C (1998). *Implementing the TRIPs Agreement; General context and implications for developing countries*. Penang, Malaysia: Third World Network: De Boeck & Larcier (2001). *Pharmaceutical patents and accessibility to drugs, German velasquez, essential drugs and globalisation*. Revue Internationale de Droit Economique. A new way to get medicines to the poor; www.spicyip.com; May 13, 2008

Counterfeit medicines, Fact sheet N°275; Revised 14, WHO/. available on http://www.who.int/mediacentre/factsheets/fs275/en/

Crabtree, A., & Rodden, T. (2007). *Hybrid Ecologies: understanding cooperative interaction in emerging physical-digital environments*. Personal and Ubiquitous Computing.

Crevoisier, O., & Maillat, D. (1991). Milieu, industrial organizations and territorial production system: towards a new theory of spatial development. In Camagni, R. (Ed.), *Innovation Networks: Spatial Perspectives*. London: Belhaven.

Cross, R., & Parker, A. (2004). *The Hidden Power of Social Networks. Understanding how work really gets done in organizations*. Boston, MA: Harvard Business School Press.

Dakhli, M., & De Clerq, Q. D. (2004). Human capital, social capital, and innovation: a multi-country study. *Entrepreneurship & Regional Development*, *16*, 107–128. doi:10.1080/08985620410001677835

Danneels, E. (2002). The Dynamics of Product Innovation and Firm Competences. *Strategic Management Journal*, *23*, 1095–1121. doi:10.1002/smj.275

Day, G. (2000). Assessing Future Markets for New Technologies. In Day, G., Schoemaker, P., & Gunther, R. (Eds.), *Wharton on Managing Emerging Technologies*. New York: John Wiley & Sons.

Deep, M. (2008). *Data protection: Emerging trends*Accessed on 3rd March 2008.

Denison, E. F. (1962). United States economic growth. *The Journal of Business*, *35*(2), 109–121. doi:10.1086/294483

Department of Science and Technology (DST). (2005). *'About the System' Science and Technology System in India.(*available at http://dst.gov.in/stsysindia/about-sys.htm, accessed January 8, 2006)

Di Benedetto, C. (1997). Identifying the Key Success Factors in New Product Launch. *Journal of Product Innovation Management, 16*, 530–544. doi:10.1016/S0737-6782(99)00014-4

Di Masi, J. A., Hansen, R. W., & Grabowski, H. G. (2003). The price of innovation: New estimates of drug development costs. *Journal of Health Economics, 22*, 151–185. doi:10.1016/S0167-6296(02)00126-1

Diamond v. Diehr, 450 U.S. 175, 185 (1980). *Excluded from such patent protection are laws of nature, natural phenomena.*

DiMasi, J. (2001). *Backgrounder: A Methodology for Counting Costs for Pharmaceuticals R&D. Centre for the Study of Drug Development.* Boston: Tufts University.

Dimminaco A.G. v. Controller of Patents, Designs & Trade Marks, (2002) I.P.L.R. July 255,269 (Calcutta H.C.)

DNDi submission to WHO IGWG (15th Nov 2006). Building a global strategy for policy change in neglected disease research. Geneva.

Dolfsma, W., Chaminade, C., & Vang, J. (2008). Knowledge dynamics out of balance: Knowledge biased, skewed and unmatched. *Research Policy, 37*, 1657–1658. doi:10.1016/j.respol.2008.08.007

Doloreux, D. (2002). What should we know about regional systems of innovation. *Technology in Society, 24*, 243–263. doi:10.1016/S0160-791X(02)00007-6

Doole, I. (2005). *Understanding and Managing Customers* (1st ed.). Harlow: Pearson Education Ltd.

Dosi, G. (1988). Sources, procedures, and microeconomic effects of innovation. *Journal of Economic Literature, XXVI*, 1120–1171.

Dosi, G. (1984). *Technical change and Industrial Transformation: the theory and the application to the semiconductor industry.* New York: MacMillan Press LTD.

Dosi, G., Freeman, C., Nelson, R., Silverberg, G., & Soete, L. (1988). *Technical change and economic theory.* London: Pinter Publishers.

Dougherty, D., & Hardy, C. (1996). Sustained product innovation in large, mature organizations: overcoming innovation-to-organization problems. *Academy of Management Journal, 39*(5), 1120–1153. doi:10.2307/256994

Doz, Y. (1996). The Evolution of Cooperation in Strategic Alliances: Initial Conditions or Learning Processes? *Strategic Management Journal, 17*, 55–83. doi:10.1002/smj.4250171006

Drews, J. (1999). *Quest of Tomorrow's Medicines.* New York: Springer-Verlag.

Drucker, P. (1993). *The Post-Capitalist Society.* New York: Harper Business.

Drucker, P. (1998). The future that has already happened. *The Futurist, 32*(8), 16–18.

Drucker, P. (1985). *Innovation and Entrepreneurship.* New York: Harper and Row.

DTI - Department of Trade and Industry. (1998). Our Competitive Future: Building the Knowledge Driven Economy. London: Competitiveness White Paper DTI.

Dunning, & Narula. (1996). Foreign direct investment and government: Catalysts for economic restructuring. London.

Dunning, J. (1979). Explaining changing pattern of international production. In defence of Eclectic theory," 41 Oxford Bulletin of econometric statistics, 269-275.

Dussauge, P., & Garrette, B. (1995). *Les alliances entre firmes concurrentes: Évolutions, conséquences pour les partenaires et impact sur la concurrence. Les cahiers de recherche. Paris: Chambre de commerce et d'industrie de Paris. Dussauge, P. & Garrette, B. (1999): Cooperative Strategy. Competing Successfully through Strategic Alliances.* Chichester, UK: Wiley.

Egbetokun, A. A., Siyanbola, W. O., Sanni, M., Olamade, O. O., Adeniyi, A. A., & Irefin, I. A. (2009). What drives innovation? Inferences from an industrywide survey in Nigeria. *International Journal of Technology Management, 45*(1/2), 123–140. doi:10.1504/IJTM.2009.021524

Eicher, T., & Turnovsky, S. (2001). Transition dynamics in non-scale models. *Journal of Economic Dynamics & Control, 25*, 85–113. doi:10.1016/S0165-1889(99)00070-6

Encaoua, Guellec & Martinez. (2006, November). Patent systems for encouraging innovation: Lessons from economic analysis. *Research Policy, 35*(9), 1423–1440. doi:10.1016/j.respol.2006.07.004

Englander, S., & Gurney, A. (1994). Medium-term determinants of OECD productivity. *OECD Economic Studies, 22*, 49–109.

Ernst & Young. (2003). *Resilience: Americas Biotechnology Report 2003*. Retrieved from www.ey.com.

Erramouspe, M. (1996). Staking Patent Claims on the Human Blueprint: Rewards and Rent-Dissipating Races. *43 UCLA L. Rev., 961*, 988–990.

European Commission. (2009). *European Innovation Scoreboard: Comparative Analysis of Innovative Performance*. Enterprise and Industry, Innometrics.

European Commission. (2005). Measuring progress towards a more sustainable Europe – Sustainable development indicators for the European Union. Luxembourg.

European Commission. *Union, Regional Policy*. Inforegio, Retrieved from http://europa.eu.int European Union. Retrieved from http://www.europa.eu.int

European Union. (2006*). Regional Innovation Scoreboard 2006*.Retrieved from http://www.proinno-europe.eu/inno-metrics.html

Evenson, R. E., & Westphal, L. E. (1995). Technological change and technology strategy. In Behrman, J., & Srinivasan, T. N. (Eds.), *Handbook of Development Economics*. Amsterdam: Elsevier Science.

Fageberg, J. (1995). User-producer interaction, learning and comparative advantage. *Cambridge Journal of Economics, 19*, 243–256.

Fageberg, J. (1987). A technology gap approach to why growth rates differ, in *Research Policy, 16*, 87-99.

Fageberg, J.(1994). Technology and international differences in growth rates, *Journal of Economic Literature,* September, XXXII (3).

Fagerberg, J. (1988). a. International competitiveness. *The Economic Journal, 98*, 355–374. doi:10.2307/2233372

Fagerberg, J., & Verspagen, B. (2002). Technology-gaps, innovation-diffusion and transformation: an evolutionary interpretation. *Research Policy, 31*, 1291–1304. doi:10.1016/S0048-7333(02)00064-1

Fagerberg, J. (1988) b. Why growth rates differ. In Dosi, G., Freeman, C., Nelson, R.R., Silverberg, G., Soete, L. (Eds.), *Technical Change and Economic Theory (pp.432-457),*London: Pinter, London.

Fernández, H., Palacios-González, E., & García-Díaz, V. G-Bustelo, B.C.P., Sanjuan-Martínez, O., & Lovelle, J.M.C. (2009). SBPMN - An Easier Business Process Modeling Notation for Business Users. *Computer Standards & Interfaces*.

Findlay, R. (1978). Relative backwardness, direct foreign investment, and the transfer of technology: S simple dynamic model. *The Quarterly Journal of Economics, 92*(1), 1–16. doi:10.2307/1885996

Florida, R. (1995). Towards the learning region. *Futures, 27*, 527–536. doi:10.1016/0016-3287(95)00021-N

Foray, D. (1996). Diversité, sélection et standardisation: Les nouveaux modes de gestion du changement technique. Revue d'Économie Industrielle, N° 75.

Forrest, J., & Martin, M. (1992). Strategic Alliances Between Large and Small Research Intensive Organizations. Experiences in the Biotechnology Industry. *R & D Management, 22*(1), 42–53. doi:10.1111/j.1467-9310.1992.tb00787.x

Foss, N. (1997). *Resources, Firms and Strategies*. Oxford, UK: Oxford University Press.Foss, N. (2000). Equilibrium vs Evolution in the Resource-based Perspective: The Conflicting Legacies of Demsetz and Penrose. In Foss, N. & Robertson, P. (Eds.), *Resources, Technology and Strategy. Explorations in the Resource-based Perspective*. Routledge Advances in Management and Business Studies.

Freeman, C. (1995). The national systems of innovation in a historic perspective. *Cambridge Journal of Economics*, *18*, 5–24.

Freeman, C. (1987). *Technology Policy and Economic Performance: Lessons from Japan*. London: London Printer Publishers.

Freeman, C. (1994). The economics of technical change: a critical survey. *Cambridge Journal of Economics*, *18*, 463–514.

Freeman, C., & Lundvall, B. (1988). *Small countries facing the technological revolution*. London: Pinter.

Freeman, C., & Soete, L. (1997). *The Economics of Industrial Innovation*, 3rd EditionLondon:Pinter.

Fu, D. (1993). *Analysis on Foreign High Technology and Its Industry Policy*. Bejing: Earthquake Press. Amsden, A., Tschang, T., et al. (2001). Do *Foreign Companies Conduct R&D in Developing Countries?* Asian Development Bank Institute Working Paper. Du, D., et al. (2003). *Internationalization of MNEs' R&D and reply strategy research*. The Subject Report of Shanghai Municipal Science & Technology Commission.

Funke, M., & Strulik, H. (2000). On endogenous growth with physical capital, human capital and product variety. *European Economic Review*, *44*, 491–515. doi:10.1016/S0014-2921(98)00072-5

Furman, J. L., Porter, M. E., & Stern, S. (2002). The determinants of national innovative capacity. *Research Policy*, *31*, 899–933. doi:10.1016/S0048-7333(01)00152-4

Gai, W., Zhang, H., & Lu, W. (2004). Compare and Analysis on International Typical High-Tech Industry Cluster and Experience Inspire. *China Soft Science*, *2*, 102–108.

García-Díaz, V., Fernández-Fernández, H., & Palacios-González, E. G-Bustelo, B. C. P., & Lovelle, J. M. C. Intelligent traceability system of Cabrales cheese using MDA TALISMAN. In *Proceedings of the 2008 International Conference on Artificial Intelligence*. July 14-17, 2008, Las Vegas, Nevada, USA

Gardiner, B., Martin, R., & Tyler, P. (2004). Competitiveness, productivity and economic growth across the European regions. *Regional Studies*, *38*(9).

Gaudin, J.-H. (1988). Structures internes des entreprises conjointes et intérêts privés. In Jacquemin, A. & Remiche, B (Eds), Coopération entre entreprises. *Entreprises conjointes stratégies industrielles et pouvoirs publics*. Paris: De Boeck université.

Gerhardsen, T.I.S. (12 March 2007), "Thailand Presents Report On Compulsory Licensing Experience," Intellectual Property Watch.

Geroski, P., Machin, S., Van, R., & Geroski, J. (1993). Innovation and profitability. *The Rand Journal of Economics*, *24*(2), 198–211. doi:10.2307/2555757

Gertler, M. (2004). *Manufacturing Culture: The Institutional Geography of Industrial Practice*. Oxford: Oxford University Press.

Gerybadze, A., & Reger, G. (1999). Globalization of R&D: Recent Changes in Management of Innovation in Transnational Corporations. *Research Policy*, *28*(2/3), 251–274. doi:10.1016/S0048-7333(98)00111-5

Glover, B. & Bhatt, H (2006). RFID Essentials (Theory in Practice). *O'Reailly Media, Inc.*

Goch, S. (2002). Betterment without airs: social, cultural and political consequences of the deindustrialization in the Ruhr Area. *International Review of Social History*, *3*, 91–99.

Goffman, E. (1963). *Stigma*. Englewood Cliffs, NJ: Prentice-Hall.

Goldemberg, J., Lehmann, D., & Mazursky, D. (2001). The Idea Itself and the Circumstances of Its Emergence as Predictors of New Product Success. *Management Science*, *47*(1), 69–84. doi:10.1287/mnsc.47.1.69.10670

Gordon, R. (2000). Does the New Economy measure up to the great inventions of the past? *The Journal of Economic Perspectives*, *14*, 49–74. doi:10.1257/jep.14.4.49

Gordon, R. (2003). *Hi - tech innovation and future productivity growth: Does supply create its own demands?*Manuscript, Northwestern University.

Grabher, G. (1993). *The Embedded Firm: on the Socio-Economics of Industrial Networks*. London: Routledge.

Granovetter, M. (1973). The strength of weak ties. *American Journal of Sociology*, *78*, 360–380. doi:10.1086/225469

Greider, W. (1997). *One World, Ready or Not: The Manic Logic of Global Capitalism*. New York: Simon & Schuster.

GRI. (2009) *Sustainability Reporting Guidelines*. Retrieved 20.07.2009 fromhttp://www.globalreporting.org.

Griliches, Z. (1992). The search for research-and-development spillovers. *The Scandinavian Journal of Economics*, *94*, S29–S47. doi:10.2307/3440244

Griliches, Z. (1980). R&D and the productivity slow down. *The American Economic Review*, *70*, 2.

Grossman, G. M., & Helpman, E. (Eds.). (1991). *Innovation and Growth in the Global Economy*. Cambridge, MA: MIT Press.

Grossman, G. M., & Helpman, E. (1994). Endogenous innovation in the theory of growth. *The Journal of Economic Perspectives*, *8*, 23–41.

Groucutt, J. (2005). *Foundations of Marketing* (1st ed.). New York: Palgrave Macmillan.

Gu, S., & Lundvall, L. (2006). Policy learning as a key process in the transformation of China's innovation systems. In Lundvall, B.-A., Intakumnerd, P., & Vang, J. (Eds.), *Asian Innovation Systems in Transitions*. Cheltenham: Edward Elgar.

Gulati, R. (1998). Alliances and Networks. *Strategic Management Journal*, *19*, 293–317. doi:10.1002/(SICI)1097-0266(199804)19:4<293::AID-SMJ982>3.0.CO;2-M

Gulati, R. (1999). Network Location and Learning: The Influence of Network Resources and Firm Capabilities on Alliance Formation. *Strategic Management Journal*, *20*, 397–420. doi:10.1002/(SICI)1097-0266(199905)20:5<397::AID-SMJ35>3.0.CO;2-K

Hagedoorn, J. (1993). Understanding the Rationale of Strategic Technology Partnering: Interorganizational Modes of Cooperation and Sectorial Differences. *Strategic Management Journal*, *14*, 371–385. doi:10.1002/smj.4250140505

Hagedoorn, J. (1995). A Note on International Market Leaders and Networks of Strategic Technology. *Strategic Management Journal*, *16*, 241–250. doi:10.1002/smj.4250160307

Hagedoorn, J., & Schakenraad, J. (1994). The Effect of Strategic Technology Alliances on Company Performance. *Strategic Management Journal*, *15*, 291–309. doi:10.1002/smj.4250150404

Hamel, G., & Prahalad, C. (1993). Strategy as Stretch and Leverage [March-April.]. *Harvard Business Review*, 75–84.

Hamel, G. Prahalad. C.K. (1994*). Competing for the future*(pp.231-232). Boston:Harvard Business School Press.

Hansen & Oetinger. (2001). Introducing T-Shaped Managers: KM's Next Generation [March.]. *Harvard Business Review*, 106–116.

Hardin, G. (1968). The Tragedy of the Commons. *Science*, *162*, 1243–1248. doi:10.1126/science.162.3859.1243

Harmaakorpi, V., & Pekkarinen, S. (2002). Regional Development Platform Analysis as a tool for regional innovation policy. *Proceedings of the ERSA 2002 Conference*, Dortmund, Germany

Harrison, J., Hitt, M., Hoskisson, R., & Ireland, R. (2001). Resource Complementarity in Business Combinations: Expending the Logic to Organizational Alliances. *Journal of Management*, *27*, 679–690. doi:10.1177/014920630102700605

Hassink, R., & Shin, D. H. (2005). Guest editorial 'The restructuring of old industrial areas in Europe and Asia'. *Environment & Planning A*, *37*, 571–580. doi:10.1068/a36273

Hati, M. (2006). *Does the new intellectual property regime impede innovation in developing countries: A case study of the Indian pharmaceutical industry*. http//www.globelicsindia2006.org/Moutushi%20 hati.pdf, accessed on Feb17th 2008-05-29

Heller, M., & Eisenberg, R. (1998, May 1). Can Patents Deter Innovation? The Anticommons in Biomedical Research. *Science*, *280*(5364), 698–701. doi:10.1126/science.280.5364.698

Helmi, S. (2008). *Turun ammattikorkeakoulun yhteiskuntavastuuraportti – Käytännön tekoja Varsinais-Suomen työelämälle*. Turku: Turun ammattikorkeakoulu.

Hill, M., & Fee, L. K. (1995). *The Politics of Nation Building and Citizenship in Singapore*. London: Routledge. doi:10.4324/9780203424438

Hoffman La Roche Ltd Vs. Cipla ltd(2008);

Hollanders, H. (2006). *European Regional Innovation Scoreboard, Past, Present and Future, 2nd MLP Regional Benchmarking Workshop*. MERIT – Maastricht University.

Hospers, G. J. (2002). Industrial heritage tourism and regional restructuring in the European Union. *European Planning Studies*, *10*, 397–404. doi:10.1080/09654310220121112

Hospers, G. J. (2004). *Regional-Economic Change in Europe: a Neo-Schumpeterian Vision*. Münster, London: LIT.

Housel & Bell. (2001). *Measuring and Managing Knowledge* (International Edition). New York: McGraw-Hill.

Howells, J. (2005). Innovation and regional economic development: a matter of perspective? *Research Policy*, *34*, 1220–1234. doi:10.1016/j.respol.2005.03.014

Hu, Z., & Khan, M. S. (1997). Why China growing so fast. *IMF Staff Papers*, *44*, 103–131.

Huangsong, L., & Zan, Z. (2005). Cause research on new round economic growth. *Management World*, *11*, 152–153.

Huggins, R., & Izushi, H. (2007). *Competing for Knowledge. Creating, connecting, and growing*. London, New York: Routledge.

I.A 642/2008 IN CS (OS) 89/2008; High Court Of Delhi http://www.vidhiglobe.com/Data_Protection_1_.pdf

Iansiti, M., & Levien, R. (2004). *The Keystone Advantage*. Cambridge, MA: Harvard Business School Press.

Illeris, S. (1996). *The Service Economy: a Geographical Perspective*. Chichester: Wiley.

Ilori, M. O. (2006). *From Science to Technology and Innovation Management. Inaugural Lecture Series, Faculty of Technology, Obafemi Awolowo University, Ile-Ife, Nigeria*. Oxford, UK: Obafemi Awolowo University Press. Ltd.

Ilori, M. O., Adeniyi, A. A., Oyewale, A. A., Sanni, S. A., & Irefin, I. A. (2002). Developing a manufacturing-based economy in Nigeria through science and technology. *Technovation*, *22*(1), 51–60. doi:10.1016/S0166-4972(00)00054-7

Ilori, M. O., Oke, J. S., & Sanni, S. A. (2000). Management of new product development in selected food companies in Nigeria. *Technovation*, *20*, 333–342. doi:10.1016/S0166-4972(99)00147-9

Industrie Canada. (2000). Profil économique du secteur de la biotechnologie canadienne. Équipe de recherche et d'analyse. *Direction générale des sciences de la vie*. Ottawa.

Info-communications Development Authority of Singapore. (2006). *Innovation. Integration. Internationalisation. Report of the iN2015 Steering Committee*. Singapore: Info-communications Development Authority of Singapore.

Ingham, M. & Mothe, C. (2000). La Coopération en recherche et développement. Les déterminants de l'apprentissage organisationnel. *Revue française de gestion*, 107, 71-82, Janvier-février.

Inkpen, A. (1996). Creating Knowledge Trough Collaboration. *California Management Review*, *39*(1).

Investment Commission Report. (2006). *Investment strategy for India*. Ministry of Finance Government of India.

IResearch Co. Ltd. (2007). *The report of current situation of Internet Cafés in China*. Shanghai: Shanghai iResearch Co., Ltd

Jacobs, J. (1961). *The Death and Life of Great American Cities*. New York: Random House.

Jacquemin, A., & Remiche, B. (1988). *Coopération entre entreprises. Entreprises conjointes, stratégies industrielles et pouvoirs publics. Association Internationale de Droit Économique*. De Boeck-Wesmael.

Jassawalla, A., & Sashittal, H. (1998). An Examination of Collaboration In High-Technology New Product Development Processes. *Journal of Product Innovation Management*, *15*, 237–254. doi:10.1016/S0737-6782(97)00080-5

Jiechang, X. (2002). Positive analysis on technical progress and economic growth and the related fiscal and tax policy. *Research on Financial and Economic Issues*, *11*, 53–59.

Jinwen, C. (2004). Demonstration research on human capital and economic growth in China. *Research On Financial and Economic Issues*, *9*, 9–13.

Johnson, G., & Scholes, K. (2002). *Exploring Corporate Strategy*. Englewood Cliffs, NJ: Prentice-Hall.

Joint Admissions and Matriculation Board (JAMB). (2007).Retrieved from http://www.jamb.org.ng (April 17, 2007).

Jolly, D. (1995). Manager un partenariat technologique. *L'Expansion Management Review*, *77*, 72–82.

Jolly, V. K. (1997). *Commercializing New Technologies: Getting from Mind to Market*. Boston: Harvard Business School Press.

Jones, C., & Williams, J. (1999). *Too much of a Good Thing? The Economics of Investment in R&D. I. Cambridge*. MA: National Bureau of Economic Research.

Jones, R. (2005). *The 2005 Economic Survey of Korea. Presentation at the OECD Breakfast Series, Washington, DC, 30 November 2005*. Paris: OECD.

Jones, L. E., & Manuelli, R. (1990). A convex model of equilibrium growth: Theory and policy implications. *The Journal of Political Economy*, *98*, 1008–1038. doi:10.1086/261717

Jorde, T. M., & Teece, D. J. (1990). Innovation and cooperation: Implications for competition and antitrust. *The Journal of Economic Perspectives*, *4*(3), 75–96.

Jorgenson, D. W., Gollop, F. M., & Fraumeni, B. M. (1987). *Productivity and US Economic Growth*. Cambridge, MA: Harvard University Press.

Jorgenson, D. W., & Stiroh, K. (2000). Raising the Speed Limit: US Economic Growth in the Information Age. *Brookings Papers on Economic Activity*, (1): 125–211. doi:10.1353/eca.2000.0008

Kadushin.C (2004). Basic *Network Concepts. The Culture and Commerce of Publishing*. New York.

Kahal, S. (2001). *Business in Asia Pacific: Text and Cases*. Oxford, UK: Oxford University Press.

Kaplan, R. S., & Norton, D. P. (2001). *The Strategy-Focused Organization, How Balanced Scorecard Companies Thrive in the New Business Environment*. Boston, MA: Harvard Business School Press.

Kaplan, R. S., & Norton, D. P. (2004). *Strategy Maps, Converting the Intangible Assets into Tangible Outcomes*. Boston, MA: Harvard Business School Press.

Kasper, H. (2006). *Services Marketing Management: A strategic perspective* (2nd ed.). West Sussex: John Wiley & Sons Ltd.

Katz, D., & Kahn, R. (1996). *The Social Psychology of Organizations*. New York: Wiley.

Katz, J. (Ed.). (1987). *Technology generation in Latin American manufacturing industries*. Macmillan.

Kedrova, J. (2004). *Measuring Productivity,* Federal Reserve Bank of Dallas Expand Your Insight, June 2004, Free Enterprise, Measures of Productivity, Krugman, P. (1991). Increasing returns and economic geography. *The Journal of Political Economy*, *99*, 483–499.

Keller, W. (1996). Absorptive capacity: on the creation and acquisition of technology in development. *Journal of Development Economics*, *49*, 199–227. doi:10.1016/0304-3878(95)00060-7

Kent, S. (2002). *Model-Driven Engineering. Proceedings of the Third International Conference on Integrated Formal Method.* (pp. 286-298). London, UK: Springer.

Kettunen, J. (2006). Strategic planning of regional development in higher education. *Baltic Journal of Management*, *1*(3), 259–269. doi:10.1108/17465260610690917

Kettunen, J. (2008). A conceptual framework to help evaluate the quality of institutional performance. *Quality Assurance in Education*, *16*(4), 322–332. doi:10.1108/09684880810906472

Kettunen, J., & Kantola, I. (2005). Management information system based on the Balanced Scorecard. *Campus-Wide Information Systems*, *22*(5), 263–274. doi:10.1108/10650740510632181

Kettunen, J., & Kantola, M. (2006). The implementation of the Bologna Process. *Tertiary Education and Management*, *12*(3), 257–267.

Kettunen, J., & Luoto, L. (2008). Cooperation between universities and ICT enterprises. In Zhao, F. (Ed.), *Information Technology Entrepreneurship and Innovation* (pp. 277–292). Hershey, PA: IGI Global.

Keyala, B. K. (2000). TRIPs patent system; implication for health care and pharma industry. *Centre for the study of global trade system of development*. New Delhi, India.

Khalil, T. (2000). *Management of Technology: The Key to Competitiveness and Wealth Creation* (International Editions). New York: McGraw-Hill.

Killing, P. (1988). Understanding Alliances: The Role of Task and Organizational Complexity. In Contractor, F., & Lorange, P. (Eds.), *Cooperative Strategies in International Business*. Lexington, MA: Lexington Books.

King, N. (1992). Modelling the innovation process: an empirical comparison of approaches. *Journal of Occupational and Organizational Psychology*, *65*(2), 89–100.

Kitson, M., Martin, R., & Tyler, P. (2004). Regional Competitiveness: An Elusive yet Key Concept? *Regional Studies*, *38*(9), 991–999. doi:10.1080/0034340042000320816

Klenow, P., & Rodrı′ guez-Clare, A. (1997). Economic growth: A review essay. *Journal of Monetary Economics*, *40*, 597–617. doi:10.1016/S0304-3932(97)00050-0

Klevorick, A. K., Levin, R. C., Nelson, R. R., & Winter, S. G. (1995). On the sources and significance of interindustry differences in technological opportunities. *Research Policy*, *24*, 185–205. doi:10.1016/0048-7333(93)00762-I

Koput, K. (1997). A chaotic model of innovative search: some answers, many questions. *Organization Science*, *8*(5), 528–542. doi:10.1287/orsc.8.5.528

Korres, G. (1996). *Technical change and economic growth: an empirical analysis of EEC countries*. London: Avebury Press.

Korres, G. (2008). *Technical change and economic growth: Inside to the Knowledge Based Economy*. London: Avebury Press.

Kotler, P., Filiatrault, P., & Turner, R. (2000). Le management du marketing. Gaetan Morin éditeur. Montréal. Canada.

Kotler.P, Jain. D, Maesincee.S (2002). Marketing Moves: A New Approach to Profits, Growth and Renewal. *Harvard Business School Press*.

KPMG Corporate Finance inc. (2002). Overview of Selected Issues of the Biotechnology Industry in Four Countries. Centre québecois de valorisation des biotechnologies, Ministère de l'Industrie et du Commerce, Gouvernement du Québec, Investissement Québec, BioQuébec.Retrieved from www.kpmg.com.

KPMG Report on Indian Biotechnology industry (2006). 18, India.

Kruss, G. (2006). University-industry links: lessons from South Africa. Retrieved from Science and Development Network (SciDev.net) - http://www.scidev.net/dossiers/index.cfm?fuseaction=dossierreaditem&dossier=13&type=3&itemid=542&language=1 on December 3, 2007.

Kuhlen (2003). *Change of Paradigm in KM-Framework for the Collaborative Production and Exchange of Knowledge*. University of Konstanz, Germany.

Lagendijk, A., & van Houtum, H. (2001). Contextualising regional identity and imagination in the construction of polycentric urban regions: the cases of the Ruhr Area and Basque Country. *Urban Studies (Edinburgh, Scotland)*, *38*, 747–768. doi:10.1080/00420980120035321

Lall, S. (2003). Indicators of the relative importance of IPRs in developing countries. [New York: Elsevier.]. *Research Policy*, *32*(9), 1657–1680. doi:10.1016/S0048-7333(03)00046-5

Lall, S. (1987). *Learning to industrialize. The acquisition of technological capability by India*. New York: Macmillan.

Lall, S. (1992). Technological Capabilities and Industrialisation. *World Development*, *20*(2). doi:10.1016/0305-750X(92)90097-F

Lall, S. (1991). Human Resources, Technology and Industrial Development in Sub-Saharan Africa. In Chhiber, A., & Stanely, F. (Eds.), *Economic Reform in Sub-Saharan Africa*. Washington, DC: World Bank.

Lanjouw, J. (2000). *Do patent matter? Evidence after GATT.* NBER working chapter no. 7495, National bureau of Economic Research, Cambridge.

Larsen, K., & Salter, A. (2004). Searching high and low: waht types of firms use universities as a source of innovation. *Research Policy*, *33*, 1201–1215. doi:10.1016/j.respol.2004.07.004

Le Gall, J-M. (1999). L'apprentissage communicationnel» *Revue française de gestion.* 130, 31-44.

Lee, B. Y. (2007). Singapore's e-Government – Sharing Our Experience. Speech by Dr Lee Boon Yang, Minister, Information, Communications and the Arts at the 4th Ministerial e-Government Conference on "Reaping the Benefits of e-Government" at Feira Internacional de Lisboa, Lisbon, Portugal, 20 September.

Lengyun, H., & Wen, Y. (2007). Economic growth: FDI spillover effect. *Economic Review (Federal Reserve Bank of Atlanta)*, *5*, 72–76.

Lerner, J., & Merges, R. (1997). The Control of Strategic Alliances: An Empirical Analysis of Biotechnology Collaborations. National Bureau of Economic Research. NBER Working Paper Series No 6014. Cambridge, MA.

Lerro, A., & Carlucci, D. (2007). Intellectual Capital and Regions: Origins, Theoretical Foundations and Implications for Decision-makers. *International Journal of Learning and Intellectual Capital*, *4*(4), 357–376. doi:10.1504/IJLIC.2007.016333

Levinthal, D. (1991). Random walks and organizational mortality. *Administrative Science Quarterly*, *36*, 397–420. doi:10.2307/2393202

Lewis, J. (1990). *Partnerships for Profit*. London: Structuring and Managing Strategic Alliances. Collier Macmillan Publishers.

Lindtner, S., Nardi, B., et al. (2008). A *Hybrid Cultural Ecology: World of Warcraft in China*. CSCW 08, November, San Diego, USA.

Lintz, G., Müller, B., & Schmude, K. (2007). The future of industrial cities and regions in central and eastern Europe. *Geoforum*, *38*, 512–519. doi:10.1016/j.geoforum.2006.11.011

Liu, S., & Xu, S. (2002). Research Summarization on International Development of Regional Innovation System. *China Science and Technology Forum*, *5*, 33–37.

Lloyd-Ellis, H., & Roberts, J. (2002). Twin engines of growth: skills and technology as equal partners in balanced growth. *Journal of Economic Growth*, *7*, 87–115. doi:10.1023/A:1015633528122

Lorange, P., & Roos, J. (1992). *Strategic Alliances, Formation, Implementation, and Evolution*. USA: Blackwell publishers.

Love, J. H., & Roper, S. (1999). The determinants of innovation: R&D, technology transfer and networking effects. *Review of Industrial Organization, 15*, 43–64. doi:10.1023/A:1007757110963

Love, J. (2003). *Evidence regarding research and development investments in innovative and non-innovative medicines*. Consumer Project on Technology.

Lucas, R. E. (1988). On the Mechanics of Economic Development. *Journal of Monetary Economics, 22*, 3–42. doi:10.1016/0304-3932(88)90168-7

Lucas, R. E. (1990). Why doesn't capital flow from rich to poor countries. *The American Economic Review, 80*, 92–96.

Lucas, R. E. (1993). Making a miracle. *Econometrica, 61*, 251–272. doi:10.2307/2951551

Lui, T. Y. (2008). Infocomm talent as a key driver of Singapore's economic competitiveness. Speech by RAdm (NS) Lui Tuck Yew, Senior Minister of State for Information, Communications and The Arts, at the National Infocomm Scholarship 2008 Award Ceremony, 5 August.

Lui, T. Y. (2009). Speech by RAdm (NS) Lui Tuck Yew, Acting Minister for Information, Communications and The Arts, at the Official Opening Of imbX Event 2009, 16 June.

Lundvall, B. A., & Johnson, B. (1994). The learning economy. *Journal of Industry Studies, 1*, 23–42.

Luoto, L., & Helmi, S. (2008). *Corporate social responsibility report*. Turku: Summary in English. Turun AMK.

Macartney, J. (2008). China watches over internet café customers in web crackdown. *Times*. Retrieved from http://technology.timesonline.co.uk/tol/news/tech_and_web/the_web/article4958778.ece

Maddison, A. (1998). *Chinese economic performance in the long run*. France: OECD Publication.

Maddison, A. (2001). *The World Economy: A Millennial Perspective. Development Centre Studies*. Paris: OECD.

Maillat, D. (1995). Territorial dynamic, innovative milieu and regional policy. *Entrepreneurship & Regional Development, 7*, 157–165. doi:10.1080/08985629500000010

Malecki, E. J., & Varaia, P. (1986). Innovation and Changes in Regional Structure. In Nijkamp, P.Elsevier Science Publishers (Ed.), *Handbook of Regional and Urban Economics* (*Vol. I*).

Malecki, E. J. (1991). *Technology and economic development: the dynamics of local regional and national change*, (eds.) Longman Scientific and Technical.

Malerba, F., & Orsenigo, L. (2001). *Towards a history friendly model of innovation, market structure and regulation in the dynamics other pharmaceutical industry: The age of random screening*. Working Papers with number 124 CESPRI, Centre for Research on Innovation and Internationalisation, Universita' Bocconi, Milano, Italy.

Malhotra, Y. (2000). KM for E-Business Performance: Advancing Information Strategy to "Internet Time". *The Executive Journal, CRC Press, 16*(4), 5–16.

Maltz, E., & Kohli, A. (2000). Reducing Marketing's Conflict with other Functions: The Differential Effects of Integrating Mechanisms. *Academy of Marketing Science, 28*, 479–492. doi:10.1177/0092070300284002

Mansfield, E. (1991). Academic research and industrial innovation. *Research Policy, 20*, 1–12. doi:10.1016/0048-7333(91)90080-A

Mansfield, E. (1995). Academic research underlying industrial innovation: sources, characteristics, and financing. *The Review of Economics and Statistics, 77*(11), 55–65. doi:10.2307/2109992

Mansfield, E. (1998). Academic research and industrial innovation: an update of results. *Research Policy, 26*, 773–776. doi:10.1016/S0048-7333(97)00043-7

Mansfield, E. (1968). *Industrial Research and Technological Innovation: An Econometric Analysis*. New York: W.W. Norton and Co.

Maria, A., & Ramani, S. (2005, February 12-18). TRIPS and its possible impact on the biotech based segment of the Indian Pharmaceutical industry. *Economic and Political Weekly*, 675–683.

Maria, A., Ruet, J., Zerah, M. H. (2004). *Final report on Biotechnology in India.* Commissioned by French embassy.

Marinova, D., & Phillimore, J. (2003). Models of Innovation. In Shavinina, L. V. (Ed.), *The International Handbook on Innovation*. Oxford, UK: Elsevier Science. doi:10.1016/B978-008044198-6/50005-X

Marr, B., & Schiuma, G. (2001). Measuring and Managing Intellectual Capital and Knowledge Assets in New Economy Organisations. In Bourne, M. (Ed.), *Performance Measurement Handbook*. London: GEE Publishing Ltd.

Marshall, A. (1916). *The Principles of Economics*. London: Macmillan.

Martin, M., & Forrest, J. (1994). *Strategic Alliances Between Large and Small Research Intensive Organizations: Experiences in the Biotechnology Industry. Discussion Papers No 132*. Halifax: Dalhousie University, Center for International Business Studies.

Martin, P., & Ottaviano, G. I. P. (1999). Growing locations: Industry location in a model of endogenous growth. *European Economic Review*, *43*, 281–302. doi:10.1016/S0014-2921(98)00031-2

Maskus, K. (1998). The role of intellectual property rights in encouraging foreign direct investment and technology transfer. *Journal of competition and International Law*.

Matter, P. S. accessed from http://www.southcentre.org/publications/publichealth/publichealth-04.htm#P327_38330#P327_38330 on 12th Jan 2008

McCarthy, A., Garavan, T., & Toole, T. (2003). HRD: working at the boundaries and interfaces of organizations. *Journal of European Industrial Training*, *27*(2), 58–72. doi:10.1108/03090590310468886

McEvily, B., & Zaheer, A. (1999). Bridging ties: a source of firm heterogeneity in competitive capabilities. *Strategic Management Journal*, *20*, 1133–1156. doi:10.1002/(SICI)1097-0266(199912)20:12<1133::AID-SMJ74>3.0.CO;2-7

Mennel, T., Steffens, J., & Klemek, C. (2007). *Block by Block: Jane Jacobs and the Future of New York*. New York: The Municipal Art Society of New York.

Merges, R., & Nelson, R. R. (1990). On the complex economics of patent scope. *Columbia Law Review* 90, Morriset, J.P. (1999), *How tax policy and incentives affect foreign direct investment: A review*. World Bank Policy Research Working Paper No. 2509; World Bank Group - Foreign Investment Advisory Service (FIAS) NEDE PIRNIA World Bank.

Millier, P. (1997). *Stratégie et marketing de l'innovation technologique*. Paris: Dunod.

Millier, P. (1989). Le marketing des produits "High Tech". *Les éditions d'organisation*, Paris.

Mohr, J. (2000). The Marketing of High-technology Products and Services: Implications for Curriculum Content and Design. *Journal of Marketing Education*, *22*, 246–259. doi:10.1177/0273475300223008

Moinjeon, B., & Ramanantsoa, B. (1995). Comment rendre l'entreprise apprenante. *L'Expansion Management Review.*, *78*, 96–103.

Montalva, M. (2005). El Marketing Estratégico Aplicado en Chile y en Canadá para Empresas de Tecnología de Información y Comunicación. *Trend Mangement*, *7*, 130–137.

Morgan, K. (2004). The exaggerated death of geography: learning, proximity and territorial innovation systems. *Journal of Economic Geography*, *4*, 3–22. doi:10.1093/jeg/4.1.3

Morgan, K., & Neuwelaer, C. (1998). *Regional Innovation Strategies: The Challenge for Less Favoured Regions*. London: Jessica Kingsley.

Moulaert, F., & Sekia, F. (2003). Territorial Innovation Models: A Critical Survey. *Regional Studies*, *37*(3), 289–302. doi:10.1080/0034340032000065442

Muller, A., & Valikangas, L. (2002). Extending the boundary of corporate innovation. *Strategy and Leadership*, *30*(3), 4–9. doi:10.1108/10878570210427891

Munos, B., (2006). *Can open source R&D reinvigorate drug research*? Nature Review drug discovery.

Nair, M.S. (21/09/2004). Data exclusivity: *The Indian perspective*. Lex Orbis; Mondaq Publication.

Nandan, K. (2001-2002). *Operation successful but patient dead; MSc thesis in Economics and social history*. Oxford, UK: University of Oxford.

National Biotechnology Development Strategy Policy. (2007). Department of Biotechnology, Ministry of Science and Technology; Govt of India, available at http:// www.dbtindia.gov.in/ biotechstrategy.htm.

National Universities Commission (NUC). (2007). *List Of Nigerian Universities And Years Founded*.Retrieved from http://www.nucnigeria.info/universities.htm (accessed April 17, 2007)

Nelson, R., & Winter, S. (1982). *An Evolutionary Theory of Economic Change Harvard*. Cambridge, MA: University Press.

Nelson, R. (1994). Why do Firms Differ and How Does It Matter. In Rumelt, R., Schendel, D., & Teece, D. (Eds.), *Fundamental Issues in Strategy*. Boston: Harvard Business School Press.

Nelson, R. R. (1993). *National Innovation Systems: A Comparative Analysis*. New York.

Nerkar, A., & Roberts, P. (2004). Technological and Product – Markett Experience and the Success of New Product. Introduction in the Pharmaceutical Industry. *Strategic Management Journal*, *25*, 779–799. doi:10.1002/smj.417

Ng, P. T. (2009). Examining the Use of New Science Metaphors in Learning Organisation. *The Learning Organization*, *16*(2), 168–180. doi:10.1108/09696470910939224

Ng, E. H. (2008). Opening Address by Dr Ng Eng Hen, Minister for Education and Second Minister for Defence, at the International Conference on Teaching and Learning with Technology at the Suntec Convention Hall, 5 August.

Niko Partners. (2008). *PC Games and Console Hardware Sales Surge in CHINA'S Booming Economy*. Retrieved from http://www.nikopartners.com/npr_050208.pdf

Niosi, J. (1994). *New Technology Policy and Social Innovation in the Firm*. London: Printer Publishers Inc.

Niosi, J. (2003). Alliances are not Enough Explaining Rapid Growth in Biotechnology. *Research Policy*, *32*, 737–750. doi:10.1016/S0048-7333(02)00083-5

Niosi, J., & Bas, T. G. (2001). The Competencies of Regions – Canada's Clusters in Biotechnology. *Small Business Economics*, *17*, 31–42. doi:10.1023/A:1011114220694

Nonaka, I., & Takeuchi, H. (1995). *The Knowledge-Creating Company*. New York: Oxford University Press.

Nonaka, I., Takeuchi, H., & Umemoto, K. (1996). A Theory of Organizational Knowledge Creation. *International Journal of Technology Management*, *11*(7/8), 832–845.

Nonaka, I. (1994). A dynamic theory of organisational knowledge creation. *Organization Science*, *5*(1), 14–37. doi:10.1287/orsc.5.1.14

Nonaka, I., & Takeuchi, H. (1995). *The Knowledge-creating Company – How Japanese Companies Create the Dynamics of Innovation*. New York: Oxford University Press.

Norris, K., & Vaizey, J. (1973). *The Economics of Research and Technology* (Cater, C., Ed.). London: George Allen and Unwin Ltd.

OECD. (1981). *North/South Technology Transfer Adjustment Ahead. Organisation for Economic Co-operation and Development (OECD)* (pp. 17–19). Publication.

OECD. (2003). *The Sources of Economic Growth in OECD Countries*. Paris: OECD.

OECD. (2005). *Economy - wide Indicators of Productivity*, Retrieved from www.oecd.org

Ohmae, K. (1990). *The Borderless World*. New York: Harper Collins.

Ohmae, K. (1995). *The End of the Nation State: The Rise of Regional Economies*. New York: The Free Press.

Oliner, S. D., & Sichel, D. E. (2001). The resurgence of growth in late 1990s: Is information technology the story? *The Journal of Economic Perspectives*, *14*, 3–22. doi:10.1257/jep.14.4.3

Olkkonen, R. (2001). The Network Approach to International Sport Sponsorship Arrangement. *Journal of Business and Industrial Marketing*, *16*(4), 309–329. doi:10.1108/EUM0000000005503

Organization for Economic Cooperation and Development Report. (2004). *Patents and innovation: Trend and policy challenges*. OECD Publication.

Ott, R. (2004).Applying the patent to living material in India. 2 Okla J.L. & Tech. 17.

Owen-Smith, J., & Powell, W. W. (2004). Knowledge networks as channels and conduits: the effects of spillovers in Boston biotechnology community. *Organization Science*, *15*, 2–21. doi:10.1287/orsc.1030.0054

Oyebisi, T. O., Ilori, M. O., & Nassar, M. L. (1996). Industry-academic relations: an assessment of the linkages between a university and some enterprises in Nigeria. *Technovation*, *16*(4), 203–209. doi:10.1016/0166-4972(95)00049-6

Oyelaran-Oyeyinka, B. (1997). *NNEWI: An Emergent Industrial Cluster in Nigeria*. Ibadan, Nigeria: Technopol Publishers.

Oyelaran-Oyeyinka, B., Laditan, G. O. A., & Esubiyi, A. O. (1996). Industrial innovation in Sub-Saharan Africa: the manufacturing sector in Nigeria. *Research Policy*, *25*(7), 1081–1096. doi:10.1016/S0048-7333(96)00889-X

Oyelaran-Oyeyinka, B. (2005). *Translating Research into Innovation in Nigeria: A Critical Perspective.* Paper Presented at the Seminar organized by the Nigerian Academy of Engineering, May 17-19, Abuja, Nigeria

Oyelaran-Oyeyinka, B. (2006). *Don't make a fetish out of R&D spending*. Retrieved from Science and Development Network (SciDev.net) - http://www.scidev.net/Opinions/index.cfm?fuseaction=readOpinions&itemid=543&language=1 on December, 12, 2007.

Oyewale, A. A. (2005). *Addressing the research-industry linkage impasse in Nigeria: The critical issues and implementation strategies*.Paper presented at Globelics-South Africa, 2005. www.globelics2005africa.org.za/papers/p0032/Globelics2005_Adesina%20Ayobami%20Oyewale.pdf (accessed December 2, 2006).

Palacios-González, E., Fernández-Fernández, H., & García-Díaz, V., G-Bustelo, B.C.P., Lovelle, J.M.C., & Sanjuan-Martínez, O. (2008). General Purpose MDE Tools. *International Journal of Artificial Intelligence and Interactive Multimedia*, *1*(1), 77–88.

Papageorgiou, C. (2003). Imitation in a non-scale R&D growth model. *Economics Letters*, *80*(3), 287–294. doi:10.1016/S0165-1765(03)00140-X

Papageorgious, C., & Perez, F. S. (2006). Dynamics in non-scale R&D growth model with human capital: Explaining the Japanese and South Korean development experiences. *Journal of Economic Dynamics & Control*, *30*(6), 901–930. doi:10.1016/j.jedc.2005.03.005

Parker, T. (2001). *Biotech companies call for patent reform.* Managing Intellectual Property. TRIPS and Public Health, Paper submitted to TRIPS Council, 29th June 2001, IP/C/W/296

Pasher, E., & Shachar, S. (2005). The Intellectual Capital of the State of Israel. In Bounfour, A., Edvinsson, L. (2005, Eds.), *Intellectual Capital for Communities*, Boston: Elsevier Butterworth-Heinemann.

Pavitt, K. (1984). Sectoral patterns of technological changes: towards a taxonomy and a theory. *Research Policy*, *13*, 343–347. doi:10.1016/0048-7333(84)90018-0

Pavitt, K., & Soete, L. (1982). International differences in economic growth and the international location of innovation. In Giersch, H. (Ed.), *Emerging Technologies: The Consequences for Economic Growth, Structural Change and Employment* (pp. 105–133). Tübingen: Mohr.

Penrose, E. (1959). *The Theory of the Growth of the Firms*. White Plains, NY: M.E. Sharpe, Inc.

Perez-Sebastian, F. (2000). Transitional dynamics in an R&D-based growth model with imitation: Comparing its predictions to the data. *Journal of Monetary Economics*, *42*, 437–461. doi:10.1016/S0304-3932(99)00055-0

Peteraf, M., & Bergen, M. (2003). Scanning Dynamic Competitive Landscapes: A Market-Based and Resource-based Framework. *Strategic Management Journal*, *24*, 1027–1041. doi:10.1002/smj.325

Pharma (2001). India: A cautionary tale on critical importance of IP protection. Retrieved from at http://www.Phrma.org/publications/documents/backgrounders/2001-03-01.202/html

Pike, A., Rodriguez-Pose, A., & Tomaney, J. (2006). *Local and Regional Development* (1st ed.). London: Regional Studies Association-Routledge.

Piore, M., & Sabel, C. (1984). *The Second Industrial Divide*. New York: Basic Books.

Pisano, G., & Mang, P. (1993). Collaborative Product Development and the Market for Know-how: Strategies and Structures in the Biotechnology Industry. *Research on Technological Innovation. Management and Policy*, *5*, 109–136.

Pohl, W., & Ponthöfer, L. (2002). *Innovationsraum Ruhrgebiet*. Berlin: Cornelsen.

Policet, A., & Noël, A. (1994). *Les alliances: une stratégie de développement pour les nouvelles PME technologiques? Cahier de recherche. Centre d'études en administration internationale*. CETAI.

Porter, M. E. (1990). *The Competitive Advantage of Nations*. Basingstoke: Macmillan.

Poster, M. (1990). *The Mode of Information: Poststructuralism and Social Context*. Cambridge, MA: Polity.

Postman, N. (1993). *Technopoly: The Surrender of Culture to Technology*. New York: Vintage.

Powell, W., Koput, K., & Smith-Doerr, L. (1996). Interorganizational Collaboration and the Locus of Innovation: Networks of Learning in Biotechnology. *Administrative Science Quarterly*, *41*, 1–16. doi:10.2307/2393988

Praveen, A. (2005). Healthcare biotechnology firms in India: Evaluation, structure and growth; *current science, 89*(3).

Projekt Ruhr. (2002). *The Ruhrgebiet*. Essen: Projekt Ruhr.

Prusak, L. (1996). *The knowledge advantage*. Strategy and Leadership, March-April.

Prusak, L. (1997). *Knowledge in Organisations*. Woburn, MA: Butterworth-Heinemann.

Prusak, L., & Davenport, T. H. (1998). *Working Knowledge* (p. 5). Boston: Harvard Business School Press.

Psacharopoulos, G. (1994). Returns to investment in education:A global update. *World Development*, *22*(9), 1325–1343. doi:10.1016/0305-750X(94)90007-8

Pulic, A. (2005). Value Creation Efficiency at National and Regional Levels: Case Study-Croatia and the European Union. In Bounfour, A., Edvinsson, L. (Eds.) *Intellectual Capital for Communities*. Boston, MA: Elsevier Butterworth-Heinemann, Rondè, P. & Hussler, C. (2005). Innovation in regions: What does really matter? *Research Policy*, 34, 1150-1172.

Qiong, H., & Zheng, W. (2006). Study on the location factors of overseas R&D investment in China. *China Soft Science*, *7*, 113–120.

Raak, A., & Paulus, A. (2001). A Sociological Systems Theory of Interoganizational Network Development in Health and Social Care. *Systems Research and Behavioral Science Syst. Res.18, 207-224.*New York: John Wiley & Sons Ltd.

Radal, C. (2000). Le nouveau rôle du Marketing dans l'industrie pharmaceutique. Retrieved from http://www.edoceo.net/articles/le_role_marketing_indu_pharma_me.htm

Rank hovis Mc Dougall case (1978). FSR, 588

Rashmi, R. (Sept 2007). Indian biopharmaceutical industry in the era of globalization: Path for novel drug industry. In *Genetically Modified Organisms: Emerging Law and Policy in India.* New Dehli, India:TERI Press.

Ratzan, S. (1998). *Mad Cow Crisis: Health and the Public Good.* New York: NYU Press.

Rebelo, S. (1991). Long run policy analysis and long run growth. *The Journal of Political Economy, 99*, 500–521. doi:10.1086/261764

Reddy, P. March 31, 2008; Foreign pharma may adopt differential pricing; Spicy IP; Accessed at http://spicyip-india.blogspot.com/2008/03/foreign-pharma-may-adopt-differential.html

Reddy, S. (2006, May). *The cost of India complying with world intellectual property rights: effect on the and access to drugs.* PhD Thesis: Berkeley University.

Report, U. N. C. T. A. D. (2004). The Biotechnology Promise Capacity-building for Participation of Developing Countries in the Bioeconomy, UNCTAD/ITE/IPC/2004/2,Geneva

Right to Adequate Food. (2002). Retrieved May 1, 2009, from Food and Agriculture Organization of the United Nations: http://www.fao.org/righttofood/principles_en.htm

Roberts, J. (2000). Developing New Rules for New Markets. *Academy of Marketing Science, 28*, 31–44. doi:10.1177/0092070300281004

Roche vs NATCO: India's First "Doha Style" Compulsory License? January 16, 2008 http://spicyipindia.blogspot.com/2008/01/roche-vs-natco-indias-first-doha-style.html accessed on February 18 2008

Romer, P. M. (1990). Endogenous technological change. *The Journal of Political Economy, 98*(5), 71–102. doi:10.1086/261725

Romer, P. M. (1986). Increasing Returns and Long-Run Growth. *The Journal of Political Economy, 94*, 1002–1037. doi:10.1086/261420

Romer, P. M. (1987). Growth based on increasing returns due to specialization. *The American Economic Review, 77*(2), 56–62.

Romer, P. M. (1990). Endogenous Technological Change. *The Journal of Political Economy, 98*, 71–102. doi:10.1086/261725

Romer, P. M. (1990). Human Capital and Growth: Theory and Evidence, *Carnegie-Rochester Conference Series on Public Policy, 32*, 251-86.

Romijn, Henny &Albaladejo, Manuel (2000) *Determinants of Innovation Capability in Small UK Firms: An Empirical Analysis.* QEH Working Paper 40

Roos, J., Edvinsson, L., & Roos, G. (1998). *Intellectual Capital: Navigating in the New Business Landscape.* New York, NY: New York University Press.

Rose, C. M. (1998). *The several futures of property: Of cyberspace and folktales, emission trades and ecosystems.* 83 MINN. *L. REV., 129*, 179.

Rothaermel, F., & Deeds, D. (2004). Exploration and Exploitation Alliances in Biotechnology: A System of New Product Development. *Strategic Management Journal, 25*, 201–221. doi:10.1002/smj.376

RVR. (2006). *The Ruhrgebiet: Facts and Figures.* Essen: RVR.

RVR. (2009). *RVR Online Zahlen & Statistik.* Essen: RVR.

Sachs, J. D., & Woo, W. T. (1997). Understanding China's economic performance.*NBER Working Paper Series, (Working Paper 5935).*

Sachwald, F. (1994). *Les défis de la mondialisation, innovation et concurrence. Institut français des relations internationales.* Paris: Masson.

Salter, A., & Martin, B. (2001). The economic benefits of publicly funded basic research: a critical review. *Research Policy, 30*, 509–532. doi:10.1016/S0048-7333(00)00091-3

Samara, A. (2007). Framework for Discovering KM Forces: The Fifth Element. *Journal of KM Practice, 8*(1).

Sanchez Medina, A. J., Melian Gonzales, A., & Garcia Falcon, J. M. (2007). Intellectual Capital and Sustainable Development on Islands: An Application to the Case of Gran Canaria. *Regional Studies, 41*(4), 473–487. doi:10.1080/00343400600928327

Sandager, I., & Andersen, B. 2008. *Complexity and learning regions: learning regions as complex adaptive systems*. In Doyle, Adams, Tibbitt E Welsh (eds.), *Building stronger communities:Connecting research, policy and practice*. Plymouth, England: Latimer Trend.

Sassen, S. (2000). Digital Networks and the State: Some Governance Questions. *Theory, Culture & Society, 17*(4), 19–33. doi:10.1177/02632760022051293

Saxenian, A. (1994). *Regional Advantage: Culture and Competition in Silicon Valley and Route 128*. Cambridge, MA: Harvard University Press.

Schiuma, G., & Lerro, A. (2008). Knowledge-based Capital in Building Regional Innovation Capacity. *Journal of Knowledge Management, 12*(5), 121–136. doi:10.1108/13673270810902984

Schiuma, G., Lerro, A., & Carlucci, D. (2007). The Knoware Tree and the Regional Intellectual Capital Index: An Assessment within Italy. *Journal of Intellectual Capital*, 9(2), 283–300. doi:10.1108/14691930810870346

Schönert, M. (1995). *Umweltwirtschaft als Hoffnung für altindustrielle Regionen? Chancen und Risiken für die Regionalentwicklung*. Working Paper 1995/141. University of Dortmund: Institute for Spatial Planning.

Schroeder, R. G., Van de Ven, A. H., Scudder, G. D., & Polley, D. (1989). The development of innovation ideas. In Van de Ven, A. H., Angle, H. L., & Poole, M. (Eds.), *Research on the management of innovation: the Minnesota studies*. New York: Harper & Row.

Schumpeter, J. A. (1934). *The theory of economic development*. Cambridge, MA: Harvard Economic Studies.

Schumpeter, J. A. (1942). *Capitalism, socialism and democracy*. New York: Harper.

Schutte, F., & van der Sijde, P. (2000). *The university and its region*. Enschede: Twente University Press.

Sebestyen, G., Nedevschi, S., Saplacan, G., Cerghizan, M., Todor, N., & Rusu, M. Towards a Traceability Solution on the Food Supply Chain. *The 8th International Conference on Technical Informatics*. June 5-6, 2008. Timisoara, Romania.

Shan, W., Walker, G., & Kogut, B. (1994). Interfirm Cooperation and Startup Innovation in the Biotechnology Industry. *Strategic Management Journal, 15*, 387–394. doi:10.1002/smj.4250150505

Shane, S. (1992). Why do some societies invent more than others? *Journal of Business Venturing, 7*, 29–46. doi:10.1016/0883-9026(92)90033-N

Shane, S. (1993). Cultural influences on national rates of innovation. *Journal of Business Venturing, 8*, 59–73. doi:10.1016/0883-9026(93)90011-S

Shanklin, W., & Ryans, J. (1984). *Marketing High Technology. Washington, DC: LexingtonBooks*. USA: Heath and Company.

Shanklin, W., & Ryans, J. (1993). Organizing for High-tech Marketing. *Harvard Business Review, 62*, 164–171.

Sharma.A (2005). *Social Networks*. University of Texas, Austin.

Sheehan, J., & Wyckoff, A. (2003). Targeting R&D: Economic and Policy Implications of Increasing R&D Spending. *STI Working Paper 2003/8*, OECD, Paris.

Shelda, D. (2006). *Knowledge Management*. New York: John Wiley & Sons Australia, Ltd.

Sherman, D., Souder, W., & Jenssen, S. (2000). Differentiel Effects of the Primary Forms of Cross Functional Integration on Product Development Cycle Time. *Journal of Product Innovation Management, 17*, 257–267. doi:10.1016/S0737-6782(00)00046-1

Shipilov, A., Rowley, T., & Aharonson, B. When do Networks Matter? (2006). A Study of Tie Formation and Decay. In Baum J.A.C, Dobrev, S. and Arjen van Witteloostuijn (Eds.), *Advances in Strategic Management* (Vol.23. June. Skyrme. D (1998). *Commercializing Organizational Knowledge*. ArkGroup Asia.Com.

Shipilov. A. V (2005). Should you bank on your network? Relational and positional embeddedness in the making of financial capital. *Strategic Organization, 3*(3).

Sieverts, T. (2005). *Zwischenstadt: Zwischen Ort und Welt, Raum und Zeit, Stadt und Land* (3rd ed.). Basel: Birkhäuser.

Silverberg, G., & Verspagen, B. (1995). Long term cyclical variations of catching up and falling behind. An evolutionary model. *Journal of Evolutionary Economics, 5*, 209–227. doi:10.1007/BF01198304

Singham, S.A. (2000). Competition Policy and the Stimulation of Innovation: TRIPS and the Interface Between Competition and Patent Protection in the Pharmaceutical Industry. 26 BROOK. J. INT'L L.363, 388-89.

Smedlund, A., & Poyhonen, A. (2005). Intellectual Capital Creation in Regions: A Knowledge System Approach. In Bounfour, A., Edvinsson, L. (Eds.), *Intellectual Capital for Communities*. Boston, MA: Elsevier Butterworth-Heinemann., Storper, M. & Scott, A.J. (1992, Eds). *Pathways to Industrialization and Regional Development.* London: Routledge.

Smith, A. (1937). *The wealth of nations*. New York: Random House.

Soboyejo, W. (2006). *A network of excellence for African development*. Retrieved from Science and Development Network (SciDev.net) - http://www.scidev.net/dossiers/index.cfm?fuseaction=dossierreaditem&dossier=13&type=3&itemid=548&language=1&CFID=1495413&CFTOKEN=24816639 on June 29, 2007.

Solow, R. (1956). A Contribution to the Theory of Economic Growth, Quarterly Journal of Economic Growth. *The Quarterly Journal of Economics, 50*, 65–94. doi:10.2307/1884513

Solow, R. (1957). Technical change and the aggregate production function. *The Review of Economics and Statistics, 39*, 312–320. doi:10.2307/1926047

Souitaris, V. (2003). Determinants of technological innovation: Current state of the art, modern research trends and future prospects. In Shavinina, L. V. (Ed.), *The International Handbook on Innovation*. Oxford, UK: Elsevier Science. doi:10.1016/B978-008044198-6/50035-8

Srinivas, K. R. (2006), *TRIPs, access to medicines and developing nations: Towards an open source solution.* Working paper no.248; Indian Institute of Management, Bangalore.

Stanton, W. J., Etzel, M. J., & Walker, B. J. (1994). *Fundamentals of Marketing* (10th ed.). New York: McGraw Hill, Inc.

Statistique Canada. (2000): Comment expliquer la croissance rapide parmi les entreprises canadiennes de biotechnologie? *Bulletin de l'analyse en innovation.* 88-003-XIF. Ottawa.

Steiss, A. W. (2003). *Strategic management for public and nonprofit organizations*. Boca Raton, Fla: CRC Press and Taylor and Francis.

Stephan, P., & Everhart, S. (1998). The Changing Rewards to Science: The Case of Biotechnology. *Small Business Economics, 10*, 137–151. doi:10.1023/A:1007929424290

Stern, S., Porter, M., & Furman, J. (2000). *The determinants of national innovative capacity*, NBER Working Paper 7876, National Bureau of Economic Research.

Sternberg, R. (2000). Innovation Networks and Regional Development – Evidence from the European Regional Innovation Survey (ERIS): Theoretical Concepts, Methodological Approach, Empirical Basis and Introduction to the Theme Issue. *European Planning Studies, 8*, 389–407. doi:10.1080/713666420

Stiglitz, J. (1996). Some lessons from the East Asian miracle. *The World Bank Research Observer, 11*(2), 151–177.

Storper, M. (1995). Competitiveness policy options: the technology-regions connection. *Growth and Change*, (Spring): 285–308. doi:10.1111/j.1468-2257.1995.tb00172.x

Suliman Al-Hawamdeh & Thomas L. Hart. (2002). *Information and Knowledge Society*.New York: McGraw Hill Education (Asia).

Sveiby.K.E (2001).What is KM? Tapscott.D (2005). *Part II: Strategy in the Age of Collaboration, New Paradigm Learning Corporation (NPLC)*. Canada.

Tay, R. (2009). *Fostering a Vibrant Infocomm Ecosystem*. Speech by RAdm(NS) Ronnie Tay, Chief Executive Officer, Infocomm Development Authority of Singapore, at the Singapore Infocomm Technology Federation 2009 ICT Business Outlook Forum at the Raffles City Convention Centre, 18 February.

Teece, D. J. (1998). Capturing Value From Knowledge Assets: The New Economy, Markets for Know-How, and Intangible Assets. *California Management Review*, *40*(3).

Teece, D. J., Pisano, G., & Shuen, A. (1997). Dynamic capabilities and strategic management. *Strategic Management Journal*, *18*(7), 509–533. doi:10.1002/(SICI)1097-0266(199708)18:7<509::AID-SMJ882>3.0.CO;2-Z

Teece, D., & Pisano, G. (1998). The dynamic capabilities of firms: an introduction. In G. Dosi, D. Teece, G. & Silverberg (Eds), *Technology, Organization, and Competitiveness: Perspectives on Industrial and Corporate Change*. Oxford, UK: Oxford University Press.

Teo, C. H. (2009). Speech by Minister for Defence Teo Chee Hean, at Committee of Supply Debate in Parliament, 12 February.

Tharman, S. (2008). Budget Statement 2008 by Finance Minister Tharman Shanmugaratnam at the Singapore Parliament, 15 February.

The General Food Law Regulation (EC) 178/2002 (2002). *The European Parliament*. Retrieved April 10, 2009, from http://europa.eu/eur-lex/pri/en/oj/dat/2002/l_031/l_03120020201en00010024.pdf

Thomas, A., & Mueller, S. (2000). A case for comparative entrepreneurship: assessing the relevance of culture. *Journal of International Business Studies*, *31*, 287–301. doi:10.1057/palgrave.jibs.8490906

Thomas, S., & Lang, T. (2007). From Field to Office: The Politics of Corporate Ethnography. In *Proc. Of Ethnographic Praxis in Industry Conference* (pp. 78-90). 2007.

Thomke, S. (1997). The Role of Flexibility in the Development of New Products: An Empirical Study. *Research Policy*, *26*, 105–119. doi:10.1016/S0048-7333(96)00918-3

Tianxia Net café Association (2008). *The report of current situation of Internet Cafés in China*. Retrieved from http://www.txwm.com/special/research/

Tiffin, S. (1997). *Building Science Technology Innovation System in Africa*. Paper delivered at the First Regional Workshop on the Restructuring of National Science and Innovation Systen in Africa Regional Programme for Technology Management (REPTEM) Lagos; May 12 – 16, 1997.

Tolosa, J. B., García-Díaz, V., Sanjuan-Martínez, O., Fernández-Fernández, H., & García-Fernández, G. (2009). Towards Meta-Model Interoperability of Models and Transformation Models. *International Symposium on Distributed Computing, Artificial Intelligence, Bioinformatics, Soft Computing and Ambient Assisted Living*(LNCS 5518, pp.315-322) June 10-12, 2009 Heidelberg, Germany: Springer.

Toyoma.K (2005). The Science of Social Networks. *Microsoft Research India*.Indian Institute of Science.

Tura, T., & Harmaakorpi, V. (2005). Social Capital in Building Regional Innovative Capability. *Regional Studies*, *39*(8), 1111–1125. doi:10.1080/00343400500328255

Ulijn, J., & Weggeman, M. (2001). Towards an innovation culture: what are its national, corporate. marketing and engineering aspects, some experimental evidence. In Cooper, C., Cartwright, S., & Early, C. (Eds.), *Handbook of Organizational Culture and Climate* (pp. 487–517). London: Wiley.

UNESCO (1969). *The measurement of scientific and technological activities*. United Nations publications.

United Nations Development Programme (UNDP). (2005). *Human Development Report 2005*. Geneva: UNDP.

United Nations Development Programme (UNDP). (2002). *Human Development Report*, Nigeria 2000/2001 Millennium Edition, 80p

United Nations Educational Scientific and Cultural Organisation (UNESCO). (1982). *An Introduction to Policy Analysis in Science and Technology*. Paris: UNESCO.

United States Supreme Court (June 16, 1980). 447 U.S. 303, 206 USPQ 193

University Companies Association (UNICO). (2005). UK University Commercialisation Survey. Financial Year 2004. Cambridge, UK.

Urban, G., & Hauser, J. (1993). *Design and Marketing of New Products*. Englewood Cliffs,NJ: Prentice Hall. van Birgelen, M.; De Ruyter, K. & Wetzels, M. (2001). Conceptualizing and Isolating Cultural Differences in Performance Data in International High-Tech Industrial Markets. *Industrial Marketing Management*, *30*, 23–35.

Users and Customers Organization. *OCU*. Retrieved June 14, 2009, from: http://www.ocu.org van der Linden, F.J. & Schmid, K. (2007). Software Product Lines in Action. *The Best Industrial Practice in Product Line Engineering*. New York: Springer.

Valentas, K. J., Levine, L., & Clark, J. P. (1990). *Food Processing Operations and Scale-up*. New York: Marcel Dekker Inc.

Van Dijk, T. (1996). Patent Height and Competition in Product Improvements. *The Journal of Industrial Economics*, *44*, 151–167. doi:10.2307/2950643

Vanessa Bradford Kerry. Kelley Lee (24 May 2007), "TRIPS, the Doha declaration and paragraph 6 decision: what are the remaining steps for protecting access to medicines?" Globalization and Health, 3:3.Wilson Tim, "In defense of patents," Times of India, New Delhi, April 26, 2008

Vang, J., & Asheim, B. T. (2006). Regions, absorptive capacity and strategic coupling with high-tech TNCs: lessons from India and China. *Science, Technology & Society*, *11*(1), 39–66. doi:10.1177/097172180501100103

Verspagen B. (1992). *Uneven Growth between Interdependent Economies*.Data Wyse:Maastricht Press.

Von Hayek, F. (1948). *Individualism and Economic Order*. Chicago: University of Chicago Press.

von Hippel, E. (1986). Lead Users: A Source of Novel Product Concepts. *Management Science*, *32*(7), 791–805. doi:10.1287/mnsc.32.7.791

von Hippel, E. (1995). *The sources of Innovation*. Oxford, Uk: Oxford University Press

Von Zedtwitz, M., & Heimann, P. (2006). Innovation in Clusters and the Liability of Foreignness of International R&D. In Carayannis, E., & Campbell, D. (Eds.), *Knowledge Creation, Diffusion, and Use in Innovation Networks and Clusters: A Comparative Systems Approach across the United States, Europe, and Asia* (pp. 101–122). New York: Praeger.

Weber. M (2002). Standardization in Knowledge Management – Towards a Common. *KM Framework in Europe, KM European Forum*.

Webster, F. (1995). *Theories of the Information Society*. London: Routledge.

Wei, L., Xianchun, X., & Zhizhou, C. (2004). China economic growth from long term tactic. *Management World*, *7*, 6–15.

Weingast, B. R. (1992). *The Economic Role of Political Institutions. Working*. Institute for Policy Reform.

Wenger, E. (1998). *Communities of practice. Learning, meaning and identity*. Cambridge, UK: Cambridge University Press.

Wennekers, S., & Thurik, R. (1999). Linking entrepreneurship and economic growth. *Small Business Economics*, *13*, 27–55. doi:10.1023/A:1008063200484

Wigg.K.M (1999). What future KM users may expect. *Journal of KM, 3*(2),, West Yorkshire, England: MCB University Press

Williams, J. (1994). Why do companies Differ and How does it Matter. In Rumelt, R., Schendel, D., & Teece, D. (Eds.), *Fundamental Issues in Strategy*. Boston: Harvard Business School Press.

Woiceshyn, J., & Hartel, D. (1996). Strategies and Performance of Canadian Biotechnology Firms: An Empirical Investigation. *Technovation*, *16*, 231–243. doi:10.1016/0166-4972(95)00066-6

World Bank. (1997). *China 2020: development challenges in the new century*. Washington, D.C.: The World Bank.

World Health Organization (2003). *Shaping the future.* World Health Report.

Xiaohua, Q., Jingping, Z., & Donghua, W. (2006). Factor analysis and prospects of China's economic growth. *Economic Research Journal*, *5*, 4–12.

Xiaolu, W., & Gang, F. (2000). *Sustainability of China's economic growth.* Beijing: Economic Science Press.

Xiaoping, Y. (2006). Staggered analysis on China's economic growth. *Market Modernization*, *62*, 200–201.

Yeo, C. T. (2000). Speech by Mr Yeo Cheow Tong, Minister for Communications and Information Technology, at the launch of eCelebrations Singapore at Suntec City, 1 March.

Yeoh, B., & Chang, T. C. (2001). Globalising Singapore: Debating Transnational Flows in the City. *Urban Studies (Edinburgh, Scotland)*, *38*(7), 1025–1044. doi:10.1080/00420980123947

Yergin, D. A., & Stanislaw, J. (1998). *The Commanding Heights: The Battle between Government and the Marketplace that is Remaking the Modern World.* New York: Simon & Schuster.

Yingxing, H. (1999). Basic meaning on transformation economic growth mode. *Management World*, *4*, 15–24.

Yongyuan, H., & Zhiyong, L. (2004). Effects of Different Types of Human Capitals on Economic Growth. *Population & Economics.*, *2*, 55–58.

Yoshino, M., & Rangan, U. S. (1995). *Strategic Alliances*. Boston: Harvard Business School Press.

Yuling, W., & Shiying, Z. (2001). Studying on decomposition of technological innovation effects spillover. *China Soft Science*, *8*, 53–56.

Yuming, Y. (2002). Analysis and calculation on TEP of China and provinces. *Economist*, *3*, 115–121.

Zahra, S., & George, G. (2002). Absorptive Capacity: A Review, Reconceptualization, and Extension. *Academy of Management Review*, *27*, 185–203. doi:10.2307/4134351

Zheng, W., Ma Cuifang. (2003). A geographical investigation into knowledge spillovers between regions. *Acta Geographica Sinica*, *58*(5), 773–780.

Zheng, W., & Qiong, H. (2007). An analysis to the possibility of regional new economy industry development in China. *Science Research Management*, *28*(2), 105–111.

Zhong, J. (2001). *System Analysis on World Silicon Models*. Bejing: China Society Science Press.

About the Contributors

Patricia Ordóñez de Pablos is professor in the Department of Business Administration and Accountability, at the Faculty of Economics of The University of Oviedo (Spain). Her teaching and research interests focus on the areas of strategic management, knowledge management, intellectual capital measuring and reporting, organizational learning and human resources management. She is Executive Editor of the International Journal of Learning and Intellectual, the International Journal of Strategic Change Management and Editor in Chief of International Journal of Chinese Culture and Management.

Rongbin W.B. Lee is the Chair Professor and Director of the Knowledge Management Research Centre, of The Hong Kong Polytechnic University. He is currently a council member of the Hong Kong Productivity Council and a member of the assessment panel of the Innovation and Technology Fund of the Hong Kong SAR Government. He established the Knowledge Solution Laboratory, the first of its kind in Hong Kong, and pioneered the research and practice of knowledge management in various organisations in manufacturing, trading, public utilities, various government departments and healthcare. Professor Lee and his team have launched Asia's first online MSc Programme in Knowledge Management. His research interests include manufacturing strategy, knowledge management, organisational learning and intellectual capital-based management.

Jingyuan Zhao is a post-doctoral researcher and senior lecturer in Postdoctoral Center & School of Management, Harbin Institute of Technologies (China). Her Ph.D. is in Management Science and Engineering in Chinese Academy of Sciences (CAS) and University Science and Technologies of China (USTC). Dr. Zhao's expertise is on regional innovation management, high-tech industry cluster, knowledge management, technologies diffusion, organization learning. She serves as a Guest Editor for several international journals and is an Invited Reviewer for China' state-run newspaper West Times to provide comments on the economy.

* * *

Vicente García Díaz is a computer engineer, PhD student in the Department of Computer Science at the University of Oviedo. His research interests include Model-Driven Engineering, Domain-Specific Languages, Project Risk Management, Software Development processes and practices and Web Engineering.

Copyright © 2011, IGI Global. Copying or distributing in print or electronic forms without written permission of IGI Global is prohibited.

B. Cristina Pelayo García-Bustelo is a Lecturer in the Computer Science Department of the University of Oviedo. PhD from the University of Oviedo in Computer Engineering. Her research interests include Object-Oriented technology, Web Engineering, eGovernment, Modeling Software with BPM, DSL and MDA.

Oscar Sanjuán Martínez is a Lecturer in the Computer Science Department of the University of Oviedo. PhD from the Pontifical University of Salamanca in Computer Engineering. His research interests include Object-Oriented technology, Web Engineering, Software Agents, Modeling Software with BPM, DSL and MDA.

Juan Manuel Cueva Lovelle is a Mining Engineer from Oviedo Mining Engineers Technical School in 1983 (Oviedo University, Spain). PhD from Madrid Polytechnic University, Spain (1990). From 1985 he is a Professor at the Languages and Computers Systems Area in Oviedo University (Spain). ACM and IEEE voting member. His research interests include Object-Oriented technology, Language Processors, Human-Computer Interface, Web Engineering, Modeling Software with BPM, DSL and MDA.

Jingyuan Zhao is a post-doctoral researcher and senior lecturer in Postdoctoral Center & School of Management, Harbin Institute of Technologies (China). Her Ph.D. is in Management Science and Engineering in Chinese Academy of Sciences (CAS) and University Science and Technologies of China (USTC). Dr. Zhao's expertise is on regional innovation management, high-tech industry cluster, knowledge management, technologies diffusion, organization learning. She serves as a Guest Editor for several international journals and is an Invited Reviewer for China' state-run newspaper West Times to provide comments on the economy.

Antonio Lerro is a Research Fellow in Business Management at University of Basilicata, where he joins the "Centre for Value Management". Antonio held his PhD at University of San Marino. He joins also Cranfield School of Management – UK - as Visiting Research Fellow. His research centres on the management and the measurement of the Intellectual Capital, innovation and change management, clusters, regional development and regional innovation systems. He is regular speaker at national and international conferences and author of over twenty academic and practitioner papers.

Giovanni Schiuma is a Professor at the University of Basilicata, Italy, where he coordinates as Scientific Director the Centre for Value Management. Giovanni is Visiting Fellow at the Cranfield School of Management and Adjunct Professor at Tampere University of Technology. Giovanni's research, teaching, and consulting focus on linking knowledge assets and organization behaviour to performance management and organization value creation. Giovanni is Co-Editor in Chief of the internationally recognised Measuring Business Excellence (Emerald publisher). His primary research interests focus around the following areas: knowledge asset & intellectual capital management, performance management systems, innovation and change management, organizational behaviors, organizational learning, industrial district and local development.

Asleena Helmi has been in the academic field for approximately 10 years in various institutions of higher learning in Malaysia. Prior to joining Swinburne University of Technology, she taught a variety of subjects which include entrepreneurship, management information systems and knowledge management at Multimedia University, Cyberjaya. Apart from teaching, she participates in consultancies and presented research papers at the Harvard Asia Project for Asian and International Relations in 2008 and has published in numerous local and international journals. She earned a distinction masters award majoring in Electronic Commerce from the School of Business at Curtin University of Technology, Western Australia. Her area of research interests are knowledge management, e-commerce and international business.

Zainal Ariffin Ahmad, PhD is an Associate Professor and has more than 15 years of teaching, training and consultancy experiences at Northern Illinois University and Universiti Sains Malaysia. He teaches in postgraduate programmes and supervises doctoral and master candidates in the areas of human resource management, organizational behavior, quality, effective teaching and consultancy. He is an active researcher and has completed nine national research projects to date and co-written 4 books and various articles in national, regional and international journals and conferences. He has conducted seminars, workshops and in-house training for academic, public sector and corporate clients on such as Siemens, Otis, Intel and Motorola. He is a member of the International Association of Applied Psychology (IAAP).

Daisy Kee is a lecturer of management at Universiti Sains Malaysia. Her areas of interests are in HRM, OB, gender and leadership. Her current program of research focuses on HRM and LMX. She holds a PhD in Business and Management from International Graduate School of Business, University of South Australia. She was the secretary of Management Case Study Journal, Australia (2004-2006). She has been voted one of the Most Outstanding Malaysian university students in South Australia by former South Australia Governor Sir Eric Neal (2006). She earned her MBA from School of Management, Universiti Sains Malaysia. She was awarded Dean's List for being one of the top MBA students (2003).

Gert-Jan Hospers is Professor of City and Regional Marketing in Europe at the Radboud University Nijmegen and Researcher of Economic Geography at the University of Twente, both in The Netherlands. He is interested in the identity, image and development of old industrial areas in Western-Europe, such as Twente, the North East of England, the Øresund Region and the Ruhrgebiet. Recently, he published – together with Paul Benneworth – The Role of Culture in the Economic Development of Old Industrial Regions (LIT, 2009).

Tomas Gabri Bas, PhD in Technology Management and Marketing of University of Quebec (Canada). His research project focus on biotechnology management, technological innovation, knowledge in natural resources and public policies. He has a MSc in Environment of University of Montreal (Canada). He is professor of technology and innovation management at the University Adolfo Ibáñez (Chile) and consultant for World Bank, United Nations, IDRC, Statistics Canada, etc. Honors: "Bright Minds - Brain World Class". One of the 10's most influential researchers of Chile in Economics and Management (according to international citation index ISI).

Rakhi Rashmi is working as Research Fellow with CNRS GREDEG, University of Nice, France, A PhD candidate of Centre for the study of Law Governance, Jawaharlal Nehru University, New Delhi, who has the degrees of LLB (Delhi University), MBL (National Law School of India University, B'lore), MFM (Pondicherry University.

Pak Tee Ng is an Associate Professor at the Policy and Leadership Studies Academic Group, National Institute of Education (NIE), Nanyang Technological University, Singapore. He read Mathematics at Cambridge University [BA(Hons), MA] and subsequently Management at Leicester University [MBA] and Bradford University [PhD]. His main areas of teaching, research and consultancy are Learning Organisation, Change Management, Knowledge Management, Quality Management, Innovation, Complexity, Leadership, Coaching and Education Policies. He is the author/editor of several books and numerous journal articles, book chapters and conference papers. He is also a journal editor and an editorial board member of several other journals.

Qiong He, Lecturer in School of Mining and Safety Engineering, Hunan University of Science and Technology (China). Postdoctoral in School of Economics, Xiamen University (China), Her PhD is in Management Science and Engineering in Chinese Academy of Sciences and University Science and Technology of China. Dr. He's expertise is on regional economic and management, resource and environment management.

Qun Ren got her Master degree in International business Finance from Bournemouth University in 2005. Currently, she is a PhD researcher in the Business school of Bournemouth University, UK. Her research interests focus on business strategy in fast-paced competitive IT industries, pricing in Internet services and national cultural issues in management.

Philip Hardwick is a Professor of Economics and Director of Postgraduate Research at Bournemouth University. His research interests concern the econometric analysis of the relative performance of financial services firms, in terms of economies of scale and scope, and cost, profit and revenue efficiencies. He has published widely in academic journals in finance and economics, including Oxford Economic Papers, Applied Financial Economics, the Journal of Management Studies, Abacus, the Journal of Risk and Insurance, the Journal of Business Finance and Accounting and the Journal of Banking and Finance.

A.A. Egbetokun holds a Bachelor's degree in Mechanical Engineering and a Master's degree in Technology Management from the Obafemi Awolowo University, Ile-Ife, Nigeria. He has also worked as a Research Assistant in the university and as a Mechanical Engineering Intern in the automobile industry. He belongs to the Nigerian Institute of Management, the Nigerian Institution of Mechanical Engineers, the International Society for Professional Innovation Management and the US-based National Association for Community College Entrepreneurship. As a Senior Research Officer with the National Centre for Technology Management, Obafemi Awolowo University, he is actively involved in policy research, evaluation and review especially in the areas of innovation, entrepreneurship, SMEs and technological capabilities. In the last two years, his research has focused on clusters and innovation systems; science, technology and innovation indicators as well as student entrepreneurship. His articles have appeared in the International Journal of Technology Management, the World Review of Entrepreneurship Management and Sustainable Development, the International Journal of Learning and the International Journal of Learning and Intellectual Capital.

W.O. Siyanbola obtained his PhD in Solid State Physics from the University of Sussex, Brighton, UK in 1991. Since 2001 he has been closely involved in Science, Technology and Innovation (STI) policy research, evaluation and review. He is a Fellow of both the Materials Society of Nigeria and the Solar

Energy Society of Nigeria; and a member of the National Business Incubation Association of America, among other prestigious professional associations. For many years, he was a resource person to the National Centre for Technology Management (NACETEM), an agency of Nigeria's Federal Ministry of Science and Technology located in Obafemi Awolowo University, Ile-Ife, Nigeria. He later became the Director-General/CEO of the centre in 2006. At NACETEM, Dr. Siyanbola now oversees policy research activities in STI and is a regular speaker at local and international STI policy debates. He also supervises postgraduate students at MSc and PhD levels at Obafemi Awolowo University.

A. A. Oyewale received training as an Agricultural Economist before obtaining a Doctoral degree in Technology Management (during which he won a United Nations University Institute for New Technologies (UNU/INTECH) Ph.D. Internship Fellowship Award in 2000) at the Technology and Policy Development Unit (TPDU) of the Obafemi Awolowo University, Ile-Ife, Nigeria. Before joining the National Centre for Technology Management (NACETEM) in 1996, Dr. Oyewale has worked as a business consultant to several organisations. A thorough-bred researcher with interests in intellectual property and the management of National Innovation Systems, Dr Oyewale has published papers in both local and international journals. He is presently the Head of the Training and Research Department of NACETEM.

Mauri Kantola is the Manager of Educational Services of the Turku University of Applied Sciences in Finland. He was previously the Quality Manager of the Turku University of Applied Sciences. He holds a MA (Soc.) from the University of Turku in Finland.

Juha Kettunen is the Rector of the Turku University of Applied Sciences in Finland and Adjunct Professor of the University of Jyväskylä in Finland. He was previously the Director of the Vantaa Institute for Continuing Education of the University of Helsinki and Director of the Advanced Management Education Centre of the University of Jyväskylä. He holds a PhD from the University of Bristol in the UK, and a DSc from the University of Jyväskylä in Finland.

Satu Helmi is the Regional developmemt coordinator of the Turku University of Applied Sciences in Finland. She holds a M.A. (Educ.) form the University of Turku in Finland.

Aikaterini Kokkinou holds a first-class PhD Degree from the University of Aegean, a first-class Master of Science Degree in Business Analysis and Finance from the University of Leicester, a first-class Graduate Degree in Economic Sciences from the University of Crete, as well as a first-class Graduate Degree in Public Administration from the National School of Public Administration, Greece. During the last 5 years, Dr. Aikaterini Kokkinou worked for the Public Debt Management Agency, Ministry of Finance, Greece. She is currently a researcher in the Department of Economics, University of Glasgow, United Kingdom. During her career, Dr. Aikaterini Kokkinou acquired a solid background in the fields of economic growth and development, productivity and efficiency analysis, entrepreneurship, innovation activities, regional growth, F.D.I. activities, competitiveness, convergence and European Union development economics. She has also collaborated in research projects, has presented her work in a number of international conferences and has published in journals, books, research papers, and conference proceedings.

George M. Korres holds a doctorate degree in Economic Sciences (Economics of Innovation) from University of London (QMW) and M.Sc. from Essex University. He is currently an Associate Professor at the University of the Aegean, Department of Geography. Over the last 10 years, during both his academic and professional career, he has gained an in depth experience in the fields of economics, business and politics in the area of European Economics and Innovation Systems. His main areas of interest and specialization include: European Integration, Social Policy and Social Inclusion, Economic and Innovation Policies, EU enlargement and the Euro-Mediterranean partnership. He has collaborated with many European universities, Institutions, NGOs as well as academia. George Korres has also been involved in various trans-European and international projects and he has published widely in this area in many International Journals, such as Applied Econometrics, Regional and Sectoral Economic Studies, Southwestern Economic Review, Management Sciences and Regional Development e.t.c.. He has also published a number of books with international publishers, such as MacMillan-Palgrave Press, Springer, Ashgate Publishers etc.

Index

Copyright © 2011, IGI Global. Copying or distributing in print or electronic forms without written permission of IGI Global is prohibited.

J

K

L

Q

R

S

T

Breinigsville, PA USA
17 October 2010
247133BV00004BI/3/P